DEGREES KELVIN

A TALE OF GENIUS, INVENTION, AND TRAGEDY

DEGREES KELVIN

A TALE OF GENIUS, INVENTION, AND TRAGEDY

David Lindley

Joseph Henry Press
Washington, D.C.

Joseph Henry Press • 500 Fifth Street, N.W. • Washington, D.C. 20001

The Joseph Henry Press, an imprint of the National Academies Press, was created with the goal of making books on science, technology, and health more widely available to professionals and the public. Joseph Henry was one of the founders of the National Academy of Sciences and a leader in early American science.

Any opinions, findings, conclusions, or recommendations expressed in this volume are those of the author and do not necessarily reflect the views of the National Academy of Sciences or its affiliated institutions.

Library of Congress Cataloging-in-Publication Data

Lindley, David, 1956-
  Degrees Kelvin : a tale of genius, invention, and tragedy / David Lindley.
      p. cm.
  Includes bibliographical references.
  ISBN 0-309-09073-3 (hbk.)
  1. Kelvin, William Thomson, Baron, 1824-1907. 2. Physicists—Great Britain—Biography. I. Title.
  QC16.K3L56 2004
  530'.092—dc22

                                  2003022885

Permission: The Syndics of Cambridge University Library in order to quote from the Kelvin and Stokes collections.

Printed in the United States of America.

CONTENTS

ACKNOWLEDGMENTS

R esearch for this book was done mostly at the Bodleian Library in Oxford, especially the Radcliffe Science Library, and at Cambridge University Library. I am grateful to the staff at both institutions for their help. I particularly thank Adam Perkins and his staff at the Scientific Manuscripts Collections at Cambridge. Last minute assistance from the Niels Bohr Library of the American Institute of Physics is much appreciated.

My agent, Susan Rabiner, helped shape the story into a more purposeful tale than the amorphous mass it might otherwise have been. Jeff Robbins at the Joseph Henry Press encouraged me to untangle some knots in the original manuscript and professed to be not unduly disturbed that I didn't quite make the deadline. Chris Butcher scanned and tweaked several of the images reproduced here. Robert Fairley kindly provided copies of Jemima Blackburn's watercolor of the Thomson brothers and Helmholtz and the striking photograph of Kelvin as an elderly man. Thanks to all.

For all kinds of other moral and practical support during a couple of peripatetic years, including but not limited to places to stay; rides to and from airports; use of the old blue Toyota; Internet connections; assorted computer peripherals plus technical assistance; beer, pizza, and bridge parties; bibliophilic companionship in Hay-on-Wye; excuses to go sight-seeing; distracting e-mails and phone calls; and a variety of opportunities to think about something other than this book, my thanks go to Liz Pennisi and Matt Butcher, Hellen Gelband, Karen Hopkin, Stephen Lindley, Bob Shackleton and Cathy Mattingly, Christine Mlot, Damaris Christensen, and Kay Behrensmeyer and Bill Keyser.

I want lastly to thank Michael Nauenberg for his review of the manuscript, which made me rethink some of my opinions, especially of thermodynamic history. Professor Nauenberg and I still don't entirely agree, but I hope our differences are honorable. The history of science is a branch of history, after all; definitive conclusions are hard to come by.

INTRODUCTION

On the morning of Wednesday, May 1, 1902, students at the University of Rochester in upstate New York assembled for chapel with untypical eagerness. Attendance, supposedly mandatory, was normally sparse. By a quarter past 10 on this day, however, students had crammed expectantly into Anderson Hall, alongside a good number of interested townspeople. The place was bursting. As senior faculty members filed in to take their seats, sporadic shouts and cheers erupted from the simmering throng. But when university president Dr. Rush Rhees at last entered, a hush came over the crowd. Leaning on Rhees's arm was a slight elderly man, with thinning white hair above a prominent forehead, his face enlivened by sharp blue eyes. The visitor walked carefully, with a noticeable limp. He was more than usually frail this morning, suffering from a bout of the facial neuralgia that had afflicted him intermittently for several years now. On rare occasions the pain was bad enough to keep him in bed for a few days. The *New York Times*, reporting his arrival from England on the Cunard ship *Campania* a week and a half earlier, noted that the old man "did not appear to be in robust health." He had been helped into a chair on the dockside while customs officials inspected his baggage. But he managed a few words

1

with reporters and would have spoken more had he been less tired. Throughout his long life he had rarely been ill, and immobility irked him. The best antidote for age and pain was to keep working, to stick to his busy schedule, and especially not to let anyone down.

As Dr. Rhees and his venerable guest moved slowly to their places, the Rochester students rose to their feet in silence. But then, if we are to believe the reporter for the local *Democrat and Chronicle*, there "broke forth such a cheer as had never before resounded through Anderson Hall. It filled the college halls, overflowed out on the campus, and could have been heard half a mile away. It was a spontaneous, generous cheer, exuberant, manly and vociferous, a cheer which must have warmed the visitor's heart, much as he is accustomed to the homage of men."

The recipient of this extraordinary acclaim was not a war hero or a beloved author, not a theater star or a famous politician, but, remarkably, a scientist and a British scientist to boot. Every age has its venerated intellectuals, but rarely do they become the subject of whooping and foot stomping by crowds of university students. In this, as in so much, Lord Kelvin was one of a kind. At 77 years of age, he was no ivory-tower academic but a public figure and a celebrity on both sides of the Atlantic. A few days earlier he had attended a reception in New York for the new president of Columbia University, where he mingled with the likes of President Theodore Roosevelt and Andrew Carnegie. In Washington, D.C., he and Lady Kelvin stayed with Mr. and Mrs. George Westinghouse at their mansion on 16th Street which, heading directly north from the White House, was lined in those days with the magnificent residences of the gilded age. At grand dinner parties on successive evenings, American politicians and foreign ambassadors, as well as technical men such as Alexander Graham Bell, accepted invitations to meet the celebrated visitor. In Rochester he was the guest of George Eastman, founder of the Kodak company, of which Kelvin was a vice-president and scientific adviser. He eagerly inspected the hydroelectric power station at Niagara, which turned the energy of the cascading cataract into electricity. For a reportedly feeble old man, he swept around the northeastern United States during his three-week visit with remarkable energy and enthusiasm.

Newspapers referred to him as a noted or eminent or distinguished scientist, an appellation Kelvin disliked. He preferred the old-fashioned designation "natural philosopher." Only in the last half century had science or natural philosophy emerged from its arcane and isolated realm to become a force in public life. Terminology had an awkward, unfamiliar air. On a trip to North America five years earlier, one newspaper talked of "Lord Kelvin, the eminent electrician." In those days an electrician was not someone who came to your house to install a new outlet or fix a broken wire; the average home didn't have such marvels. Rather, an electrician was one versed in the science of electricity and magnetism, natural phenomena that had only recently begun to yield to scientific understanding and that still retained a good deal of mystery. Kelvin had indeed been a pioneer of the new science of electromagnetism, and of much else besides, but that could hardly account for his renown. The names of his equally meritorious contemporaries, men such as Faraday and Maxwell and Weber and Helmholtz, may have evoked a sliver of recognition among the nonscientific public. But these were not widely known names at the end of the 19th century, any more than they are today. Kelvin, on the other hand, was a genuine celebrity.

After the raucous Rochester students had settled themselves, the usual chapel service followed. Then university president Rhees spoke of their distinguished guest. "Lord Kelvin's visit," he began, "has called to mind his many contributions to the practical applications of science to modern needs." He mentioned the laying of the first transatlantic submarine telegraph cables some 40 years earlier, an enterprise with which Kelvin had been crucially associated. Of particular concern to Rhees's audience was Kelvin's long-standing involvement in the development of systems to generate electric power by tapping the enormous energy going to waste every second as water plunged endlessly over nearby Niagara Falls. He talked of Kelvin's countless laboratory investigations, which underpinned the work of many pioneers of electrical science and technology. "His patient study and passion for exactness have put in his abiding debt all students who follow in the path of physical investigation in which he has been so illustrious a leader."

These achievements, Rhees was careful to note, sprang from the mind of a man who was not simply an inventor but one whose prime achievements lay in the realm of pure science. He mentioned the doctrine of the conservation of energy. Kelvin had been one of those who had brought into being this profound law, which now stood "as the basis of not a few of the advancements made during the last half century in both pure and applied science."

Here was a man, in other words, who had contributed profoundly to the development of fundamental physical principles and who had in addition turned those elementary insights toward practical ends. In 1902, when Rhees lauded his visitor, the creation of mechanical devices and technological instruments according to the principles of science was not yet a routinely accepted part of ordinary life. The telephone had been around for two decades or so but was still considered a luxury. Electricity as a source of domestic power was not yet widespread. Cars were barely known, airplanes nonexistent. Technology was just beginning to impinge on the lives of ordinary men and women. It was seen almost without reservation as a boon and a blessing. Science was the harbinger of a new world of convenience, of labor-saving devices, of vast industries. It represented progress, as yet unsullied by doubts. Those who brought technology to life were rare and remarkable men. Often, like the incomparable Thomas Edison, they were men of incalculable ingenuity but no deep scientific knowledge. The true scientists, on the other hand, generally stayed aloft in their abstract realm and did not deign to come to earth. Uniquely, Kelvin existed in both spheres, as scientist and technologist, academic and entrepreneur, a philosopher and a practical man rolled into one.

When he died in December 1907, a few years after his visit to Rochester, Kelvin was buried at Westminster Abbey with all the pomp and ceremony Great Britain could muster. He was laid in a tomb alongside Isaac Newton, that unapproachable icon of pure science. The proximity seemed just: surely the names Kelvin and Newton would live on forever in the same exalted rank.

***

Not exactly. Everybody has heard of Newton. Few these days know of Kelvin. His name survives, for those with a little physics education, in the absolute temperature scale. The lowest temperature attainable, $-273.15°$ Celsius, is zero on the Kelvin scale. There used to be an American company, based in Detroit, that made refrigerators under the brand name Kelvinator. There is no direct connection. The company launched the line in 1918, to pay scientifically appropriate tribute to a man who had done much to develop the modern understanding of temperature and no doubt also to cash in on a still-famous name.

Celebrity is notoriously fragile, of course, but scientific reputations do not normally wax and wane according to the whims of one era or the next. Milestones in science stand forever; those who erect them gain permanent recognition. Yet in a poll conducted by the U.K. Institute of Physics in 1999, Kelvin's name did not feature in the top 10 all-time greats of physics, or even among 18 also-rans. How could a man routinely described in his lifetime as Britain's and perhaps the world's greatest scientist have become so cruelly neglected?

Even while he was alive, however, a gulf had developed between Kelvin's public persona and his reputation among researchers. In the newspapers he was scientific knowledge and brilliance personified; at scientific meetings, which he attended eagerly until the year he died, he had become something of a crank, a living fossil, a holdover from an almost forgotten era. He had reservations about the existence of atoms; he believed the earth was no more than a hundred million years old; he would not wholly accept the novelty of radioactivity. Even in his position as "eminent electrician" he had parted company from the mainstream. James Clerk Maxwell's theory of electromagnetism had by then gained universal recognition as a full account of electric and magnetic physics. Or not quite universally: Kelvin would not accept it, even though, decades earlier, his own innovative ideas had been Maxwell's first inspiration.

In his judgments on technological matters too he displayed the dogged certainty of an opinionated old man. The *Times* reporter who caught a few words with him as he disembarked in New York in 1902 asked him about the prospects of two new scientific wonders, wireless telegraphy and airships. The first, Kelvin said, "is one of the world's most remarkable inventions . . . very marvelous indeed." Of the second he declared that "they will never be able to use dirigible balloons as a means of conveying passengers from place to place. . . . It is all a delusion and a snare . . . not practicable."

Thus stands, in histories of science, the enduring image of Lord

Kelvin. An old man, out of touch with the new science of the early 20th century. An old man who said no to atoms, no to Maxwell's electromagnetic theory, no to radioactivity. A crank, in other words. In the decades after his death Kelvin's scientific reputation sank rapidly and has still not risen back to anything like its peak. His name was posthumously attached to the scientific temperature scale in 1954, and with good reason. I wonder, though, whether the average physicist today could explain in any detail what it was that Kelvin did to make this commemoration appropriate. What mostly survives, for those who know the name Kelvin at all, is the image of a crotchety, white-haired man, quick to oppose what he couldn't understand.

***

That, in any case, is how I too was ready to perceive Kelvin when I first came across him in historical context. For much of his life he fought a running battle against geologists and biologists over the age of the earth. Starting from elementary laws of energy conservation and heat loss, Kelvin declared with unwavering conviction that our planet could be no more than a hundred million years old. Most likely it was no more than 20 million years old. Geologists, who had devised increasingly respectable theories of the formation and erosion of rocks, and biologists, armed with discoveries of fossils and more recently with Darwin's theory of evolution, chafed against this limitation. Kelvin's reasoning was, at the time, not unreasonable, but he stuck to it with blind stubbornness even as new facts and ideas came to light that knocked holes in many of his arguments.

I used this episode as a cautionary tale in my book *The End of Physics*, a critical survey of the development of physics through the 20th century. Kelvin's arguments had been solid and his logic impeccable. Even so, he was wrong. The earth is now known to be 4.5 billion years old, a figure Kelvin would have found ludicrous. Tough! Kelvin struck me as the perfect illustration of a physicist inclined to lay down the law to lesser scientific disciplines, despite imperfect information and unproven assumptions. When I learned also of his attitude about atoms, electromagnetism, and radioactivity, I got a clear picture of a man who with unfailing consistency put his money on all the wrong horses. There was the curious circumstance that Queen Victoria, or rather her advisers, had seen fit to raise him to the peerage—the first scientist, as I later discovered, to be so honored—but at that time I didn't know how celebrated he had been in his day, or why. I continued to think of him as one of the numerous minor-league scientists who populate old textbooks and later sink into the footnotes.

My views began to change a few years ago as I was researching my account of the life and work of the Austrian physicist and atomic pioneer Ludwig Boltzmann. I encountered Lord Kelvin at a much earlier stage in his life, when he still went by the name he was born with, William Thomson. In the late 1840s and early 1850s he was in his 20s, and his reputation existed only among his fellow scientists. But what a reputation! This young man, only a few years out of college, had already made astonishing progress in the quest to understand the nature of heat, work, and energy, and in the parallel effort to elucidate the nature of electricity and magnetism. Both subjects were then in their infancies, and here in the middle of it all was William Thomson, a mathematical prodigy who had first published original work at the age of 16 and whose inspirations were showing others how to untangle the great puzzles of the time. Not just in Britain but among the great scientists of France and Germany he was regarded as the most promising talent to have appeared in decades. The middle of the 19th century saw the foundation of what we now call classical physics—the science of heat and light, of electricity and magnetism—and William Thomson, not yet 30 years of age, was at the heart of it, propounding ideas and principles still taught today at the core of any course on basic physics.

But this only deepened my earlier puzzlement. From my own education in physics I was familiar with a good number of notable names from the 19th century, but either I had forgotten about William Thomson completely or else I had never learned much about him in the first place. Why were his apparently fundamental contributions not part of my general knowledge? And how did the quicksilver William Thomson, agile and original, turn into the doddery and skeptical Lord Kelvin?

Gradually I began to amass references to Kelvin and his wide-ranging achievements. There was Kelvin's circulation theorem in fluid mechanics, and the Kelvin-Helmholtz timescale in the physics of stars.

The V-shaped waves created by the bow of a ship speeding through water diverge, someone told me, at the Kelvin angle. I learned from a footnote somewhere that an important proof in vector calculus known as Stokes's theorem first appeared in a letter from William Thomson to his friend George Gabriel Stokes, who later set it as an exam problem for Cambridge students, which is why his name and not Thomson's became attached to it.

Then there was his connection to the transatlantic submarine cables, for which he developed the theory of undersea signal transmission, and his involvement with the British Royal Navy in devising new navigational instruments. There was the firm of Kelvin and White, a Glasgow maker of telegraph equipment, scientific and laboratory instruments, and, toward the end of the 19th century, household electricity meters.

I came across Kelvin again in a still more unlikely setting. Lauren Belfer's 1999 novel *City of Light* is a melodrama of murder and romance set in Buffalo in 1901—the year of President McKinley's assassination there. The new hydroelectric power plant at Niagara Falls feeds electricity to the booming town and its greedy businessmen. A rag-tag group opposes the power station, fearing that those behind it want to take every drop of water from the falls and dry up one of the country's natural wonders. The leader of the activists claims that the industrialists aim to consume Niagara Falls altogether, to sacrifice it completely to capitalism. "Has not their own prophet declared this policy?" the activist declares. "Their own 'President of the International Niagara Commission,' their prophet of darkness—Lord Kelvin!"

Prophet of darkness indeed! Here was a contrast to the awestruck praise from Dr. Rhees and the wild cheering of his Rochester students!

***

But I am running ahead. Behind the name William Thomson, Lord Kelvin, lay, so it now seemed, half a dozen different people. The question remained: How did youthful brilliance turn into resistance and obstructionism? Was the aged Kelvin a disappointed man? Angry? Oblivious?

This book is my attempt to disentangle and then recombine the many elements of his life in order to resolve its mysterious, possibly tragic path: early renown, established brilliance, stubborn old age, and abrupt posthumous fall. It's a complicated tale, and a chronological catalog of his activities would make confusing reading, to say the least. Inevitably, I have had to tease and rearrange, with good intentions, I hope, although we all know where those can lead. As my apologia I can do no better than borrow Kelvin's own words. In January 1883 he delivered a lecture to the Institution of Civil Engineers in London in which he dilated on the seemingly dreary but actually, as we shall see, contentious and intricate matter of the standardization of practical units for the measurement of electrical quantities. Characteristically, he struck out with enthusiastic digressions on subjects he had intended to dispose of in just a few sentences, and his lecture waxed on beyond the allotted time. Even so, his audience detained him with questions, and he drew warm applause and thanks for his efforts. In conclusion he replied:

"I wish I could have made it more clear, and placed it before you more methodically. All I can say is, that I have done my best, and I am much obliged to you for your patience."

1

CAMBRIDGE

Seventeen years old, William Thomson arrived in Cambridge at the end of October 1841 to begin his undergraduate studies. For a few days he felt adrift and aimless. With his father James Thomson, professor of mathematics at Glasgow University, he had made the tedious trip from Scotland to eastern England: mail coach from Glasgow to Carlisle, in northwestern England, where they spent the night; another coach across the country to the eastern seaport of Hull; across the Humber estuary in a little steamboat; and finally on to Cambridge in another coach, arriving in the late evening. Professor Thomson managed to get an inside seat, but William had to perch on top until they got to Bedford, 30 miles west of Cambridge. His father stayed a couple of nights and, an academic man himself, dined at the fellows' table of St. Peter's College while his son ate with the new and returning students. Then Professor Thomson returned to Scotland to prepare for his own classes, leaving his son to start his new life. William had roamed a good deal in Scotland and had traveled to continental Europe twice, but these had been family adventures, in the lively company of his three brothers and his older sisters, Elizabeth and Anna, with a maid and often a tutor, all under the strict guidance of their father. William was an affable, gregarious young man but unused to isolation.

"Since he has left I have had very little to do, since lectures have not begun yet, & I have not got any advice as to what I should read," he wrote to Anna after his father's departure. So he idled about, getting to know the town and St. Peter's College (as it was usually called then; now it is known as Peterhouse). The grandeur of the place surprised him. "I had no idea there were such fine gardens and grounds about the Colleges," he wrote to Elizabeth. Wryly he informed Anna that "to make the time pass less heavily, I have been going out every now and then, & coming in again." He went on short walks beyond the city. Accustomed to hiking around the lonely lochs and spectacular mountains of Scotland, he found the flat, empty landscape around Cambridge disheartening, but soon he was out walking and talking with his fellow undergraduates. Conversation and burgeoning friendships pushed the dreary landscape into the background. With guidance from his sisters he learned to prepare a small breakfast in his college rooms. He asked Anna to tell him how much tea to put in his cup for breakfast. From Elizabeth he wanted to know whether he should put coffee in the water before it has boiled or after.

William's outgoing nature soon asserted itself. He began to make friends with other new students. The college tutor, Dr. Cookson, called on the young men to attend a wine party in his rooms. "I made my appearance in fear and trembling," he told Anna with comic exaggeration. He had grown up in a house where distinguished academics came and went, and knew how to behave with charm and civility. At Cookson's party he fell into awkward conversation about "college, buildings, Fitzwilliam museum" and other small matters, all of which, he allowed, "was on the whole not unedifying." His initial nervousness overcome, he was not going to be overawed by the chitchat of the dons.

After a few days like this, lectures started, and William was instantly busy. "I have got no time to be dull," he now told Elizabeth, "as I have got as much to do as I can possibly accomplish, and more besides." More rewarding than walking, he soon discovered, were the pleasures of rowing on the Cam. William knew all about boating on the clean open waters of Scotland. The narrow river in Cambridge—"an exceedingly muddy and sluggish stream," he told Elizabeth—had traditions all its own. "I adventured myself to-day for the second time in a funny (or funey, or funney), i.e. a boat for one or two people to row in. It is certainly rather a venture to go in them as we can hardly stand upright in them for fear of upsetting them." Quickly he mastered the little craft. Though tempted, he shunned traditional Cambridge rowing, in eights, in which his college was doing well. "Rowing for the races is too hard work for getting on well with reading; and besides, the men connected with the club are generally rather an idle set." Not only that, a few days earlier a foolhardy student from Queen's College had managed to drown himself trying to shoot the modest three-foot fall on the Cam.

After a week or two, William's unease dissipated entirely and he threw himself into college life. He went to his lectures, read voraciously, studied in his rooms, rowed on the Cam, tramped around the muddy fields with his new friends, and, like any smart undergraduate away from home for the first time, stayed up late talking earnestly of this and that. His already advanced knowledge of mathematics set him apart from most of his fellow students. And a secret about him soon came to light that made him the object of unconcealed awe.

Six months earlier, in May, the *Cambridge Mathematical Journal* had published a short paper correcting errors in a recent book by the Edinburgh professor of mathematics, Philip Kelland. Only the cryptic initials P.Q.R. identified the author of this concise and confident work. The editor of the *Journal* was David F. Gregory, a fellow of Trinity College. He knew P.Q.R.'s identity and was eager to meet the young man. And so a few days after his father left, William Thomson called on Gregory to discuss not only a brief addendum to his first P.Q.R. paper, which subsequently appeared in the November *Journal*, but also more advanced work that he had developed over the summer and wished to submit to Gregory's editorial scrutiny.

Back in Glasgow, meanwhile, James Thomson had a visit from Archibald Smith, son of a local sugar merchant. Smith was about to return to Cambridge where he too was a Trinity mathematician, having graduated a few years earlier. He dropped in on Thomson after being startled to learn from Gregory that the author of the P.Q.R. paper was none other than Thomson's son. He "asked your age, and was surprised you could have written it," James Thomson wrote proudly to William. Toward the end of November, a bare month after William's arrival at

Cambridge, Gregory and Smith came by his undergraduate rooms to talk about mathematics. "It was certainly a great honour for a freshman of St. Peter's to have two fellows calling on him," he wrote blithely to his father. "They staid, I suppose, nearly three quarters of an hour." But success had come so easily to William, even at this young age, that he scarcely wondered at it.

***

The previous summer, 1840, James Thomson had taken all six of his children on a holiday to Germany. The older children combined recreation with education. They studied German with a tutor and went on hikes, good for body and mind. Their father pointed out geological features and notable plants along the way. Continental expeditions were unusual for the times and, for a large family, expensive and cumbersome. James Thomson was a thrifty man and only moderately affluent, but for the improvement of his children he would spend what he must. The summer before he had taken his clan to Paris to see the sights and to improve their French. Leaving the four boys behind with a maid and tutor, he took off with his daughters to explore southern Germany and Switzerland on horseback. Elizabeth was then 20 years old, Anna 19. Young women, he told his daughters, had little opportunity to travel by themselves, and he wanted to give them this chance for adventure while he could. Back home, sober Glaswegians regarded them as brave and unusual girls and invited them out to dinner to tell of the exotic sights they had seen. The trip was doubly memorable for Anna. At a hotel in the Grindelwald in Switzerland they ran into William Bottomley, who had been a student of her father's when he had taught mathematics in Belfast, before moving to Glasgow. Anna stayed up late talking to this "delightful young man." They married in 1844.

On these trips William Thomson halfheartedly kept a diary. Before going on to Paris, the family had stayed almost a month in London, where Robert, the youngest son, underwent surgery to remove a stone. This was 19th-century surgery, before anesthetics. The boy, 10 years old, was trussed up firmly on the operating table while the doctor opened him up. Elizabeth waited at the door, alarmed by his cries and moans, and rushed in to comfort him as soon as the ordeal was over. William and his father, meanwhile, spent the morning touring the British Museum.

The danger of infection, often fatal, hovered over Robert for several days, but he was quickly on the mend. In the meantime the other children saw the sights. William and Elizabeth were up and about early, feeding the ducks in St. James's Park or exploring nearby Westminster Abbey. William recorded in his notebook a few of his impressions of London, revealing a pleasantly impish humor that largely disappeared from his later notes and correspondence. Once when he and his sister were in Hyde Park they heard that Queen Victoria was in the vicinity "and we accordingly determined to wait and see her. After waiting for about half an hour, we succeeding in catching a glimpse of the top of her bonnet, a most overwhelmingly interesting spectacle, so much so indeed, that I forget its colour." A few days later, on his own, he spotted the royal carriage again, chased after it, and was amply rewarded: "As the Queen's head was averted I had the inexpressible felicity of seeing her most gracious bonnet."[1]

On the way to Germany the following summer, William again made a few desultory notes. On May 21 the family went by steamship from Glasgow to Liverpool then to London by train and on to Rotterdam by steamer, and finally by steamboat down the Rhine to Bonn, arriving close to midnight on the last day of June. Then William put his notebook down and left the remaining pages blank.

But he had no need to write down his impressions of Germany. The few weeks he spent there remained bright in his memory until he died. There his career in physics began; there he acquired a perspective (it would be too grand to say philosophy) on mathematical science that he maintained, for good and ill, throughout his life.

***

In 1822 the Frenchman Jean-Baptiste Joseph Fourier published his great book *Théorie Analytique de la Chaleur* (Analytical Theory of Heat), one of the milestones of physical science. At that time, and for a couple of decades after, the nature of heat resisted understanding. Most natural

---

[1] But perhaps this is not irony, after all. Perhaps a glimpse of the royal headwear really was thrilling to a 15-year-old provincial boy. Or perhaps William felt that he ought to be thrilled and duly recorded that he was.

philosophers held that it must be some kind of tenuous subtle fluid. Heat flowed, after all, and could be stored up and transferred with some degree of control from one place to another. It resembled a fluid more than it resembled anything else. This putative heat-fluid acquired the name "caloric."

But if scientists didn't understand what heat was, they knew a good deal, practically speaking, about what it did. Fourier's particular genius was to construct a quantitative mathematical theory of the behavior of heat on a foundation of knowledge derived from observation and experiment, so that ignorance of the nature of heat was no impediment. This was an innovation, philosophically as well as scientifically. Mathematicians had always worked in an axiomatic way—start with principles and deduce consequences—and when they came to apply themselves to natural phenomena, they aimed to apply the same style of reasoning. But if no principles or axioms were known, then how to begin?

Direct observation, on the other hand, showed that heat flowed more readily through some substances than others, was retained by some materials longer than by others, and flowed faster where the difference in temperature was greater. Building on these elementary facts Fourier saw how to create a general theory of heat flow. Imagining a three-dimensional body with heat distributed within it, so that it was hotter in some places, cooler in others, Fourier came up with the crucial concept of the *temperature gradient*. Just as water flows downhill along the steepest incline, so heat moves in the direction along which temperature falls most sharply. In turn, the movement of heat changes the temperature pattern within the body. Ultimately heat will flow within a closed body so as to iron out all differences. When the temperature is the same everywhere, there is no gradient at any point, and heat will no longer move about.

When he was just 15, William Thomson heard the Glasgow astronomy professor John Pringle Nichol praise Fourier during his lectures. He asked Nichol whether he should read the *Théorie Analytique*. "The mathematics is very difficult," Nichol cautioned, but William got the book from the library anyway and set himself to understand it.

How easily he read and digested Fourier's theory depends on which reminiscence we are to believe. In 1903, speaking at the unveiling of a stained glass window memorial to Nichol, the 79-year-old Lord Kelvin recalled that "in the first half of the month of May 1840 I had, I will not say read through the book, I had turned over all the pages of it." But to his biographer S. P. Thompson three years later he offered a more robust account: "On the 1st of May . . . I took Fourier out of the University Library; and in a fortnight I had mastered it—gone right through it."

Either way, Fourier undoubtedly had a formative influence on William Thomson's scientific thinking. He saw that the great virtue of the Frenchman's work was that, despite his title, the book did not in fact provide a theory or explanation of heat. As Fourier proclaimed: "Primary causes are unknown to us; but are subject to simple and constant laws, which may be discovered by observation, the study of them being the object of natural philosophy. . . . The object of our work is to set forth the mathematical laws which [heat] obeys. . . . I have deduced these laws from prolonged study and attentive comparison of the facts known up to this time. . . ."

In other words, it is not necessary to understand the true nature of a physical phenomenon; instead, one observes and measures how it behaves and devises mathematical laws accordingly. That such an analysis yielded powerful and general results struck Thomson with the force of youthful revelation. Fourier's words became a mantra to him, the bedrock of his view of physics. He acquired a lifelong detestation of speculation or metaphysics. Any scientific proposal must be grounded in a combination of established principles and empirical facts, and must yield mathematically rigorous results. Fourier's magnum opus was some 400 pages long, but his new theory occupies only a small part of the book. The bulk of the treatise consists of numerous calculations of heat flow in all sorts of geometries (blocks, rods, spheres, and so on) to show the versatility and universal validity of his methods. It is not so implausible after all that William may have "mastered" the volume in a couple of weeks: the elementary ideas, once grasped, make evident good sense, and the mathematically acute young man would not have found Fourier's repetitious calculations forbidding.

Not everyone embraced Fourier. Philip Kelland, the Cambridge-educated professor of mathematics at Edinburgh, published in 1837 his own *Theory of Heat*, in which he claimed to find contradictions and inconsistencies sufficient to invalidate many of Fourier's findings. A couple of days before the Thomsons left Glasgow for their trip to Germany, a copy of Kelland's treatise came into William's hands. He was, he later recalled, "shocked to be told that Fourier was mostly wrong. So I put Fourier into my box, and used in Frankfort [sic] to go down to the cellar surreptitiously every day to read a bit of Fourier." Surreptitiously, because he was supposed to be learning German, but when his father discovered that William was avoiding one lesson in order to delve into something much deeper, he could hardly be displeased. And any displeasure evaporated when William jumped from his seat one day and declared abruptly, "Papa! Fourier is right, and Kelland is wrong!"

Kelland stumbled not over Fourier's theory itself but over a novel method he used to solve the equations of heat flow. The trick involves the construction of what are now called Fourier series. The vibration of a violin string furnishes a classic example. The fundamental musical note derives from the oscillation of the string as a whole—its ends fixed, the center point moving up and down with a fixed period. But then there are the higher harmonics: the center point stationary and the two halves moving in opposite directions, one going up while the other is going down; then the string divided into thirds, quarters, and so on. At any moment, the shape of the violin string represents the sum of a series of simple waves with successively smaller wavelengths.

Any smoothly varying mathematical function defined over some finite length can be likewise represented as the sum of an infinite series of waves with suitably chosen amplitudes. In his book, Fourier used this method to solve many examples of heat flow. He might imagine, for instance, a cylindrical rod, initially at the same temperature throughout, with one end abruptly brought into contact with some body at a lower temperature. Heat flows out of the rod, and a gradient develops along it. To determine the mathematical form of this changing temperature gradient, Fourier found it easier to calculate the components of a suitable sum of waves: a Fourier series. The technique has become a standard tool in applied mathematics.

Kelland didn't understand it. Fourier was partly to blame. He constructed series in a number of slightly different but essentially equivalent ways and would jump from one to another, depending on which was more convenient for a particular problem, without always making it clear what he was up to. But William Thomson had the wit to see that Fourier reached the right conclusions despite his occasional sloppiness and showed that the Frenchman's slips and omissions were not fatal. Kelland, by contrast, simply saw the problems and stopped dead. Where Kelland displayed a pedantic sense of logic, Thomson demonstrated real insight. He showed himself more acute than Kelland and more rigorous than Fourier. He provided simple proofs of some assumptions that Fourier had made but not verified. This was bravura from anyone, let alone a boy who had celebrated his 16th birthday only a few weeks earlier.

William quickly convinced his father that indeed Fourier was right and Kelland wrong. James Thomson, a good if not original mathematician, saw that this was a substantial result. His son had provided a clear and reasoned decision in a dispute between two eminent men. His sharp analysis warranted publication in a mathematical journal.

Back in Scotland, father and son worked up a paper explaining Kelland's errors, and early in 1841 James Thomson sent it off to Gregory, hoping it could appear in the *Cambridge Mathematical Journal.* Established just four years earlier by the young mathematicians Gregory and Archibald Smith, the *Journal* aimed to provide a venue in English for the new kind of mathematical physics that the French especially were developing. It was at James Thomson's insistence that William disguised himself as P.Q.R. The letters had no particular meaning, except that they are often used as a triplet of variables in three-dimensional mathematical problems. This anonymity was ostensibly for his son's benefit, as James Thomson apparently thought it inappropriate that a boy should publish openly in a scholarly journal. He may also have wished to spare Kelland, a fellow professor, the embarrassment of having his errors pointed out by a child.

As a matter of propriety, Gregory decided Kelland should know the name of his accuser and see the paper before it went into print. So James Thomson wrote directly to Kelland and received at first a cool response: "As to the insertion of the paper in the journal I think Mr G did quite right in corresponding with you first for two reasons. 1. Because an author never gets any credit for rectifying blunders. 2. Because the plain wording of the remarks is not quite what should appear in a periodical lest it should awaken the wrath of parties concerned & the blame fall on the editor."

Kelland added some technical criticisms, and James Thomson, always ready to act the diplomat in pursuit of larger aims, agreed to remove some of the phrases that had irked the other man. He further mollified him by declaring that William's "sole object is to establish what is true, and to remove any false impressions with regard to Fourier."

Privately, to James Thomson, Gregory agreed that "the flippant manner in which Mr Kelland speaks of Fourier would deserve pretty strong terms of reprobation." To his credit, though, Kelland saw the error of his arguments and the correctness of William's and quickly agreed to publication. "I am very much pleased with it and think if he works it up well into a paper it will be most interesting," he wrote, after James Thomson had soothed him. "Send my regards to your son," he added, for his "great service to science." Although Kelland played no further role in William's life, his sister Elizabeth says that the two became good friends later on.[2]

James Thomson's tact made what could have been an awkward scientific debut into a rather smooth performance. The paper appeared in the *Cambridge Mathematical Journal* of May 1841. Though admittedly written with James Thomson's help, it displays an assured, straightforward manner. After describing briefly the problems Kelland raised in his "excellent *Treatise on Heat*," William immediately showed that Fourier's answers are right, even though some of his arguments appear patchy. "I have examined the other series given by Fourier, on this subject, and they seem all to be correct, with the exception of misprints and mistakes in transcription, which, unfortunately, are very numerous," he wrote. In one case he gave a detailed argument to show that Fourier must have done a calculation correctly, even though some of the intermediate steps printed in the *Théorie Analytique* are wrong. There is nothing apologetic or obsequious about the paper. William states his purpose, writes out his calculations, and presents his conclusions. It is an adult work.

***

During his three undergraduate years William published a dozen papers in the *Cambridge Mathematical Journal*. He read eagerly and stud-

---

[2]Robert Louis Stevenson, an Edinburgh undergraduate in the early 1870s, remembered Kelland as a "frail old clerical gentleman, lively as a boy, kind like a fairy godfather, and keeping perfect order in his class by the spell of that very kindness" (R.L.S. in *The New Amphion*, 1886).

ied hard, at first with his college tutor Cookson and then for two years with William Hopkins, a highly regarded private tutor. He was never idle. For exercise he went on the river and strode for miles about the dull Cambridgeshire landscape. His fellow undergraduates would visit his rooms, or he theirs, and they would talk mathematics, dabble in political questions or other news of the day, discuss religion, perhaps touch on Shakespeare or the classics or lighter reading. He took up the cornopean (also cornopiston, from the French *cornet à piston*), a kind of French horn, and in 1843 became a founding member, later president, of the Cambridge University Music Society. Sometimes he spent too many hours on the cornopean and regretted that he hadn't read as much mathematics as he might have. Sometimes he read for so long he needed to walk or row to refresh his mind. But always he got his work done. A lifelong habit of incessant activity took root.

Despite William's precocious ability and prodigious achievements, James Thomson suffered from a constant fear that Cambridge would seduce his son away from a rigorous intellectual path into a dissolute and purposeless life of wine parties, rowing, and the reading of light novels. Born to a poor farming family in what is now Northern Ireland, James Thomson had doggedly used his intellectual talents to build himself a sound and solid life, resisting along the way any distraction. William, by contrast, overflowed with almost casual brilliance. He devoted hours to his studies but had nervous energy to spare. Rowing, walking, and music were not distractions but essential recreations.

To his father, though, these extracurricular activities represented time and effort not applied in laying the foundation of a secure career. For some significant proportion of Cambridge students in those days, undergraduate life was devoted mainly to forming friendships and connections, learning how to comport oneself at afternoon tea parties, playing rugby or cricket, boating on the Cam, carousing and drinking in the evening—anything but studying in earnest. There were no entrance exams to the university. Anyone who had money and preferably a helpful family connection could enroll as a student.

Through most of the 19th century, the Cambridge student body fell into three roughly equal divisions. About a third were the sons (no women, of course) of the gentry, with private means, a family Oxbridge tradition, and no urgent concern to find a career or profession. These students generally left the university with an ordinary rather than an honors degree, which still bestowed on them the right to regale friends and family with stories of their time at Cambridge. Another third were poor students, clever but from meager backgrounds, who survived on scholarships or charity and struggled to live while they devoted themselves to studying. For such people Cambridge was a lifeline from poverty, to be grasped securely and never let go.

Constituting the remaining third were the children of what we would now call middle-class professionals: doctors, lawyers, clergymen, headmasters, and the like. William was of this group. James Thomson never ceased to worry that his son might through inattention or complacency slip back down the social ladder that he had so determinedly ascended. Brilliance was all very well, but what brought success in the world was the correct attitude. He bombarded William with cautions and admonitions:

> You know my views about a strict and proper economy, not merely on account of expense, but also on account of your own health and habits. At the same time, *always making moral correctness and propriety your aim above all things else*, you must keep up a gentlemanly appearance and live like others, keeping, however, rather behind than in advance.
>
> Recollect my maxim never to quarrel with a man (but to waive the subject) about religion or politics.
>
> Never forget to take every care in your power regarding your health, taking sufficient, but not violent exercise. In "your walk of life" also, you must take care not only to do what is right, but to take equal care always to *appear* to do so.
>
> Healthful and innocent exercise and amusement, I wish you, of course, by all means to take in a suitable degree; but, above all things, take care to be moderate and wise in the formation of your notions and habits.

In Cambridge he saw his son beset all around with temptations and perils that could upend his promising career and destroy his future at a stroke. Even ice skating was on the list of dubious recreations. In December 1841, learning that William had been out on the frozen river, his father let fly with a paragraph that jumped from one lively fear to the next. He was concerned for William's safety on the ice, he wrote, but hoped "farther, that it will not lead you into company that will injure or relax your moral feeling. I am sorry to hear that you have been boating—not on account of the thing itself, as I think there can be no danger, but that you may be brought into loose society, a thing that would ruin you forever. I find that Ayrton [another Scottish undergraduate] goes to no wine parties, because of the excesses and other evils to which they lead. *At present,* I do not say, you should go to none, unless with fellows; *but you should scarcely go to any others*; and if you do go, observe the strictest caution, and always tell me about any thing you find. In more advanced years, you will see that my cautions are well founded."

William replied with casual reassurance: "With regard to wine parties, I have gone to as few as I possibly could, and at any to which I have gone there has not been the least approach to excess. . . . I have given no wine parties, or indeed any parties yet, but I suppose I must return some of the invitations next term." Although he didn't row with the college team—"an idle and extravagant set," he agreed—he took to the water as often as he could, reassuring his father that "I always row by myself in a funny, (or as it is called skulling. . .) or at least go in a two-oared boat, with some friend with whom I should otherwise be walking."

William had a fine sense of what he could get away with. In February 1842 he wrote to his father with the startling news that he had, without permission, spent all of £7 on a secondhand boat, "built of oak, and as good as new." This extravagance, he claimed, actually represented fiscal astuteness, since he would no longer need to hire a boat. His father was shocked by this insubordination. "You are quite right in anticipating that I would be surprised," he wrote. "You allowed yourself to be cajoled and probably cheated. . . . Seven pounds for a tub that will hold only one person!!!" He called his son "a soft freshman" for being duped.

He threatened to make William return the boat immediately and get his money back. But with his remarkable and gifted son, the instinct to be firm ran up against a habit of indulgence. He consulted Cookson, the tutor, on the pros and cons of boating, evidently received a favorable opinion, and grudgingly paid up. The money came to William in a letter from his younger brother John, who reported the reaction of his mischievous sister Anna.

"I hope [the boat] is to your liking but it is not at all to Anna's as she would like exceedingly that it were broken up, for firewood, or employed for a *washing tub*, as, till that time she will be constantly on the look out in the obituaries for the drowning in the Cam of an extraordinarily clever, young Cantabridgian: and, if I say to her that you could surely swim across the Cam, she says that I know quite well that you might take the cramp."

A month later James Thomson had softened entirely after hearing again from Cookson, who wrote "so favourably and so kindly regarding you" that he sent another £10 for the boat and other expenses. He apologized a little for his "admonitory style" but told William that "at your period of life, and placed as you are among many persons of different characters and habits, you require to be most circumspect, and to be firm in your adherence to what is right and proper, and in resisting every advance to what is bad." Rowing itself, a vigorous and manly activity, James Thomson could not honestly object to. It was the lurking fear of loose morals and roguish young men that animated his concerns.

James Thomson's numerous, lengthy, and repetitious letters to his son make him seem cautious to a degree, humorless, and puritanical. But he was not in person as dour as all that. John Nichol, son of Professor Nichol and playmate to the younger Thomson children, recalled Professor Thomson thus: "Good-hearted, he was shrewdly alive to his interest, without being selfish, and would put himself to some trouble, and even expense, to assist his friends. He was a stern disciplinarian, and did not relax his discipline when he applied it to his children, and yet the aim of his life was their advancement. . . . He was uniformly kind to me, and I owe him nothing but gratitude."

His lowly origins colored James Thomson's personality and anxiety over his children's future. Born in 1786 on his parents' farm near the small town of Ballynahinch, County Down, he was the great-great-grandson of a John Thomson who had fled religious persecution in the lowlands of Scotland around the time of the English Civil War, at the end of which, in 1649, Oliver Cromwell had sent Charles I to the executioner's block. In Scotland, Protestants at first sided with Cromwell against the overbearing Charles, then against him as he too attacked Scottish political and religious traditions. Many Scots fled to the northern counties of Ireland.

Agnes King, daughter of William's sister Elizabeth, says that the Thomson family was of "the fine old stock of Scottish Covenanters." These were adherents to the King's Covenant, which James VI of Scotland (later James I of England and father of Charles I) had signed in 1580 in formal renunciation of the Pope and the Catholic Church. The religious ramifications in Scotland of the Civil War in England verge on the incomprehensible (as well as the Covenanters, there were the Protesters, the Remonstrancers, the Resolutionists, and others), which is one reason the repercussions linger in Northern Ireland to this day.

Some of the Thomsons moved on to America. Others, James's ancestors, stayed where they were even after Scottish affairs had quieted but continued to think of themselves as ancestrally Scottish. Over the years their religious ferocity abated into mainstream Presbyterianism. In 1798, when James Thomson was a boy, he witnessed bloodshed at the Battle of Ballynahinch, when English soldiers put down a brief Irish rebellion inspired by the recent French and American revolutions. His family provided food for the rebels, but the insurgents were quickly and easily defeated.

James Thomson acquired a fierce disgust of religious favoritism and sectarianism. He devoted his considerable talents and self-discipline to the furtherance of his own life, and though he was firm in his principles he always aimed to resolve difficulties by diplomacy rather than protest.

James received a little education from his father but went on to teach himself from books and later enrolled at a local Presbyterian school, taking in the standard improving diet of classics and mathematics. He was unusually bright, but even more remarkably diligent. While still taking higher classes at the country school, he served as assistant teacher to the lower forms. Later he became a master at the school in the summer months and for six years sailed to Glasgow every autumn to attend university there, which ran for a single long session from November to May. He obtained his M.A. in 1812 and two years later became a teacher of mathematics, geography, arithmetic, and bookkeeping at the Belfast Academical Institution. The following year he became professor of mathematics. In 1817 he married Margaret Gardner, daughter of a merchant family, whom he had met in Glasgow. He built a house in Belfast and started a family. Elizabeth came first, in 1818, then Anna in 1820, James in 1822, and William on June 26, 1824. Three younger children followed: John in 1826, Margaret in 1827, and Robert in 1829.

He began writing textbooks, not only on elementary arithmetic but also on geography, astronomy, and more advanced mathematical subjects, including algebra, geometry, trigonometry, and differential and integral calculus. He had a way of bringing mathematics alive through illuminating examples, and his books became standards in many schools and colleges. They sold well enough to substantially increase his income, and some remained in print for decades. As late as 1880 his sons James and William together edited the 72nd edition of his *Treatise on Arithmetic in Theory and Practice*. In the preface (written for the 23rd edition of 1848), James Thomson senior decried the aridity of teaching by rote memorization, by which mathematics, "peculiarly fitted to call forth and improve the reasoning powers, is degraded into a dry exercise in memory."

The liveliness and effectiveness of his teaching brought him renown. He lavished the same care and attention on the education of his children. He rose at four in the morning to work on his textbooks, taught at the institution during the day, and in the evening tended to his offspring. He began by reading to his children from the Bible and the classics, and introduced them, the girls as well as the boys, to arithmetic, geography, botany, and other elementary subjects. He resorted to tutors only for music, dancing, and French. As the children grew older, he would read to them from newspapers and magazines of current events and encourage them to comment on both style and substance. The older boys, James and William, he particularly encouraged in mathematics, and both proved quick—William quicker than James, James more thorough than William. At the ages of 8 and 6, the boys took a few classes at the Belfast Institution and took the top two prizes. In a presentiment of what was to become a common pattern, William came first ahead of his older brother. Elizabeth scrubbed and washed the wriggling William and dressed him in white trousers, black jacket, and tie, leaving James to dress himself similarly, then proudly marched them off to the institution to receive their awards.

James Thomson's devotion to his children redoubled after his wife died in May 1830, having never regained her health following Robert's birth. During their mother's decline, Elizabeth recalled, James Thomson strained to keep his grief to himself and present only a sturdy figure to his children. Once, unseen, she saw him emerge from their mother's bedroom and was "frightened to see my beautiful father, so tall and strong, standing outside the door pressing his head against the wall." The moment passed. He collected himself and went back to his wife's bedside.

That very evening the five oldest children were summoned to their father's study. "He was sitting there alone, at the side of the fire," Elizabeth recalled years later. "As the little troop came into the room, he opened his arms wide, and we ran into them, and he clasped us all to his heart. I was the tallest, and his head dropped on my shoulder, and he said with a choking voice 'You have no Mamma now.' He held us a long time so; his whole breast heaving with convulsive sobs. Then he gathered the two little ones, William and John, on his knees, and kept his arms tight round us all,—his head resting on the cluster of young heads pressed closely together; and there we remained in silence and darkness, except for the glow of the dying embers, till at last the nurse came and asked leave to put us to bed."

His youngest daughter, Margaret, had never been well and died the following year, not quite four years old. From this time on James Thomson was "both father and mother to us, and watched over us continually," Elizabeth said, although she herself, as the oldest child, took on a maternal role and was always a more serious girl than playful Anna. To William, not yet six, the death of his mother only briefly darkened the happy progress of his childhood. He recalled nothing of her in later years.

The following year, 1832, the clan moved to Glasgow where James took up the professorial position he held until he died. He and his family got off to a difficult start. That year cholera raged through the city, as it periodically did, bursting out of the slums to threaten the whole population. The university at that time was near the old center of Glasgow, in an area that had been engulfed by cheap tenements for the inrushing factory workers while the more affluent Glaswegians drifted west. The mushrooming industrial cities of Great Britain all had their share of disease and squalor and drunkenness and crime, but Glasgow was among the worst. Many years later, when the university had moved to a new site, those who had known it in the old days could afford to let nostalgia color their recollections. As one long-serving professor recalled many years later:

> There was something in the very disamenities of the old place that created a bond of fellowship among those who lived and worked there. . . . The grimy, dingy, low-roofed rooms; the narrow, picturesque courts, buzzing with

student life, the dismal, foggy mornings and the perpetual gas; the sudden passage from the brawling, huckstering High Street into the academic quietude, or the still more academic hubbub of those quaint cloisters, into which the policeman, so busy outside, was never permitted to penetrate . . . the roar and the flare of the Saturday nights, with the cries of carouse or incipient murder which would rise into our quiet rooms . . . these sharp contrasts bound together the College folk and the College students, making them feel at once part of the veritable populace of the city, and also hedged off from it by separate pursuits and interests.

The Thomson family took up residence in one of the 11 faculty houses forming a tight quadrangle known as College Court. It was a "dingy old place," John Nichol remembered. They hunkered down for weeks until the cholera had burned itself out. Elizabeth recalled with a shudder the dead-cart taking bodies away at all hours. On top of this James Thomson discovered that his regular salary was far less than he had expected. Instead his income came largely from fees collected from the students who attended his classes. Few came at first to hear the new professor, and family legend records that his Glasgow position, far from solidifying Thomson's entry into the professional classes, cost him money for the first year.

But for James and William, the arrival in Glasgow marked the beginning of their intellectual lives. Huddled in College Court, they made the acquaintance of other academic families, notably that of John Pringle Nichol, who introduced William to Fourier a few years later. John Nichol, the professor's son, remembered Elizabeth and Anna as "both clever, good talkers and sketchers." One of them (he diplomatically doesn't say which) was "very pretty." In sketches done by Elizabeth around this time both girls look charming, though the artist gives herself a slight edge. With the four boys they formed "a pleasant and happy group," according to Nichol.

William and James began to sit in on their father's classes. If their fellow students were surprised to see an 8-year-old in their ranks, they were astonished when the professor posed a difficult question that left the class silent except for the small fair boy who jumped up from his seat pleading, "Do, papa, let me answer!" He had always been a blessed child, so it seemed to Elizabeth. He was a bonny baby, fair-haired and blue-eyed. In Ireland a local artist had borrowed him one day as a model for an angel, suitably adorned in frills and ribbons. As a 2-year-old he was once discovered sitting on the floor, staring at his reflection in a mirror and cooing to himself, "P'itty b'ue eyes Willie Thomson got!"

James Thomson doted on this most adorable of his sons, as Elizabeth records in a curious passage from her memoir: "William was a great pet with him—partly, perhaps, on account of his extreme beauty, partly on account of his wonderful quickness of apprehension, but most of all, I think, on account of his coaxing, fascinating ways, and the caresses he lavished on his 'darling papa.' When our father came in he would run to him, and jump about him like a little dog, exclaiming, 'Oo's nice good pretty papa, oo's nice good pretty papa,' and when his father stooped to greet him, the child would fling his arms about his neck and smother him with kisses, and stroke his cheeks endearingly. He had not words adequate to express his affection, and tried every conceivable way to make it felt. And this was not occasional demonstration; it was his constant habit, and had been from infancy. Sometimes the others thought there was a little affectation in this, especially when he used baby language after he could speak quite well; and we laughed at him, but he never heeded."

This odd behavior, Elizabeth claims, excited no jealousy or resentment among the other siblings. William was a sweet-natured child. His siblings were proud of their beautiful and bright brother and pleased that he brought such obvious happiness to their recently bereaved father. In 1834, when William was 10 and James 12, they enrolled formally as Glasgow University students and frequently won the top prizes in their classes, in classics as well as mathematics. Most often, as in Belfast, William came first, James second. Nonetheless, William did not become spoiled or vain. His tutors and fellow undergraduates at Cambridge recalled him as a charming and sociable young man. "A most engaging boy, brimful of fun and mischief, a high intellectual forehead, with fair, curly hair and a beauty that was almost girlish," recalled one contemporary years later.

Consciously or not, William learned through his childhood how to use his charm and his father's affection to get his own way. After his unauthorized purchase of the "funny," he frequently reminded his father how favorable rowing was for his health and therefore also his studies. "I have been reading moderately, and skulling a good deal in this vacation, so that every one tells me I am looking much better than I did some time ago. Today, just before Hall, I returned from a skull of fourteen miles . . . and I am not in the least tired, but I shall be in excellent condition for reading in the evening," he reported, and a few days later added, "I find that I can read with much greater vigour than I could when I had no exercise but walking, in the inexpressibly dull country round Cambridge." About this time he recorded his weight as 8 stone 10 pounds (122 pounds) in his rowing jersey.

These little reminders paved the way for William telling his father, in May 1842, that he had used more of his tuition money to buy out his partner's half-share in the boat. His fellow undergraduate, he explained, hadn't rowed very much and when he did had damaged the boat and broken an oar, so that as before William's new expenditure was in truth an economy. "I am sure you will perfectly approve of that way of spending the money since I have found the skulling, after two or three months trial, to be most beneficial to my health and reading," he confidently asserted. His father grumbled, then paid up.

***

William's reading focused, of course, on the study of mathematics, the exception being an irksome examination colloquially known as the "little-go," which all honors students had to pass in their second year in order to demonstrate at least a passing acquaintance with Latin and Greek authors as well as works of a general religious or philosophical flavor.[3] For this William boned up on a section of the Aenead and a little Xenophon, recording in his diary that he had been practicing on the cornopean a good deal "to relieve my head from the seediness concomitant upon littlego subjects." Despite some fear that the classics would trip him, he easily negotiated the little-go. Then it was all mathematics. For the second and third years he studied mainly with his private tutor, or coach, William Hopkins. He had two goals. He wanted first to learn as

---

[3]The physicist J. J. Thomson, an undergraduate in the 1870s, told of a Greek grammar written especially for the little-go, "which contained a long list of words which were irregular to the point of impropriety . . . not one half of which my classical friends had ever come across." The time spent in these studies, he says, "was utterly wasted" (Thomson, 1936, p. 35).

much advanced mathematics as he could, to build on what he had learned from his father and by his own initiative and develop a wide-ranging and systematic command of the subject. James Thomson had his doubts about Cambridge mathematics—there was an excessive reverence for Newton and a consequent resistance to the new ideas coming mainly from France—but even so, it represented the pinnacle of mathematical attainment in Great Britain. There was no doubt in his mind that a Cambridge mathematical education was what his son needed, but still, there was some anxiety about entrusting the youngster's ripening brilliance to hands other than his own.

The second goal was to become "wrangler" for his year. This is the undergraduate who gets the highest marks in the mathematics honors examinations—the tripos, as Cambridge exams are still known, not from any tripartite nature of the subject examined but from the precarious three-legged stool on which examinees in olden days had to sit. Candidates for the mathematical tripos sat (in the 19th century at ordinary chairs and desks) a grueling series of eight lengthy papers undertaken over a period of six days. Nervous and sweating examinees were expected to spill out, in coherent manner, the most arcane and involved elements of the subject they had digested over the previous years. The top man was senior wrangler, the second junior wrangler; positions were reported down to 10th or even 20th wrangler. The London *Times* published the list. Being wrangler was a moderate sort of national honor as well as a university distinction.

Like all examinations but to an extraordinary degree, the competition for wrangler was a test of genuine knowledge, power of recall, concentration, nerves, and handwriting. Over the years it had acquired the qualities of a ritual, like the compulsory figures section of the Olympic ice-skating competition in which contestants must perform prescribed moves and jumps with precision and control. Technical mastery rather than originality or flair was the key. The well-coached wrangler candidate knew his essential mathematics but also knew how to write out stock answers to standard questions as concisely and rapidly as possible, in order to do as many problems as he could in the time available. It did not help, in the heat of the exam, to start thinking of more profound or comprehensive solutions than the one the examiners were looking for.

Nor was there any reward in perceiving valuable generalizations or wider implications of a narrowly defined answer. Compact, tidy handwriting was an asset. William Thomson wrote in a large though readable scrawl.

In the middle decades of the 19th century, because good teaching was rarely a high priority of the colleges, the system of mathematical coaching developed to a fine degree in response to the demands of wrangling. As a married man, Hopkins could not be a college fellow, but in the end he acquired a far greater reputation, not to mention a better income, as a coach than he would have done in a formal appointment. The sporting analogy is apt: A good coach secured high places in the exams for his pupils and thus attracted the better students in subsequent years. Hopkins charged £72 per student per year for twice-weekly sessions, might coach 10 or 12 students in total, often took additional classes during the long summer vacation, and thereby easily earned £800 a year or more—considerably more than the typical college fellows and a solid upper-middle-class income.

As with coaching for gentlemen who wished to row, the aim was to develop strength and stamina and the ability to perform reliably and repetitively strange motions that neither body nor mind would take up naturally. Coaches were generally men who had placed well in the wrangler competition in years past and who had a knack not only for training young men in the same art but also for predicting from one year to the next the questions that were likely to come up. Problems that would test the wranglers-to-be could come only from certain advanced branches of mathematics, yet they had to be solvable in the allotted time. An apt question, like a nifty crossword clue, was a praiseworthy construction in itself.[4]

Peter Guthrie Tait, another Scottish student who became a close friend of William Thomson, was senior wrangler in 1852. As one who had survived with distinction a difficult and painful ordeal, he was later scathing about the Cambridge system. "College Tutors and Lecturers take

---

[4]A modest theory: the popularity of cryptic crossword puzzles in England, especially among graduates of the older universities, testifies to the continuing influence of antiquated Oxbridge educational philosophy. The skillful deployment of arcane knowledge in a wholly artificial manner and in a deliberately inappropriate context—this is the key to solving cryptic crosswords.

but small part in the process of education," Tait told the students and faculty of Edinburgh University in 1866, where he was a professor. "Private Tutors, 'Coaches' there, 'Grinders' we should call them, eagerly scanning examination papers of former years, and mysteriously finding out the peculiarities of the Moderators and Examiners under whose hands their pupils are doomed to pass, spend their lives in discovering which pages of a text-book a man ought to read, and which will not be likely to '*pay*'. The value of any portion as an intellectual exercise is never thought of; the all-important question is—*Is it likely to be set?* I speak with no horror of, or aversion to, such men; I was one of them myself, and thought it perfectly natural, as they all do. But I hope such a system may never be introduced here."

The wrangler system, over the years, acquired a patina, a hushed mystique into which all new aspirants must be inculcated. J. A. Fleming, a Cambridge undergraduate in the late 1870s who became a pioneer of the pretransistor electronics industry, recorded in appropriately flat prose his experience of studying for the mathematical tripos. The student would visit his tutor, Fleming recounted, "at an appointed time, and the 'coach' gave him an examination paper of questions and supplied a hint or two as to how the questions were to be solved, and also marked certain chapters or parts in a text-book to be read. . . . Then the student went back to his own room and tackled the paper of questions, and read as requested. The next day we took our results to the coach, who noted successful answers, and gave a further hint as to the solution of unsolved problems. The coaches had great experience in forecasting the kind of question likely to be put in Tripos exams, and it would have been quite impossible to obtain a high place without their aid."

One can easily imagine this exchange taking place in a sepulchral silence, sheets of carefully annotated paper gliding back and forth across a polished table to the accompaniment of restrained gestures and indications, as if it were part of a training program for novices in some secretive and highly regulated religious order. The greatest proportion of senior and junior wranglers, as it happens, went on to careers as ministers in the Church of England. Mathematics afforded few professional opportunities, beyond a few Oxbridge fellowships and a clutch of professorial chairs at the four Scottish universities—that or schoolmastering. In any case advanced mathematical training was seen more as a kind of general strengthening of the mind than it was as preparation for a life of work with numbers and equations. The same was true of the classical tripos. Students of the classics learned by heart great stretches of Caesar and Cato and Ovid and Horace, and acquired the ability to ad lib a suitable Latin ode on any occasion. Men trained in this way were regarded as having intellects honed to less demanding tasks, such as running Her Majesty's government and directing the course of empire.

William Hopkins was a superlative mathematical coach. Merely seventh wrangler in 1827, he had by 1849 coached 17 senior wranglers and 44 top three places; his pupil E. J. Routh, senior wrangler in 1854, became a coach with 27 senior wranglers to his credit. Those who gained the top handful of places each year could, if they wished, find a pleasant college fellowship or work as a coach and then fashion a career producing more wranglers. Thus did wrangling perpetuate itself over the generations, and as often as not it was the competent but less imaginative men who went on to become fellows and coaches, while those with some other ability besides that of writing out long mathematical answers at great speed took up other careers.

Hopkins was an exception to this dreary practice. He had genuine scientific ambitions. He pioneered the application of quantitative mathematical methods in geology, which had until then been largely a descriptive science, like botany. He analyzed the earth's orbital motion, its rigidity and interior dynamics, the movement of glaciers, and most notably the distribution of heat within the earth. To his pupils he brought not just the tools for doing well at the tripos but also a deep appreciation of the nature of scientific problems and the use of mathematics in solving them. In the middle of the 19th century, the relevance of mathematics to science in general, as opposed to a few highly specific areas of physics, was by no means commonly accepted. Hopkins saw how rational analysis could be brought to bear on all manner of questions—but in the case of his exceptional pupil William Thomson, this was a superfluous lesson.

Even before he began his coaching with Hopkins, William had published a third paper in the *Cambridge Mathematical Journal* of February 1842. Where his first two papers, correcting Kelland's misimpressions of Fourier, were works of mathematics, his third was recognizably a piece of physics, and a sophisticated piece at that. Written in Scotland in August

1841, before he arrived in Cambridge, it owed nothing to Cookson, Hopkins, or his other tutors or advisers. In it he described an analogy between the heat flow formulation of Fourier and the way an electric field spreads across space between charged objects. The Frenchman Charles Coulomb had established in 1785 that the electric force between two charges, like the gravitational force between two masses, decreased in proportion to the square of the distance between them. Subsequently the self-taught English genius Michael Faraday, son of an impoverished London blacksmith, had devised an alternative portrayal of electrical interactions, employing what he called "lines of force." In Faraday's vivid imagination (he knew no mathematics to speak of and substituted an acute, largely pictorial way of understanding physical phenomena), forces between electric charges were conveyed along curved lines, something like intangible elastic strings; these lines, moreover, repelled each other and so distributed themselves as economically as possible through space, creating a tension that we now recognize as the electric field.

These were vague notions, and William Thomson remained skeptical for some time of Faraday's powerful but allusive insights. Nonetheless, he proved in his 1842 paper that, with physical quantities suitably redefined, he could use Fourier's mathematics of heat flow to portray the geometry of Faraday's lines of force. This was more than mathematical cleverness; it hinted that electric force "flowed" through space just as heat flowed through matter. It turned out, moreover, to be a powerful way of analyzing electric forces. Coulomb's inverse square law was fine for dealing with the simple case of two discrete electric charges but became intractable if one wanted to investigate more complex geometries—the force between an electrically charged sphere and a flat plate, for example. Fourier's treatment of temperature distributions became in William's adaptation an equally general way of dealing with distributions of electric charge. The method (as subsequently developed by Thomson and others) is still taught today. For so young a man to have devised it when both heat and electricity were so poorly understood was a remarkable step.

A fourth paper for the *Cambridge Mathematical Journal* was the first that William produced from Cambridge. Again it displayed sharp physical insight. The pattern of temperature within some body, Fourier had shown, would always become more uniform as time passed. Conversely, William realized, temperature must become less uniform, more irregular, as one turned the clock backward. He proved a striking result. A pattern of temperature in some object cannot in general be calculated backwards in time without limit, because mathematically impossible distributions of heat arise. To turn it around, a heat distribution existing at the present moment can be the outcome of an initial arrangement that existed only some finite time ago. This was a straightforward mathematical demonstration, but it was not long before important physical applications came to William's mind. Most notably, he applied this reasoning to the present state of heat within the earth (so far as it was known) and reached the conclusion that the earth could not have an unlimited past. This seems uncontroversial today; in the middle of the 19th century it was not. Even the application of mathematical reasoning to "cosmological" questions such as the origin of our planet struck many Victorian minds as close to sacrilegious.

While he was cramming relentlessly for the wranglership, William kept up a remarkable rate of publication. The subjects he had broached—the flow of heat, the geometry of electric fields, and the mathematical parallels between the two—formed the seeds of work that he developed much more deeply in the first part of his scientific career. It all rested on Fourier's principle of formulating mathematically sound arguments relating to observed phenomena. As his friend P. G. Tait remarked many years later, "Fourier made Thomson."

***

One of William's fellow undergraduates, reminiscing years later, reported that the startlingly accomplished young man was being touted as a senior wrangler just days after his arrival at Cambridge. William himself suffered doubts from time to time. For a few months in early 1843, halfway through his second year, he kept an intermittent diary of his undergraduate routine and habits, and more interestingly of his fears and anxieties.[5] A student from Germany named Ludwig Fischer aroused concern: "I must read very hard and try to be at least as well prepared as he

---

[5]This notebook is the only truly personal record of William Thomson's that has survived, and like his earlier diaries of the visits to Europe, it is for the most part half-hearted and desultory. Possibly he wrote other diaries, but I suspect not. Except during these anxious months of early 1843, interior rumination was not his thing.

is," William noted on February 15. He felt no lack of intellectual fire-power but worried that his incomplete education before Cambridge left him at a disadvantage. A week later, returning from a coaching session with Hopkins along with Fischer and another undergraduate, Hugh Blackburn, who came from a well-to-do Scottish family, he "was mali-ciously glad to find that Fischer had not done all the problems. Blackburn had got solutions for all, but nobody had given interpretations except myself." But his confidence swung up and down. "I am beginning gradu-ally to be violent in my apprehensions regarding Fischer since we have started Mech[anics]," he wrote a couple of weeks later, but just three days after that he reported to his father that "Fischer does not get on quite so well with the statical problems, as he did in Geom[etry] of 3 Dim[ensions], and if he continues so when we come to Dynamic, I shall not be so much afraid."

February and March of 1843 saw William, for the only time in his life, doubting the feasibility of a career in mathematics. In conversations with friends the idea came up that he might take up the law for a profes-sion, as the Glasgow mathematician Archibald Smith eventually did. "I have pretty nearly determined to go to the Chancery bar, if something else do not succeed, though I cannot get over the idea of cutting math-ematics," he confided to his diary on February 19 (employing a now obsolete subjunctive). A month later he jotted down the same dreary thought. He attended a court hearing in Cambridge, to get a feel for the thing, and was impressed by a lawyer's eloquent speech, less so by the somnolence of the judge.

This was a low time. He worried about Fischer beating him, ago-nized over his future, and from time to time was homesick and (a few months short of his 19th birthday!) nostalgic over past triumphs. A diary entry of March 14 finds him melancholy indeed. Looking over issues of the *Cambridge Mathematical Journal* containing his youthful work, he "spent an hour at least in recollections. I had far the most associations connected with the winter in w^h I attended the natural phil[osophy class, in Glasgow] and the summer we were in Germany. I have been thinking that my mind was more active then than it has been ever since and have been wishing most intensely that the 11th of May 1840 would return. I then commenced reading Fourier, & had the prospect of the tour in Germany before me."

In this febrile mood, William succumbed to another temptation: literature. "Blackburn I find is a great man for reading Shakespere, besides Beaumont & Fletcher, Ben Jonson, &c," he recorded on March 24, and a little later he sampled the dangerous fruit himself: "looked over some of Shakespere's poems, and have just seen enough to make me wish to see more." He even set aside his extracurricular study of the French mathematician Poisson for Richard III and Henry IV. A modest taste for Shakespeare was no cause for alarm, but Fischer was dabbling in materials far more perilous: "I went to see Fischer and found him reading [Goethe's novel] *Wilhelm Meister*. I got him to read me over some in German." Later that evening William went to bed but couldn't sleep and got up again. "I have been looking out of the window, and have got back my journal to endeavour to fix my impressions. The moon is shining brightly on the mist w^h lies on the meadow (like the like the [sic] silvery clouds we saw from Ben Lomond. I have been looking out of the window for a long time, and listening to the distant rushing of waters, the barking of dogs, and the crowing of cocks."

Now he was hooked. He picked up Fanny Burney's 1778 novel *Evelina*, a racy, breathless tale of society and manners, and a best-seller in its day. Evelina is a young woman whose mother died giving birth to her and whose high-society father, so she believes, has disowned her. She is raised by a loving but limited country parson, who reluctantly lets her be introduced into society by better-connected acquaintances. It's a ripping yarn, full of snobbery and disdain and gentlemen behaving badly. Inevitably, a young noble, Lord Orville, eventually marries Evelina, although not until she has survived various scrapes and perversely misunderstood his intentions.

Aspects of the tale may have struck a nerve with William Thomson. Like Evelina, he was a young provincial full of talent and promise. Where Evelina was trying to make her way in the beau monde of London and Bath, William was momentarily struggling in a Cambridge society quite different from the Glasgow world he grew up in.

More particularly, Evelina's guardian, the Reverend Villars, may have put William in mind of his own father. Villars worries that Evelina, introduced to the brilliance and excitement of society, would develop the taste for a life ultimately forbidden to her because of her lack of means and connection. Villars observes: "A youthful mind is seldom totally free from ambition; to curb that, is the first step to contentment, since to diminish expectation, is to increase enjoyment" and later "I would fain guide myself by a prudence which should save me the pangs of repentance."

Evelina is caught between Villars's cautionary admonitions and the excitement of new adventures with people of wealth and *ton*; she had to contend with "gentlemen" who, thinking her a naïve country wench, were eager to deceive and use her badly, as the euphemism had it. William's predicament was hardly so dire, but as his father constantly reminded him, Cambridge abounded in wealthy, idle young men who stood ready to draw unworldly young people into loose living and depravity.

Burney's novel is a genuine page-turner, and William's eagerness to stay up late and find out how it all ends is easy to understand. But this tale of the innocent abroad evidently awoke unfamiliar feelings. One evening after reading for a while, "I spent a long time looking at the sheep, and listening [sic] the birds, whose singing *filled* the air. . . . I got to bed with a very strange feeling."

The next day he had the usual round of walks and conversations with friends, then returning late picked up the book again: "On Sunday night, after I was left alone, I read Evelina till 2^h 20^m, when I finished it (the 1st novel I have read for 2 or 3 years)."

Evelina's tale, after a series of implausible coincidences and discoveries, ends in unalloyed happiness. Her true aristocratic parentage is revealed and her fortune restored; she reconciles with her father, who had been deceived into raising another young girl as his daughter; she marries the noble Lord Orville; the salacious Sir Clement Willoughby is sent packing. But as Burney sharply relates, the revelation that Evelina is a blue blood after all causes all the previously disdainful society women to embrace her as one of their own. William Thomson could expect no such absurd denouement. If he felt at all uneasy in Cambridge, out of his depth with people whose style and manners were strange to him, unsure of his future, he could expect no miraculous lifeline. The blithe passage of his youth and the easy renown he had achieved in Glasgow paled a little before the contemplation of his new environment and the future it promised. If he were to prosper it would have to be entirely his own doing, and for a few months that burden troubled him.

The day after he had closed the book on Evelina's thrilling adventures he went out to scull for a while, came back to his rooms, answered a letter from his father, then "went to Fischer. I find he has been reading Goethe to a great extent."

Wilhelm Meister, like Evelina, is a young person seeking a path in the world. He falls in with an itinerant set of actors but soon finds them shallow and vain and wonders where his future lies. Searching for his vocation is a solemn task for young Wilhelm: "What mortal in the world, if without inward calling he take up a trade, an art, or any mode of life, will not feel his situation miserable?" After a series of mishaps and adventures and coincidences even more implausible than Evelina's, Wilhelm gives up his dreams of the stage and begins to see a more responsible and worthwhile future. "To thine own self be true" is the message, with the implicit assumption that one's own true self will turn out to have merit.

Early in May, James Thomson wrote that he would be coming down to London over the summer because Robert again needed surgery to remove stones. The prospect of a visit from his father overjoyed William: "I hope most intensely that you will come here, instead of waiting in London to meet me." What passed between father and son is unrecorded. From London at the end of May, James Thomson wrote warmly of his visit: "With my trip to Cambridge I have been much gratified. I am glad to say that what I saw and heard of you was very satisfactory. Your success in your studies, and in making the most valuable of all acquisitions— character, has afforded me great pleasure." He was pleased with the circle of friends his son had formed and, clumsily but earnestly, offered William encouragement at rowing: "Tell me about your races. You see that though I consider it necessary you should give them up for the future, yet I feel an interest in them so far as you are concerned."

William broke off his diary while he was in Scotland for the summer. Interpretation is difficult not only because of its sparseness but also because pieces of it have been carefully torn out—sometimes a line or two, sometimes a paragraph, sometimes whole pages. The diary came to the Cambridge physicist Joseph Larmor in this condition after its author's death.

When William returned to Cambridge, on the portentous date of Friday the 13th of October, so too his anxieties returned. He tried his diary again, but even more sporadically than before. On Sunday he recorded a few casual remarks, then something more personal, of which only this remains, part of a page that was incompletely torn away:

> my thoughts have been a tissue of . . .
> may take places. I shall never . . .
> them here. When I was writing my journal, I endeavoured to
> keep some of them a secret from myself.

The next day there is a little more: "During last week I have been rather unsettled and not applied myself to reading nearly as much as I could have wished. The idleness however did not depend upon external circumstances as I have been in my room almost every night at or before 7, but partly to my having Wilhelm Meister. . . ." The rest of the page is torn off. When it resumes he mentions reading mathematics with Blackburn, and there is more about boats, but "I was very little interested about the race."

The date of this entry is unclear. In late 1843, though he resumed his studies with a vengeance, he also kept up with rowing. At the end of November William wrote cheerily to his sister Anna that although he had to catch up on a lot of work for Hopkins "I am practising now everyday for a great skulling race w^h will take place on Tuesday. The winner will get a cup of about fifteen guineas value as well as the honour of holding a pair of silver skulls in his hands for a year. I don't however aspire to such an honour, and I shall be very well satisfied if I come in second or third. Blackburn and I went on very regularly with Faust till James [his brother, who visited Cambridge briefly] came, but since that we have been rather interrupted. I have very seldom time now to take out my cornopean, but after the skulling is over, I mean to miss going down to the river one day at least in the week, and to have some practice on the cornopean."

His father and sisters, learning of his renewed interest in rowing, became fearful again, and William wrote back in his usual reassuring way. Anna replied this his letter "containing all your reasons for having joined the boat races . . . has one good effect at least—that of convincing us all that you are a most excellent logician." Despite his protestations he came first in his sculling race and won the Colquhoun Cup. To the end of his life he remained immensely proud of his athletic triumph. Fifty years later, when he was supposed to be replying to letters congratulating him on being raised to the peerage, he fell to reminiscing and declared that getting the cup was "better than winning in an examination."

He ended his diary for good with an undated half page, perhaps written after his victory on the Cam. He described his latest work with Hopkins, then announced—to himself? to his father's shade?—that from now on he would keep only a private mathematical journal. He concluded: "I need not stop to commemorate anything about boating last term on this, as I suppose I shan't forget it in a hurry. I shall at least remember everything worth remembering about it."

There is an echo here from Wilhelm Meister, who after giving up his pretensions for the stage, mused that he had stayed with the actors "longer than was good: on looking back upon the period I passed in their society, it seems as if I looked into an endless void; nothing of it has remained with me." Did William feel the same away about the time he had spent reading plays and poetry and novels? If, after he closed his 1843 diary, he suffered any further lack of confidence, it stayed securely within his own mind. No one would ever remark that he was a thoughtful, reflective, introspective man. He developed no taste for deep literature. These were childish things, and by the end of that tortured year he had put them all away. A diary entry from April 12 betrays, rather comically, a sense that William didn't have the stuff of dark yearning in his soul. After staying up late reading Shakespeare one evening he remarked: "This is a beautiful moonlight night, and my rooms are quite romantic. If I were only sentimental enough to enjoy it, I might lose a great deal of time looking out of the window."

He went on, during the rest of his life, to fill some 150 notebooks, bound in green, with mathematical calculations, jottings and ideas on science, drafts of papers and letters, occasional philosophical reflections—but nothing personal.

A sharp mind may make its course along many channels. William Thomson's mathematical prowess showed itself early, but at Cambridge, for a few months, as he turned 19, a different sensibility briefly awoke. By the end of 1843, however, these intrusive and unsettling feelings had been put to rest. Thereafter William worked every waking hour at science and showed only a passing and conventional interest in music or art. On the rare occasions, much later in life, when he read a novel, it would usually be a seafaring tale. He liked Beethoven, Mozart, and Weber and was especially fond of the Waldstein, one of Beethoven's perkier piano sonatas. He told one of his nieces off, light-heartedly, for playing Grieg on the piano while he was in the next room. Too modern! He didn't particularly care for Wagner, not so much because of the music but on account of the silliness of the plots.

***

The following summer, 1844, William Hopkins organized an extra session of mathematical coaching at Cromer, on the East Anglian coast overlooking the North Sea. "This is a very pleasant place, for England, and especially for Norfolk, w^h is rather remarkable for its dullness," William reported to his father. He lodged with Fischer and Blackburn in an old house near the cliff-top, which "cannot survive another winter, I think, unless great care is taken to protect the cliff below it with a wall." Jeopardy College, they called it. The students saw Hopkins every other morning; William saw him alone the intervening mornings. They scrutinized old exam papers. In the afternoons they read or went bathing in the sea. James Thomson complained about the expense—"Your lodgings are surely unnecessarily fine. For what you pay, we could have good lodgings on the Clyde for no inconsiderable family"—but paid up anyway, even when William had to write again a couple of weeks later asking for more money.

Whether studying or enjoying seaside amusements or both, William failed to write enough letters to his siblings, bringing a rebuke from his youngest brother, Robert: "I think you might write a *little* oftener. . . . John is the only one in this house you deign to write to except Papa & that when you want money." William was a seasoned student now. For all his father's scolding, he knew he could rely on getting £5 or £10 to sustain him, even if he had to plead, apologize, and ask twice. He was learning his independence, studying with his undergraduate friends, becoming closer acquainted with Hopkins, making plans to travel for a while after the summer session with his friend Blackburn.

He did not neglect his mathematics. Hopkins wrote to reassure James Thomson that William was not idling. "I am happy to say that he has given me entire satisfaction. His *style* is very much improved, and though still perhaps somewhat too *redundant* for examination in which the time allowed is strictly limited, it is very excellent as exhibiting the capacious-ness of his knowledge as well as its accuracy. I consider his place as quite certain at the *tip-top*. . . . I hope we shall be able to send him forth with such a character as few are able to carry with them from the University."

Back in Cambridge after a short break in Glasgow, William girded himself for the final run at the wrangler competition. He had now completed three years and was in his tenth Cambridge term. Honors exams took place early in the new year, and candidates had one last chance to study and cram before the grueling test came upon them. Expectation grew. William should be an exception among senior wranglers, so great by now was his reputation. "To have him come out as a common place senior wrangler," Hopkins had written to James Thomson from Cromer, would be "a grave disappointment."

By the middle of December, William's tutor, Cookson, was writing in the same vein: "Your son is going on extremely well. He says that his health & spirits are good though he is perhaps a little more pale than usual. . . . We fully expect him to be first and indeed it would be a great disappointment to all his friends and a great surprise to the University if he were not—I do not know of any candidate whom he has any reason to fear."

The candidate himself was not so sure. There may have been giddiness in Glasgow, but William urged calm. He wrote to Aunt Agnes Gall, his late mother's sister who was housekeeper for James Thomson's family: "I do not feel at all confident about the result, but I am keeping myself as cool as possible, and I think I shall not be very much excited about it when the time comes. One thing at least I am sure of is that if I am lower than people expect me I shall not distress myself about it, and if any of you lose any money on me I shall consider it your own fault for giving odds."

To his father a few days before the ordeal began he wrote: "The prospect is of course rather terrible, as all the three year's [sic] course of Cambridge reading is for the one object of getting a good place, so that in that respect anyone's whole labour may be lost very easily. I hope however that if I do not get as high as people expect, that it will not be much disappointment, as I think my time will still not have been quite thrown away. . . . I am sure that many others will be quite as well prepared and I am determined to be satisfied whatever may be the result, and I hope you will not be disappointed if I do not succeed well. . . . You need not be in any fear about my health as I never have been better than I am now."

**

On top of the pressure of the wrangler competition, another battle lurked in the back of William's mind, the result of a plot cooked up by his father with William's not wholly enthusiastic assent. William Meikleham, the Glasgow professor of natural philosophy, was old and ailing. He had been a professor since 1799 and was 70 years old when William started at Cambridge. Though not an original scientist, Meikleham had been a good teacher and, like James Thomson, a devotee of the modern French style in mathematical science. In 1839, however, his health had begun to fail, and his lectures were delivered by substitutes, first Nichol, the astronomy professor, and then David Thomson (no relation), a young Cambridge graduate. But as long as he breathed, Meikleham remained professor. The question of his successor discreetly arose, with James Thomson taking a close interest: He wanted a lucid teacher, a modernist, a scientific man in the new fashion, and from time to time he quizzed William about some of the Cambridge fellows he had encountered. Gregory was a possibility, but he died unexpectedly in 1844. Archibald Smith might serve, but he had exchanged his Trinity fellowship for a more lucrative career as a London lawyer.

As time went by and Meikleham clung to life, another possibility came to James Thomson's mind: his own son. In 1841, when William started at Cambridge, that idea was absurd. A couple of years later, with William's reputation rising not merely as a likely wrangler but as an increasingly prominent contributor to the *Cambridge Mathematical Journal*, James saw ever more clearly before him the prospect of his son as his professorial colleague. He found ways to throw out the suggestion to some of his closer Glasgow associates (including the professor of medicine, who regrettably for our story was also named William Thomson[6]).

William's attitude toward his father's scheming is hard to judge. James

---

[6]Along with James Thomson, William Thomson the elder, and David Thomson, all unrelated, there were also the brothers Thomas Thomson, professor of chemistry, and Allen Thomson, professor of anatomy.

Thomson, from his earliest moments, had plotted out his life with a view to security and income. William, raised in a comfortable household and aware of his own talents, worried less about the future. Something would turn up; he would find a way, as he always had thus far. If he did well as wrangler, a Cambridge fellowship was his. He conceived of spending some time in Paris, to make the acquaintance of the French scientists whose work and style he so admired. Like the young researcher today, he could imagine spending a few years here and there before settling into a permanent position. His father, on the other hand, knew how rarely a secure post came open. If the Glasgow chair went to another young man, it could be closed out to his son for the rest of his life. In the whole of Great Britain no more than a handful of comparable positions existed. So he kept the fire gently alive and wrote to William of the ups and downs of Meikleham's health.

Toward the end of 1843, when William still had another 18 months at Cambridge ahead of him, James Thomson delivered worrying news: "I am sorry to say that Dr Meikleham has a second attack of his distemper, and that, though he *may* yet get over it, he is considered to be in a most precarious state. I wish he were spared for two or three years longer. In such things, however, we have no controll [sic]." He asked William for his opinion of Gregory or even Hopkins. The great concern at Glasgow, he explained, was that Cambridge men were thought too abstract, too mathematical, too *superior*. They wanted a practical man, who knew a little bench science, and could teach to a less exalted student body than attended Cambridge.

William made only passing and noncommittal responses to these overtures. Next April his father reported a conversation with Dr. William Thomson, the medical man, concerning Meikleham's successor: "I felt . . . I ought to mention to him my views regarding you. . . . He was naturally quite struck with the idea of your youth, &c.; but he received the proposition as favourably as could be expected. He asked about your *experimental* acquirements, particularly in chemistry . . . and he said that a mere mathematician would not be able to keep up the class." James Thomson advised his son to find a way of doing some laboratory chemistry and relayed Dr. Thomson's opinion that although there was no doubt of William's being "an accomplished analyst in mathematical and physical science, yet it would operate much against you . . . should you not be able to give evidence of your acquaintance with the manipulations, to a certain extent, of experimental philosophy. Turn the whole matter carefully in your mind, and write to me soon about it."

William replied that he could not possibly do chemistry experiments in his college rooms and promised vaguely that he would look for some alternative. But his father, the bit between his teeth, galloped on regardless. More than ever he reminded his son of the necessity of mature and circumspect behavior: "What you have to do, therefore, is to make character, general and scientific, so as to justify the Lord Rector, the Dean, and the other electors who usually act with me, in supporting you—a matter of difficulty on account of your youth." Warning of one of the senior Glasgow faculty members "who, as to private affairs, is more nearly omniscient that any one I have known," he reminded his son yet again of the perilous associations of rowing: "Avoid boating parties in any degree of a disorderly character, or any thing of a similar nature; as scarcely any thing of the kind could take place, even at Cambridge, without his hearing of it."

These events unfolded during the same spring that William fell under the influence of *Evelina* and *Wilhelm Meister*. Career planning was not uppermost in his mind. After his father's visit to Cambridge, he reapplied himself to his studies. Jockeying for the Glasgow chair occupied his father's attentions far more than his own. In August James Thomson reported that "Dr. Meikleham has had another attack—a very bad one. He has weathered it, and is pretty well again. In all probability, he will not survive another."

That spring William spent a week or so in London, staying with his older brother James, who was then apprenticed to an engineering firm in Millwall. He also visited Archibald Smith, who despite taking up the law advised William to stick with science. Smith "seems to be getting on very well, and I think now has no idea of giving [the law] up, though he says that he thinks of all lives that of a professor must be most enviable," William told his father. "He said that I should not go to the bar, and when I said that I might not be able to get anything else, he answered that if Dr. Meikleham would live a little longer, I might be appointed his successor."

Just a couple of weeks earlier William had lamented to his father that "three years of Cambridge drilling is quite enough for anybody." Eager to apply himself to real science rather than formal study, William finally began to see the attractions of a professorial appointment beside his father and for the first time began to write as if it were his ambition and not just his father's dream. In April Meikleham suffered yet another setback, falling in his room and bashing his head on the fireplace. William responded to this news with less than commendable concern: "For the project we have it is certainly much to be wished that he should live till after the commencement of next session." By the middle of May, James Thomson reported, Meikleham was "greatly changed. He is silent—vacant—and seems to notice little of what is going on around him. . . . I shall be much surprised indeed if his chair be not vacant before the beginning of next session."

Against the odds, however, Meikleham held on, reduced almost to a vegetable state, clearly incapable of teaching, but professor of natural philosophy still. William studied at Cromer over the summer and returned to Cambridge for the final assault on the mathematical tripos. Meikleham continued to loom silently over his life. Dr. William Thomson kept up with advice to his younger namesake, suggesting he try writing a popular lecture on some scientific subject to prove that there was more to him than rarefied mathematics. "A Cambridge education did not *always* give the power of easy expression or of commanding the attention of an audience," the older man had remarked to James Thomson, with an ironic smile.

As William readied himself for the exams, disquieting news emerged from St. John's, one of the larger colleges and a traditional leader at wrangling. They had a student, Stephen Parkinson, who displayed an extraordinary capacity for absorbing old exam answers and blurting them out again at high speed. Parkinson crammed until he was gray, and as the examinations unfolded, the Johnians felt confident enough to place bets on their man.

On January 1, William dashed off a brief note to his father to say that he had cruised through the first two papers with time to spare but afterwards thought of things he had forgotten to put in. His tutors also wrote to keep James Thomson abreast of the drama. Hopkins reported that William "is going through his examination with vigour and cheerfulness, and in good health, and as far as I can judge has, in the majority of his papers, done himself justice. . . . My confidence that he will be senior wrangler has always been very strong. and I can only say that it remains unshaken." Cookson sent on one of the exam papers with the confident assertion that "I think that he cannot fail to do almost every question in this paper."

But on January 5, 1845, a rest day in the middle of the examination, William wrote with mixed news: "I have been getting on very well with the examination, and certainly quite as well as I expected. . . . However, I hear the Johnian has been getting on exceedingly well and so I must not be too confident. . . . Yesterday I was told I was the only man who did not look seedy with the examination." A week later, when the trial was over but the outcome still unknown, he reported that "the Johnians are talking confidently of their hero. . . . I have not been making myself anxious on the subject."

The competition, always an exciting event on the Cambridge calendar, became unusually intense that year and was the subject of much college gossip. A firsthand account comes from Charles Arthur Bristed, an American student who wrote a memoir of his five years among the natives:

> This present year, however, one of the Small College men [i.e. William Thomson] was a real Mathematical genius, one of those men who . . . are said to be 'born for Senior Wrangler,' while the Johnians were believed to be short of good men. . . . But now their best man [Parkinson] suddenly came up with a rush like a dark horse, and having been spoken of before the Examination only as likely to be among the first six, now appeared as a candidate for the highest honors. [R. L. Ellis, an examiner] was one of the first that had a suspicion of this, from noticing on the second day that he wrote with the regularity and velocity of a machine, and seemed to clear everything before him. And on examining the work he could scarcely believe that the man *could* have covered so much paper with ink in the time (to say nothing of the accuracy and performance) even though he had seen it written out under his own eyes. By-and-by it was reported that the Johnian had done an inordinate amount of problems, and then his fellow-collegians began to bet odds on him for Senior Wrangler. But the general wish was for the Peterhouse man, who, besides the respect due to his celebrated scientific attainments (he was known to the French Mathematicians by his writings while he was an Undergraduate), had many friends among both reading and boating men, and was very popular in the University.

Parkinson's style of automatic writing was not for Thomson. Despite coaching, he could not prevent his mind from running away with him. He would always see fascinating implications beyond the immediate scope of the question, or a way of solving the problem that might turn out a little cleverer than the approved way. Or not: How he could tell until he tried? He saw more than the typical examinee and was duly penalized.

When students and faculty gathered at the Senate House on January 17 to see the results posted on the notice board, it was William Thomson's friends who were downcast. He was junior wrangler, Parkinson senior. Bristed's tale resumes:

> The unexpected award of the Senior Wranglership was the great surprise of the year, and the subject of conversation for some time. It was said that the successful candidate had practiced writing out against time for six months together, merely to gain pace, and had exercised himself in problems till they became a species of bookwork to him. . . . The Peterhouse man, who, relying on his combined learning and talent, had never practiced particularly with a view to speed, and perhaps had too much respect for his work to be in any very great hurry about it, solved eight or nine problems leisurely on each paper, some of them probably better ones than the other man's, but not enough to make up the difference in quantity.

William's prior hesitance now allowed him a perverse kind of victory over his father. "You see I was right in cautioning you not to be too sanguine about my place. . . . The only thing I feel in the least degree about it is that it may make it more difficult to succeed in getting the professorship in Glasgow. . . . I hope you will not think I have misspent my time here. I feel quite satisfied that I have spent as much time on reading and preparation as I could consistently with higher views in science. The Johnians give themselves up to one object, and it is fair that they should have their reward."

He wrote again the following day: "I hope by this time you have recovered from the shock of what I am afraid you must have considered very unpleasant intelligence. . . . Ellis tells me, and does not hesitate to tell others of his friends, that even without previous opinion, he could see by my papers that I am better than Parkinson, but that I fell short in quantity."

His father's guarded reply would only partly have assuaged him: "The place you have got at the examinations is an excellent one, and you and all of us ought to be well satisfied with it. In point of *name* the next higher place would have been desirable; but coupling with your place all the distinctions that you can claim, we can and will make out a good case for you."

From his sister Elizabeth he got warmer reassurance: "I must confess that the unlooked for result of the examination has somewhat disappointed me; but Papa says he thinks you have the *character* of Senior Wrangler notwithstanding, and he trusts you will maintain it. . . . I was very sorry in reading your letter which arrived Sunday when I came to the part where you say you are afraid Papa will think you have misspent your time at Cambridge. He does no such thing, he is very proud of his son and not in the slightest degree less pleased with him since the small humiliation he has met with."

It was James Thomson's perennial weakness not to be wholehearted in any judgment until he had obtained authoritative support. From Hopkins he soon received a lengthy and unreserved testament to his son's abilities: "I confess that your son's not being senior wrangler is to me a very great disappointment. I can assure you however that the circumstance has not affected in the slightest degree the high opinion in which I hold both his talent and acquirements. . . . The fact of your son being *second* is perfectly explicable without lessening the conviction that in the high philosophic character both of insight and knowledge, he is decidedly *first*. . . . While others are simply *answering* a question, he will often be writing a *dissertation* upon it. . . . One of the examiners . . . told me he thought it highly probable that while your son would be hereafter building up for himself a *European* reputation, his opponent might be scarcely known beyond the bounds of the University. . . . Your son bears his disappointment extremely well, better I think in fact than his friends."

This assessment was acute: while Thomson went on to build a reputation greater even than Hopkins imagined, Parkinson remained in Cambridge for the rest of his life, achieving little except the induction of further generations of diligent men into the realms of wranglerdom. James Thomson seized on Hopkins's letter, not only to dissolve any personal disappointment but to fire up again his project of bringing William to Glasgow. He made sure the letter came to the attention of senior Glasgow faculty and joyfully told William: "Hopkins's letter has done you *great* good here. The man that wrote it has a *heart* and a *head.*" Professor Buchanan read the letter in the Thomsons' dining room and, said James, "exclaimed 'that is the kind of man we should have!' I do not see any meaning to put on this except what at once occurred to myself and what will readily occur to you."

William had one final ordeal before concluding his student career. The Smith's prize examination came after the honors tripos and was reckoned to be a deeper test of scientific understanding than the mad race to be wrangler. Each year students tried to answer a number of long questions on mathematical physics rather than pure mathematical methods. The Smith's prize rewarded analytical thought more than trained rapidity of response. It too had its own mystique and pressures. Shortly after Thomson's time, the Smith's prize examination was overseen ("invigilated," to use the preferred Cambridge term) by his close friend Professor George Gabriel Stokes. The exam took place at Stokes's house in Cambridge, in the dining room. Mrs. Stokes and the children took their meals in the kitchen for the few days the exam lasted, and tiptoed about while the young students huddled at the dining table, scratching their pens at the papers while Stokes, a taciturn man at the best of times, said nothing except presumably, "Turn the paper over!" and then, some hours later, "Gentlemen, put down your pens." The imagination easily supplies Stokes's even breathing, a clock ticking somewhere in the quiet house, hushed childish voices and padded feet beyond the firmly closed dining room door. Stokes's daughter recorded years later an occasion when two young men, having survived the morning session, felt unable to face the afternoon and ran away through the garden after eating the lunch Mrs. Stokes had prepared. An unfortunate incident, Stokes commented ruefully, as the two men had actually been doing rather well. Thereafter he made sure the garden gate was locked before he started the exam.

In 1845, William handily beat Parkinson for the Smith's prize. Cookson conveyed the good news to James Thomson: "I have seen your son, who is overjoyed. . . . Some of the papers for the Smith's prize examination were of a more difficult nature than those [for the tripos] and required a more profound & philosophical view of the subjects. It is to this that your son's success may be attributed."

Overjoyed he may have been, but William wasted no time in self-congratulation. Meikleham was hardly speaking, barely breathing, but still alive, and for the time being neither he nor his father could do anything to bring the Glasgow appointment closer. William had had his fill of reading and studying and cramming. Now it was time for his real education in science. On January 24, 1845, he wrote hurriedly to his father with the result of the Smith's prize. The next day he went to London with his friend Hugh Blackburn, and from there the two continued to Paris, where many of the greatest exponents of mathematical and experimental science lived and worked.

***

As well as Fourier's mildly controversial work on heat, published in 1822, the great classics of French mathematical science from the late 18th and early 19th centuries include Lagrange's *Théorie des Fonctions Analytiques* (1797), Laplace's five-volume *Traité de Mécanique Celeste* (1799-1825), Cauchy's *Cours d'Analyse* (1821), and Poisson's *Traité de Mécanique* (1833). Taken together, these monumental works created the modern form of differential and integral calculus and their application to mechanics and the motion of bodies. No matter what inscrutable processes of thought Isaac Newton had employed to devise these ideas in the first place, when it came to compiling the tremendous *Philosophiae Naturalis Principia Mathematica* of 1687 he reverted to a presentation that Euclid would have recognized. Far from being a system of mechanics, Newton's work strikes the modern student as a collection of geometrical exercises and trigonometric problems. This style not only made the book hard slogging, it concealed much of the mathematical structure of the underlying theory. The French mathematicians enlarged and systematized Newtonian mechanics into the sophisticated body of analytical methods that students learn today.

French scientists had also raised the art of experimental investigation to new importance. Charles Coulomb and Jean-Baptiste Biot had established laws for forces acting between electric charges and simple magnets. André-Marie Ampère measured interactions between magnets and electric currents and proposed a sophisticated mathematical theory that held sway for a decade or two. In optics, Biot and Augustin Fresnel and Dominique-François Arago investigated the polarization and diffraction of light and attempted to tie their findings into mathematical systems from the pens of Poisson and others.

The French achievements of this era became, and remain, the foundation of a "mechanical" description of nature, consisting ultimately of inanimate objects responding to elementary forces. What the French called *la physique* aimed to combine experimental investigation and mathematical sophistication into a seamless whole. In 1816 Biot published a textbook, *Traité de Physique Expérimentale et Mathématique* setting out what we would now call (to use a much abused word) a reductionist view of natural phenomena, in which forces acting on particles are at the bottom of every physical process. Of course, there were mathematicians and scientists outside France too: Euler and Gauss in Germany, Young and Herschel and Babbage in England, Brewster and Nichol in Scotland. But Paris was the birthplace and center for a view that came to influence the rising generation in Britain and eventually penetrated Cambridge too.

One man who grasped quickly the superiority of these gallic innovations was James Thomson. In 1825, writing in the *Belfast Magazine and Literary Journal*, he offered this scathing judgment: "Since the days of Newton, however, the British mathematicians have been far surpassed in several branches of science, by their neighbours on the continent [especially in] the higher and more difficult parts of pure mathematics, and in physical astronomy." Here he noted particularly the work of Lagrange and Laplace, as well as Euler, and went on: "While these brilliant achievements were crowning the efforts of the mathematicians on the continent, the men of science in Britain were wasting their time and talents, some in restoring the ancient geometry of Greece, and some in following servilely and implicitly the *manner* in which Newton presented his investigations, without being actuated by the *spirit* by which he was directed in his researches. Fond and proud of that eminent man almost to devotion, and prejudiced against his rivals on the continent, partly by feelings of national jealousy, and partly by the scientific war between the adherents of him and of Leibnitz,[7] they generally clung, even in the minutest particulars, to the methods pointed out by their great leader; and, falling behind

---

[7]Newton and Leibnitz, in the late 17th century, independently devised differential and integral calculus, but because Newton failed to publish his work except under extreme duress from Edmund Halley, an intense and bitter dispute broke out as to who

on the march of discovery, they scarcely contributed in the slightest degree, during the lapse of a century, to the advancement of science, in its higher and more difficult parts."

But as James Thomson went on to say, change was afoot. A group of reformers led by William Whewell (who brought the words "science" and "scientist" into common parlance), William Herschel (discoverer, in 1781, of Uranus), and Charles Babbage (designer of the famous "difference engine," a mechanical calculator) began to clear away the old dogmatic Newtonian lore in favor of the more flexible, systematic, rigorous, and general mathematics of the French school. Or rather, at Cambridge, this was half a reform: Cambridge slowly embraced French mathematics but failed to succumb fully to the charms of *la physique*. In an intellectual divide that remains today, at least at the level of stereotype, the French hankered after a grand overarching system in which all phenomena ultimately referred back to theories of a single, logically consistent formulation, while the pragmatic Anglo-Saxons preferred to analyze empirical matters piecemeal. The French wanted to portray mechanics, light, electricity, magnetism, and heat all as parts of a universal underlying theory of matter and forces. The British were content to come up with satisfactory accounts of each of these subjects on their own terms.

By the time William Thomson studied at Cambridge, Whewell was ironically beginning to seem like part of the old guard, resistant to further continental scientific innovations (the "despotic Whewell," William once called him in a letter to his father, because of his resistance to change in the mathematical tripos). In Paris, William quickly became acquainted with the surviving French savants of the great generation, including Biot and Cauchy. He met Joseph Liouville, editor of the *Journal de Mathématique*, who set him a problem. Like his French colleagues, Liouville had difficulty with Faraday's schematic but suggestive portrayal of the electric tension between objects as an influence carried along curved lines spreading throughout space, and thought it contradicted Coulomb's inverse-square force acting along a straight line between two charges.

---

was really first. So insistent were the English on Newton's priority that they took well over a century to accept that Leibniz's formulation was in many respects easier to use and more flexible in application, though it embodied the same mathematics.

This notion—action at a distance, as it was usually called—was a piece of Newtonian thinking taken unchanged into *la physique*. Faraday found action at a distance philosophically objectionable and believed that there must be a medium pervading space by which electric forces transmit themselves from one place to another. His lines of forces were an attempt to capture this vision.

Full development of Faraday's insights lay in the future. For the time being Liouville could see only one theory versus another: action at a distance along straight lines versus curving, elastic lines of force. As William reported to his father, "He asked me to write a short paper for the Institute explaining the phenomena of ordinary electricity observed by Faraday, and supposed to be objections fatal to the mathematical [i.e. French] theory. I told Liouville what I had always thought on the subject of those objections (i.e. that they are simple verifications) and as he takes a great interest in the subject he asked me to write a paper on it." In truth, no contradiction existed. Faraday did not dispute the magnitude of the force predicted by Coulomb's law, and for a simple case such as two charges separated by some distance, the lines of force would be perfectly symmetrical around the line joining the charges, so there would be no sideways forces. In the case of three or more bodies, William easily showed, the geometry obviously became more intricate, but Faraday's picture nevertheless predicted the same forces as Coulomb's law did.

But a significant conceptual difference existed between the two pictures. According to Coulomb, electrical forces acting in a complex arrangement of multiple charges were best imagined as the summation of independent forces acting between all the pairs. According to Faraday, the charges created a state of electric tension pervading the whole of space around them, and the force acting on any one charge arose from the electric tension where that charge resided. William showed that the pictures came to the same thing, however one looked at it, and for static arrangements of charges nothing more need be said. Faraday's depiction, vague and poorly formulated as it seemed to Liouville and the French, ultimately had more physics in it. But that was not yet apparent. For the moment William had shown with simple clarity that Faraday and Coulomb did not disagree. To do this he combined hard mathematics with an appreciation of the physical phenomenon concerned. It was a kind of problem solving at which he excelled. And he showed a sympathy for *la physique* while retaining a certain outsider's perspective.

Equally important for William's career was his stint of work in the Paris laboratory of Victor Regnault, a 35-year-old experimental scientist. At the behest of the French government, Regnault had embarked on a long project to measure the thermal properties of steam—its rate of expansion with temperature, the quantity of heat needed to raise its temperature by some amount, and so on. As the new industrial economy grew, steam power became an ever more important foundation for national prosperity. In Britain, birthplace of the practical steam engine, inventors and amateur scientists continued to develop the new technology in laissez-faire style. In France, politicians saw an opportunity to spend national revenue on a project of national importance. Efficiency in a steam engine meant economy of operation (more power from the same quantity of coal), but next to nothing was known at the time of the scientific principles behind steam power. Regnault's work aimed to establish a foundation of practical knowledge by which to improve steam engine design.

William's time in Regnault's lab introduced him not only to practical science, at which he proved adept and ingenious, but also to the implications of the theory of heat in technological matters. In his reading of Fourier he had come to know heat as an element of fundamental physics. Working long days with Regnault, he discovered heat as a source of motive power, causing gases to expand, pistons to slide, and crankshafts to revolve. The French scientists, for all their love for the rigor and elegance of higher mathematics, also insisted on the importance of empirical knowledge. William's four and a half months in Paris in 1845 turned him from an applied mathematician into a man of science. Decades later, as Lord Kelvin, he told the French Academy on receipt of an honor that France "is without doubt the Alma Mater of my scientific youth, and the source of that admiration for the beauty of Science which has enchanted and guided me throughout my career."

***

William's letters to his father from Paris expressed unabashed excitement at being inducted into the fellowship of true scientists. His father's responses were equally enthusiastic but for a less exalted reason. Meikleham ailed but lived still, and the Glasgow chair remained open. "Dr W Thomson [the medical man] and Dr Nichol are anxious that you should become acquainted as much as you can with the great men of Paris, as testimonials from them may serve you much, and it will be pleasant to make their acquaintance irrespective of this." And later: "Dr W.T. is much pleased to hear that you have got fairly into Regnault's Cabinet [de physique—his lab], and hopes you will be able to get a good testimonial from him . . . and others regarding your general knowledge of Physique, and showing that you are not merely an expert x plus y man." James Thomson, relaying the oracular advice he regularly obtained from the older Dr. William Thomson, urged his son to get promises of a testimonial from any great Parisian he happened to meet, no matter how slight the occasion. William wrote of the wonderful things he was learning and the ideas he discussed with Regnault, Liouville, Cauchy, and the rest. At length he admitted he could probably get letters from Liouville and Regnault, but he begged off asking anyone else, saying he had had insufficient contact. He hoped those two testimonials would be adequate reward, in his father's eyes, for his time in Paris.

He left France at the end of April with no clear plan beyond going back to Cambridge and picking up some coaching to make a living while he pursued all the new ideas he had absorbed in Paris. At the end of June, however, he wrote to his father that "very much contrary to my expectations" he had been elected to a fellowship at St. Peter's. His protestation seems disingenuous in the extreme: as an undergraduate he was already publishing alongside the great scientists of the day, and despite his tiny failure in being only junior wrangler, one of the examiners had reportedly declared to another, "You and I are just about fit to mend his pens."

This news brought joy to his father, tinged with wistfulness. He congratulated William on getting "forward so far at so early an age! At your age I was teaching eight hours a day at Dr Edgars,[8] and during the extra hours—often fagged and comparatively listless—I was reading Greek and

---

[8]The master of the small village school in Ballynahinch, where James Thomson got his first education.

Latin to prepare me for entering college, which I did not do till nearly two years after."

During the summer William coached undergraduates (as he had been coached only six months earlier) and came back to Glasgow in September, catching up with his family whom he had not seen for a year. There were also, of course, "important matters in consideration at present" to be advised of by his father, and no doubt a good deal of plotting and preparation went on between them. But Meikleham, neither better nor worse, clung silently on to life. William returned to Cambridge. He had "as many pupils as I would wish" and also gave lectures in college every morning at eight o'clock, which he said he enjoyed more than the coaching.

The following February James Thomson informed his son that a teacher of mathematics at the Glasgow High School was ill and would probably soon die, and wondered if William might make a move for the position. William's reply makes clear that he was beginning to establish his own life and would not go along with his father's every scheme to get him back to Glasgow. He turned down the opportunity because "I am afraid I should have to give up any thing in the way of original research. At present, at Cambridge, I can with ease make more than enough money to support myself, and when I commence receiving money for my fellowship, I think with what I receive for lecturing, I should be independent of private pupils, in pecuniary respects. The only event which could make me require more than I get at present would be marriage, and if that were ever to happen when I am here, I think I could, by private pupils, get as much, or very nearly as much money as is mentioned for the situation in Glasgow, with as little work, and that of course of a higher kind." The fellowship paid £200 a year, with rooms in college—a comfortable living.

James Thomson didn't push the matter further and in fact asked his son's advice on who else might be suitable. Instead, after a disturbing rumor came to his ears at Glasgow, he resumed his campaign for the big prize. He had heard, he wrote in early May, that William was "said not to bring down your instructions to the capacity of ordinary students. . . . Such a report may seriously injure you. . . . You *must* take care to cure the evil, if it exist; and if not, to teach so simply, clearly, & slowly, that you may be able to get decidedly good testimonials on that point. *Do attend to this above all things.*" The origin of this tattle-tale never became clear. William suspected it came from a man at Trinity but in any case wrote strenuously to say it wasn't at all true. All too cognizant of his father's concern and tenacity, he got his former examiner, now colleague, Ellis, to write to James Thomson in the same vein: "The idea, if there is such an idea, that it is the common opinion in this university that as a private tutor he advances too rapidly or 'talks over men's heads' is I verily believe, perfectly unfounded. I have never heard a syllable to that effect." Once these fears had blossomed, however, James Thomson could not let them go, and he harped incessantly on the question of his son's ability to teach at the appropriate level. He reported that Dr. William Thomson was now worrying that young William suffered from "timidity and want of effective locution" and wondered if he could work up some sort of nonmathematical lecture of a general nature to assuage fears that "your ideas and expressions are bound up in the icy chains of x's and y's, +'s and −'s." William hardly bothered make any reply to these further charges, the possibility of calming his father seemed so hopeless.

In any case, the battle was now engaged in full. Meikleham died at last, early in May 1846, and James Thomson upped again the barrage of letters and advice and instructions to his son. He wrote with long lists of the names of eminent people who might supply testimonials. He wanted proof (what it might be was unclear) of William's teaching abilities. He derived further anxiety from another rumor: It appeared that Archibald Smith had some thoughts of entering the competition for the Glasgow post. Smith had at first said he would write a letter on William's behalf but then wrote a couple of days later to say he was thinking about applying himself. William visited Smith in London and found only that he hadn't made his mind up. James Thomson, learning this, decided that some great plot or betrayal was under way. Smith wrote considerate if indecisive letters to both father and son in which little is clear except that he wasn't sure what he planned to do.

Although Meikleham's death was no surprise, except in coming abruptly after so long a postponement, William at first seemed uncertain how to react. A Cambridge fellowship was no bad thing. News of the ancient professor's demise "took me quite by surprise, as of late I have been composing myself to the idea of being fixed here for two or three years," he wrote, but went on to say he would start arranging for testimonials. He allowed himself some reservations about his father's all-out campaign, however: "I think it will not be a good plan to attempt to get too many [testimonials]. A few, from those who should know my qualifications such as Hopkins, Cookson [and other examiners] will I am sure have more influence without many others than if they were overwhelmed in a flood of testimonials from people who do not understand what is wanted for a professor of Natural Philosophy."

William could never act with sufficient vigor to convince his father he was in earnest. James Thomson mentioned the Lord Rector of Glasgow University, Rutherford, and the Dean, Maconochie: "Could you 'get at them'? Maconochie is a *vain* man and would be flattered by a letter from a great or learned man." (A letter from his brother James a couple of weeks later spelled out 'Maconochie' at the top, in large, heavy block letters, as William was in the unfortunate habit of putting two n's in the middle of the man's name). More from his father: ". . . double your efforts to procure testimonials. . . . Could you have nothing from Chasles [another of the French *physiciens*] or Gauss? . . . Do all you can. . . ." He urged him to try for Herschel and Faraday, among other notable English scientists. William had met Faraday once or twice by now, but though he offered good wishes and encouragement Faraday explained he never wrote testimonials for anyone, as he thought the whole process absurd and possibly corrupt. By mid-June James Thomson was writing, "I am afraid you are resting too quietly on your oars about testimonials. . . . I may add that the remark of one of your friends here on reading your note which has just arrived, is that it does not appear to be the note of a person who is trying to obtain a valuable appointment for life, and who is perhaps at a principal crisis in his career." William's siblings mostly stayed out of this frantic exchange, although his brother James wrote once or twice when their father was ill, to relay the latest instructions. Anna, safely away from the fray in Belfast where she was living with her new husband, William Bottomley, provided one suggestion: She told her brother to "get a beard fast so as to make you more imposing."

James Thomson took up again the question of quantity versus quality in testimonials: "Cookson &c are right in their views about the *fewness* of testimonials, were all electors as philosophical and judicious as they are themselves. [The Glasgow electors] are *small* men, however. Get therefore what testimonials you can from fellows and other respectable people. I wish you could get something from Herschel and Airy. Perhaps you could through some friends." And he added the news that old Mr. Smith, Archibald's father, had just returned from Malta, intending "you may be sure, [to] use, *without much scrupulousness*, every means in his power to forward his son's views."

This soap opera played out over the summer. James Thomson was never able to convince himself he had done enough to compensate for his son's lackadaisical attitude. If William had been wavering about giving up the comforts of his Cambridge existence, a fortuitous intervention by the Bishop of Ely may have nudged him. At the end of June he had to go to the old cathedral city of Ely, some 15 miles northwest of Cambridge, to receive formal admission to his St. Peter's fellowship. Colleges were still at least nominally religious institutions, and appointments needed an imprimatur from the local see. William traveled to Ely expecting to get some sort of document signed in absentia, but the bishop happened to be there and took care of business himself. William told him he had applied for the Glasgow post but would stay at Cambridge if it didn't come through. The bishop responded, William reported to his father, with a mild reproach. He said he "hoped I do not intend to 'fritter away my time with taking pupils here' and spoke a good deal to the effect that it would be very desirable if men who have gone through the Cambridge course could be induced to continue studying and endeavouring to make discoveries in science."

The result of this months-long campaign was a printed and bound volume of 29 testimonials, along with a separately printed letter from Liouville, which arrived late. Most of the notices were warm but routine, adverting to William's great brilliance, his achievements so far and those undoubtedly yet to come. Hopkins was careful to add a sentence about William's easy manner and amiable nature as a teacher, which led James Thomson to decide his testimonial was the best of the bunch. Cookson, the college tutor, concluded with: "He is already blessed with a reputation which veterans in science might envy, but his friends look for still greater lustre. . . . God grant that he may live and do honour to his country," which rather embarrassed William.

Liouville's remarks, transcending any chauvinism, were warm indeed: "I believe M. William Thomson is destined to attain a high rank among that stellar group [*pléiade*] of *savants* that England is justly proud to claim as her own." He added a personal note: "Continue, Monsieur, to work as you have already done for a number of years, and brilliant success will crown your efforts. . . . Your future is bright, believe me." It comes as no surprise to find that this private letter was included in the printed materials ordered up by James Thomson to bolster his son's case.

This tortured tale ends in utter anticlimax. Archibald Smith never applied for the Glasgow chair. On September 11, 1846, the faculty met and unanimously selected William Thomson to be the new professor of natural philosophy. The relentless campaign at last over, all foes, real and imaginary, put to flight, William's persevering father finally allowed himself the pleasure of unfettered joy. According to Elizabeth's husband, the Reverend David King, "a countenance more expressive of delight was never witnessed. The emotion was so marked and strong that I only fear it may have done him injury." The next day he was outwardly calmer but "every now and then a quiet, happy smile stole over his features, and he seemed quite full of enjoyment."

William took it all in stride. Success had come again. He was 22 years old. A few weeks later, when he had returned to Glasgow, Elizabeth reported that "William does not look in the slightest degree elated. He is perfectly composed. You would hardly think that it was he who had succeeded so brilliantly."

2

CONUNDRUMS

Geography made northern England the engine room of the industrial revolution. First came the mill towns, with clattering looms driven by fast-running water. In the bucolic south, placid rivers meandered across the rolling landscape, ambling between lush fields full of grazing cows. Farther north, torrents rushed in narrow channels down the steep slopes of the Pennine hills, where sheep nibbled the thin grass. In valleys where a few powerful streams came together, the mill towns blossomed. Tiny agricultural settlements quickly became home to thousands of mill workers—men, women, and children working six days a week for wages that barely kept them alive. This was the birth of the urban industrial working class.

By the end of the 18th century, steam power had begun its long ascendancy. The first steam engines appeared around 1700, and simple steam-powered pumps with a rocking action were used in a few mines not long afterward. But it took almost a century of inventions and improvements for rotary engines and, later, locomotives to become practical. Only then did the vast factories of the 19th-century industrial revolution begin to appear. Fast-flowing water from mountain streams was no longer a requirement, but steam engines needed water in bulk and coal in large quantities. Industry moved down from the hills and into the flatlands. Cities sprang up beside large rivers, where coal and raw materials could be shipped in and finished goods shipped out. Manchester, Birmingham, Leeds, and many more sprawling, smoke-belching factory cities settled on old farmlands. Ports such as Liverpool and Glasgow grew too, bringing in raw materials, especially cotton and sugar, from British possessions around the world.

The transition from fast-running water to steam as the source of motive power had a striking scientific parallel. The working of a water mill explains itself with little difficulty. Water falls, turns a wheel, then goes on its way. Ideally, no water is lost. Nor is it immediately obvious that anything is lost *from* the water. If you go a little way downstream from a mill, the water may flow just as fast as it did upstream. Engineers of the 18th century were well aware, following Newton, that a water wheel generated mechanical work. And they knew too, if only from the obvious visual appearance of the thing, that this mechanical work had its origin in the fast-moving stream. But it seemed, from casual inspection, to be an inexhaustible source.

When scientists first began to ponder the working of steam engines, the analogy with water wheels proved irresistible. Heat from a furnace turns water into steam. Steam in the cylinder pushes on a piston, producing mechanical effort. Then, with the piston at full stretch, the steam (cooled because of its expansion) is vented into the air to allow the piston to slide back. From the hot furnace to the spent steam, a quantity of heat, it appeared, fell from a high temperature to a low temperature. Heat flowed, in some ill-defined sense, through a steam engine, yielding mechanical effort as it did so. It was not obvious that any heat was "used up" in the process. Indeed, the nature of heat being so enigmatic, it was impossible to say what using up heat might mean. Where would it go? What would it become? Far more reasonable to suppose that the quantity of heat stayed the same. Only its quality—its temperature—underwent any evident change.

***

During his 1845 stay in Paris, investigating with Regnault the properties of steam, William Thomson came across an intriguing paper written in 1834 by Emile Clapeyron. Clapeyron gave a quantitative analysis relating the amount of work produced by a steam engine to the quantity of heat passing through it. More intriguing still, Clapeyron explained that his analysis was not original but rather was his attempt to put into tight mathematical form a prosy discussion he had found in a pamphlet written 10 years earlier, in 1824, by Sadi Carnot, and titled *Réflexions sur la puissance motrice du feu et sur les machines propres à développer cette puissance* (Reflections on the motive power of heat and on machines capable of developing that power, usually known by just the first half of the title). Carnot's work, contemporary with Fourier's great book, was utterly new to Thomson—and, as he quickly discovered, apparently unknown to the whole of Paris. The library at the Collège de France did not have it. Thomson went to bookstores and to the used-book stalls along the banks of the Seine asking, in his Scots-accented French, after Carnot. "*Caino?*" the merchants would respond, "*Je ne connais pas cet auteur.*" When he managed to convey the name correctly he was shown works on military engineering by Lazare Carnot and on social matters by Hippolyte Carnot. Of Sadi Carnot he found not a trace.

Lazare was Sadi's father, Hippolyte his younger brother. Lazare Carnot was a man of some technical knowledge, who had applied his talents to military matters and then to politics, in which he showed impressive agility. He served with Robespierre on the Committee of Public Safety in the revolutionary 1790s and after a period of exile had come back as a minister under Napoleon in the early 1800s and again during the "hundred days" of 1815, when Napoleon returned from Elba only to be defeated at Waterloo and sent into more distant seclusion on St. Helena. Then finally Lazare's ingenuity ran out. He went into exile in Prussia and died in 1823. His son Sadi, born in 1796, haphazardly educated, growing up in the shadow of political reverses, and with his father only intermittently present, became (according to his brother) a sullen and mistrustful man. After a brief military career, he hung about Paris in the occasional company of engineers and technical men but made little acquaintance with the world of science. Then in 1824 he produced one of the most profound and original scientific works of his or any era.

In pondering what went on inside a steam engine, Carnot's acute insight was to think in terms of a repeating cycle. For the piston to begin a new stroke, it had to come back to its starting position. Ideally, Carnot imagined, the engine would also start each cycle with precisely the same conditions of temperature and pressure in the cylinder. Each cycle was then identical and independent.

The conceptual obstacle to analyzing an engine was that complicated and interacting processes of heat transfer and work production occurred in what seemed to be a hopelessly entangled way. Carnot untangled the problem by idealizing his cycle so as to isolate the roles of heat and work. He imagined a furnace maintained at some constant high temperature, while the spent steam was expelled from the piston at some constant low temperature. He constructed a four-part cycle. First, the cylinder was charged with steam at the high temperature; second, the piston was released so that the charge of steam produced an amount of work, cooling and expanding as it did; third, spent steam was discharged at the low temperature; finally, the piston returned to the starting position.

Details aside, the trick here was that heat entered only at the high temperature and left only at the low temperature, in parts of the cycle where no work was being produced. During the work phase, on the other hand, no heat was moving in or out. By isolating the operations of an engine into distinct steps of the cycle, Carnot was able to compare heat going in to work produced and thus ponder the engine's efficiency—the amount of motive power obtained from a given quantity of heat, or rather, as he explicitly said, a given quantity of the supposed heat-fluid called caloric. He declared firmly that "the production of motive power in a steam engine is due not to an actual consumption of caloric *but to its passage from a hot body to a cold one*" (Carnot's italics). This is the mill wheel analogy. Heat passes through an engine, going from the high to the low temperature, but is not used up.

This sounds all very well, but Carnot's cycle is by design highly idealized. How can it help in understanding the working and efficiency of a real engine—an actual machine with leaky gaskets, squeaking pistons, sticky valves, and all manner of other imperfections?

Here was where Carnot's ingenuity blossomed into true originality. Because of the way he constructed it, his cycle was reversible. That is, if a certain amount of caloric moving from high to low temperature produced, via his ideal engine, a certain amount of work, then that same quantity of work, applied so as to run the engine backward, would push precisely the same amount of caloric "uphill," from the lower to the higher temperature. Now imagine, he said, a hypothetical engine that could produce a greater amount of work from the same transfer of caloric. He could use part of the work from such an engine to run his cycle backward and push all the caloric back to the higher temperature—and he would have some work left over. This would mean he could create work by shunting a finite amount of caloric from high temperature to low and back again, ad infinitum. He declared roundly: "This would be not only perpetual motion, but an unlimited creation of motive power without consumption either of caloric or of any other agent whatever. Such a creation is entirely contrary to ideas now accepted, to the laws of mechanics and of sound physics. It is inadmissible." Therefore, he concluded, the ideal cycle he had devised was the most efficient possible means to create work from transfer of caloric. No other engine could do better.

He offered one final extension of his reasoning. Engines did not have to run on steam. A Scotsman, the Reverend James Stirling, had invented an engine in which heated air pushed a piston. Now imagine, Carnot said, an ideal cycle running on air coupled to another cycle running on steam, but in reverse. If the efficiency of the air engine was greater than the efficiency of the steam engine, it would be possible as before to move heat uphill without a net expenditure of work. Therefore the efficiency of the two engines had to be the same.

To sum up, Carnot had constructed an idealized engine cycle with an efficiency that could not be surpassed by any other engine. It therefore represented a theoretical maximum. Moreover, this efficiency must be the same regardless of the engine's working substance, so it was a *universal* theoretical maximum. Carnot stated with great emphasis the connection between maximum efficiency and reversibility. If any heat was lost in the running of the engine, whether from the escape of steam or from conduction through metal parts, it would not be restored if one ran the engine backward. The efficiency must then be smaller. Although Carnot concluded his analysis by using what knowledge he could find of the properties of steam to estimate actual work production by typical engines, he could not establish a completely general formula. He did demonstrate, though, that the efficiency could depend only on the upper and lower temperatures between which the engine cycled.

Carnot's essay of 1824, starting from next to nothing, created in a single burst of originality the foundations of a new science relating heat and work. It was utterly without precedent and dense with implications. Published privately in an edition of 600 copies, the *Réflexions* sank immediately into obscurity. Hardly anyone read it. Few who did understood it. And certainly it made no impact among French scientists.

For this Carnot has to take some of the blame. He argued in words, not mathematics, yet at the end provided tables of numbers representing the work that could be got out of an ideal steam engine working at certain temperatures, using a certain amount of coal, and so on. His style of exposition was clear and emphatic in places, cryptic and obscure at some crucial and delicate points. Carnot was an early exponent of scientific writing according to the principle "say what you're going to say, say it, then say what you just said." To mathematical scientists of the French school accustomed to tight algebraic exposition, his work came across as an exercise in rhetoric, with pages of numerical results arriving seemingly out of nowhere. To engineers occupied with making their furnaces and cylinders as heat tight as possible, and worrying about seals and friction and lever arms, Carnot's abstract pronouncements about an imaginary kind of ideal cycle seemed spectacularly beside the point.

Even when Clapeyron, 10 years later, simplified some of Carnot's arguments, left out some of his shakier reasoning, and wrote down a simple mathematical statement of the main result, he failed to attract much interest. The prospective audience—natural philosophers with mathematical inclinations yet interested in working out the theory of an industrial machine—barely existed. William Thomson, however, was the perfect reader. Heat had fascinated him since he encountered Fourier. He had embraced the strategy of tying mathematical reasoning to empirical knowledge rather than abstract principles. On top of that, he had grown up in an industrial city and, with his brother James, had made toy steam engines and other machines as a child.

He failed to find a copy of Carnot's treatise in Paris, but he studied Clapeyron's paper and, during the Thomson family's summer outing at

Knock Castle on the Ayrshire coast, he told his brother James about it. Some months later James found a copy for himself and wrote to William: "The preliminary part, of wh. you told me the substance at Knock, is I think a very beautiful piece of reasoning, and of course is perfectly satisfactory."

James Thomson, the professor's oldest son, would have been celebrated as an extraordinarily bright young man had he not been blessed with an even more extraordinarily bright younger brother. He began attending his father's lectures at the age of 10, but 8-year-old William already outshone him. Almost every year William won first prize in the class, and James came second. James was perhaps a little slower than his brother, but the more telling difference was that he tended (rather like his father) to be cautious and circumspect. Sometimes he seemed more profound or rigorous; at other times he could seem merely pedantic and unimaginative.

Three sketches, from the early, middle, and late parts of his life, furnish a consistent portrait of James Thomson. John Nichol, son of the astronomy professor, recalled: "Of the sons I liked James the best. He was crotchety and apt to be sulky with those who would not enter into his crotchets; here, as far as I know, his faults end." Talking of his later career Nichol added: "I believe some of his inventions were excellent, but there was always some practical obstacle which prevented their bringing to the inventor either the fame or fortune they merited. James was an idealist in his way."

In the 1860s the German scientist Hermann von Helmholtz visited William Thomson in Glasgow and while there met James. He wrote to his wife that James "is a level-headed fellow, full of good ideas, but cares for nothing except engineering, and talks about it ceaselessly all day and all night, so that nothing else can be got in when he is present. It is really comic to see how both brothers talk at one another, and neither listens, and each holds forth about quite different matters. But the engineer is the most stubborn, and generally gets through with his subject."

When the brothers were old men, the engineer J. A. Ewing had much the same impression: "It was also, sometimes, difficult not to be impatient; for James, great as was his insight, seemed wanting in some sort of mental perspective, and had very little sense of time. There was never a flaw in his logic; it was devastatingly thorough and would tolerate no admission of even the most obvious preliminaries. Occasionally one listened to his argument as the wedding guest listened to the tale of the ancient mariner, wondering not so much when it would end as when it would really begin."

After their youthful studies in Glasgow, William did not take his degree, otherwise he could not have entered Cambridge as an undergraduate. James took his B.A. and went into a series of apprenticeships at engineering companies. There was no undergraduate engineering school to attend in those days. Technical men and inventors learned their craft on the job, sometimes picking up analytical skills along the way, sometimes not. In 1842 James started at Horseley's, a company in Walsall, close to Birmingham in the smoky industrial heartland of England. While William studied, rowed, walked, and played music among the elegant buildings and sumptuous gardens of Cambridge, James wrote to him of a harsher reality: "I have a good many warnings about taking care of my fingers among the machinery. Mr Bell's son has got 3 off his right hand and another of the pupils has just returned from London where he went to get his hand taken care of after having taken off one finger and destroyed another." Another letter (from his lodgings, delightfully named Mrs. Grim's) speaks of a boiler blowing up and killing a man. James helped draw up blueprints for iron bridges, which cost tens of thousands of pounds each; errors cost money, a factor that may have both suited and reinforced his natural inclination to extreme carefulness. But later letters complain of a lack of work coming into Horseley's (as well as a lack of letters coming from William), and by early summer of 1843 the company had failed. James returned home to Glasgow.

The following year he went down to London, with no great enthusiasm, and found after only a couple of months that the company he had become attached to was about to be sold. By the end of the year he was in Manchester, in a pleasant part of town, he reported, not at all smoky except when the wind blew wrong. As William approached his final exams, James wrote to contrast his situation: "I wish my apprenticeship was as nearly done as yours, but even when it is done, I fear I shall have no such comfortable berth to step into as that which is probably waiting for you."

Then ill health brought him down. He had often suffered through colds and infections, and on one of his walking trips in Europe he had damaged a knee that never fully healed. Early in 1845, fatigued and listless, he was diagnosed with a weak heart and returned to the family home in Glasgow where he received medical attention typical of the era. He had a "blister over my heart w^h kept me in bed for a fortnight" and afterward had "a silk cord put through my skin with the ends left out so as to cause a permanent running." Regularly he received an "infusion of digitalis" to bring down the pulse, and he was instructed to take "no animal food or spirits of any kind." A blister over the heart, it should be explained, was not an ailment but a treatment. A blister might be induced by burning or by the application of a poultice containing the dried bodies of cantharides, also known as Spanish fly or blister beetle. The irritation and subsequent infection were meant to scold the heart into working harder. A couple of years later Elizabeth was subjected to the same doctoring, and James wrote to William to say how he knew from experience that it was "really a most painful and distressing thing."

Secluded in the Thomson home in Glasgow and unable to enter on any more apprenticeships, James pursued his theoretical investigations of engineering matters. While he and William were puzzling over Clapeyron, still unable to find Carnot's original essay, their project of understanding the scientific principles of steam power came sharply up against a new and seemingly contradictory piece of information. In the summer of 1847, William went to Oxford for the annual meeting of the British Association for the Advancement of Science. This organization, founded in 1831 by a group of young reformers exasperated by the fuddy-duddies who ran the venerable Royal Society, had already established its annual meeting as the prime venue for announcing new results, sounding out one's colleagues, and keeping abreast of the activities of scientific men across Great Britain. The BA attracted amateurs, gentlemen, engineers, and academics in equal measure, in contrast to the Royal Society, which had degenerated in the early 19th century into more of a London club for aristocratic dilettantes than a scientific organization. Facing pressure from the BA and other new scientific groups, the Royal Society had by midcentury largely regained its former reputation.

At the Oxford BA meeting, Thomson met James Prescott Joule, an example of the new scientific man emerging from the industrial revolution. Born into a successful Manchester brewing family,[1] Joule had studied for a while with the great chemist John Dalton, a Mancunian who discovered strict numerical laws of proportions in chemical reactions and had gone on to propose the existence of atoms. Thereafter Joule largely educated himself in science and engineering, and had the resources to support a substantial laboratory in his own house. He was an active member of the Manchester Literary and Philosophical Society, established in 1781 as an intellectual forum for the emerging middle classes of the industrial city. The Lit & Phil, as it was known, began publishing its own scientific journal in 1785. True to its mercantile origins, the society held to a firm belief in the practicality of science. A visiting German scientist, Carl Jacobi, recalled speaking to the members of the Lit & Phil in 1842, when he "had the courage to say that it is the honour of science to be of no use, which provoked a powerful shaking of heads."

While working in the family brewery, James Joule began to do scientific experiments. His first project, perhaps motivated in part by the rapidly growing abundance of noisy, smoky steam engines in Manchester, was an investigation of electric motors (invented by Faraday in 1821) as not only a cleaner alternative but potentially a more efficient and therefore cheaper one. It seemed not impossible at the time that electromagnetic motive power might be limitless. As with the analogy to water power, it appeared that a magnetic field, appropriately arranged, could cause an armature to rotate without itself being affected. This, Joule soon discovered, was not the case. He found that electricity passing down a wire creates heat in proportion to the square of the current. He found that as electric motors were made bigger, their coils somehow developed a resistance to the applied magnetic field that was trying to turn them. This was mysterious, but a general lesson urged itself on Joule's mind. In modern idiom, you can't get something for nothing. He observed that a current passing through an electromagnet will generate a certain amount

---

[1] A truism about the California gold rush is that those who did best out of it were the hoteliers and suppliers of pick axes and panning equipment. Similarly, William Joule & Son became the biggest brewer in the English midlands, while engineering companies, such as those James Thomson apprenticed for, came and went.

of heat; he then noted that if the same magnet was part of a motor, the amount of heat generated was reduced according to the amount of work the motor did.

Joule became a stickler for measuring things accurately and convinced himself that if one effect of an electric current—heat, mechanical power, a magnetic force—were somehow reduced, whatever was lost had to show up somewhere else, in some other form and, crucially important, in an equivalent amount. These studies led him to experiment on, and measure, the conversion of mechanical energy into heat. He arranged for a falling weight to turn a magnet and measured the current generated. He forced water through narrow pipes and measured the temperature increase. He used a known force to turn a paddle in an enclosed container of water and again looked for a temperature increase. Over a period of years he satisfied himself of a fundamental principle: A certain quantity of mechanical work, when efficiently and completely transformed, always created an equivalent amount of heat. This conversion factor he named the mechanical equivalent of heat, and in the ungainly units of the day he concluded that a quantity between about 600 and 1,000 foot-pounds of mechanical effort was needed to heat one pound of water by one degree Fahrenheit.

Getting his results published proved difficult, the *Proceedings* of the Royal Society being especially resistant. In later years he joked that these rejections didn't surprise him. "I could imagine," he said, "these gentlemen in London sitting around a table and saying to each other: 'What good can come out of a town where they dine in the middle of the day?'"[2] Joule had more luck with the Lit & Phil journal and the *Philosophical Magazine*, a London journal founded in 1798 specifically to promote science of a practical, empirical nature. Even so, Joule's presentations of his findings to British Association meetings, in Cork in 1843 and in Cambridge in 1845 (which Thomson is known to have attended), attracted mainly indifference.

In 1847 in Oxford he presented his latest results, using what he re-

[2]An irregular line can be drawn roughly east to west across the middle of England. South of this division, people eat breakfast, then lunch, then dinner; to the north, they take breakfast, dinner, and tea. Incalculable social consequences follow.

garded as his most trustworthy method. By turning a paddle wheel to heat water, Joule had concluded that it took a little over 780 foot-pounds of effort to heat one pound of water through one degree Fahrenheit. In the audience was the new Glasgow professor, William Thomson, who at this time was wholly persuaded of Carnot's principle that heat passed through a steam engine unchanged in quantity, creating mechanical work as it went. A Carnot engine running in reverse, therefore, used mechanical work to *move* a quantity of heat from a low temperature to a higher one, but now here was this man Joule saying that he could use mechanical work to *create* heat. According to his own recollection 35 years later, Thomson "felt strongly impelled at first to rise and say that [Joule's conclusion] must be wrong," but "as I listened on and on, I saw that . . . Joule had certainly a great truth and a great discovery, and a most important measurement to bring forward. So instead of rising with my objection to the meeting, I waited till it was over, and said my say to Joule himself, at the end of the meeting." At a reception that evening at the elegant Radcliffe Camera, the two spoke further. "I gained ideas which had never entered my mind before, and I thought too I suggested something worthy of Joule's consideration when I told him of Carnot's theory," Thomson recalled.

Though skeptical, Thomson was impressed by Joule's modest sincerity and earnestness and by the obvious care with which he had conducted his experiments. He didn't know what to make of Joule's findings, but he saw something new and of profound significance. "Joule is I am sure wrong in many of his ideas, but he seems to have discovered some facts of extreme importance, as for instance that heat is developed by the fric[tion] of fluids in motion," he wrote to his father, telling him to tell James to look out for reports of Joule's work. He quickly developed a sympathy with Joule, who was overjoyed to find someone taking him seriously— and someone who was, by reputation, the rising star of British science. A little later Thomson sent copies of Joule's works to his brother, with the warning "I enclose Joule's papers which will astonish you." James, in his measured way, wrote back two weeks later with his interim verdict: "I certainly think [Joule] has fallen into blunders [but] some of his views have a slight tendency to unsettle the mind as to the accuracy of Clapeyron's principles."

Thomson's first meeting with Joule had an odd postscript. From the end of July to early September, Thomson traveled to Paris and then to Switzerland, meeting scientists but mainly enjoying a walking holiday. Near Chamonix, in the French Alps, he had an unexpected encounter. As he recalled it 35 years later: "Whom should I meet walking up but Joule, with a long thermometer in his hand, and a carriage with a lady in it not far off. He told me he had been married since we parted at Oxford! and he was going to try for elevation of temperature in waterfalls. We trysted to meet a few days later at Martigny, and look at the Cascade de Sallanches, to see if it might answer. We found it too much broken into spray."

Looking for a temperature increase from the top to the bottom of a waterfall was a more hopeful than plausible way of finding out the heat generated by motion. Thomson's charming tale, recounted many years later, is a fine example of his capacity for embellishment. At the time of the meeting he gave his father a simpler account. "Before leaving the St Martin road, I met, walking, Mr Joule, with whom I had recently become acquainted at Oxford. When I saw him before, he had no idea of being in Switzerland (he had even wished me to make some experiments on the temperature of water in waterfalls) but since that time had been married, & was now on his wedding tour. His wife was in a car, coming up a hill. As we were going different ways, we had of course only a few minutes to speak." In other words, Thomson had told Joule he was going to Switzerland, and Joule had asked about the feasibility of measuring the temperature of waterfalls. The long thermometer that Joule carried while his new bride waited patiently in the carriage seems to be pure invention. Thomson was fond of these occasional ornamentations. To his credit, his inventions are rarely for self-aggrandizement, just to make a good story.

***

Settling into his new life in Glasgow, Thomson continued to puzzle over the apparent contradiction between Carnot and Joule but for the moment could see no way forward. In the meantime he developed his lecture courses, pondered other scientific problems and, judging by scraps of evidence from his correspondence, enjoyed a social life. Ludwig Fischer, his old Cambridge rival, later companion, had recently come to Scotland as professor of mathematics at St. Andrews University. Having seen a note from Thomson to another Cambridge friend, Fischer wrote: "I must say I am not at all satisfied with the 'pious' wish you express at the end concerning matrimony, having hoped that your attention might have been much engaged at Oxford by certain young ladies, on whom, I learn from good authority, you have made the most favorable impression."

Earlier in the year J. B. Dykes, an undergraduate musical friend on his way to a career in the church, had responded to some sort of jokingly admonitory letter from Thomson: "Your most grave & sober counsel had, I rejoice to say, a most beneficial & salutary influence upon me, & made me there & then, on the spot, repent in dust & ashes for my sins of omission & commission, mentioned by you in your epistle & more especially that heinous sin of flirtation. I felt most keenly the force of your remarks, & that they were so very much to the point inasmuch as I felt convinced that they came from 'a party' who was quite conversant with the topics on wh: he wrote & who in his daily & nightly serenades & promenades & searches after '*them*' would have himself experienced so lately those pleasant & touching little sensations which he so wisely & properly reprehended in me. . . . Now don't you go for to flirt with any young women at Oxford remember '*them*'. . . ."

And when he took up his Glasgow professorship, another friend dashed off this warning: "Mind you don't get married before you are aware of it—you are in a *very dangerous* position now—all the prudent mammas in Glasgow will be asking you to tea—but take care!" following up two months with a rumor lacking any foundation: "There is a tremendous report afloat in Cambridge about you—viz—that you are supposed to be married. I hope you will authorize me immediately to contradict it."

These fragments give a sense of young men adopting a bluff and jocular style to hide their unworldliness. But Thomson was clearly no dry academic. The young wife of a Glasgow friend recalled: "I was asked to go to balls to chaperone him. The ballroom was a dirty trades hall badly lighted and with second rate music. William always used to ask me to take him home at 12 o'clock, but he was generally unwilling to come so soon. . . ."

Elizabeth and Anna had tried teaching their brothers to dance when they were young, though at the time William in particular "professed utter scorn." James never danced, but William evidently found a taste for it, or for the benefits it brought, and Elizabeth suspected he took private lessons as a young Glasgow bachelor, though he was careful to conceal it.

When he had returned to Glasgow in 1846, the Thomson family was intact except for Anna, who was married and living in Belfast. Elizabeth had married the Reverend David King in 1843, and they lived elsewhere in the city. The rest still lived with their father, looked after by their Aunt Agnes Gall. The two younger Thomson sons, on whose education their father had lavished less personal attention, were adequate but not outstanding scholars. John, a lively and amusing youngster, at first went into the business world, learning the ropes in a commercial warehouse in Glasgow. But in May 1844 he was pleased to write to William in Cambridge to say that he had given up his job—"regular drudgery" he called it—and planned to study medicine. He would rather be happy than make money, he explained. He did well, winning the second prize in medical studies at the end of the 1846-1847 session, which was also William's first session as professor. But studying to be a doctor was a perilous path. In April, while doing his rounds at the hospital, John caught a fever. Within a couple of days, at the age of 21, he was dead.

The following October Elizabeth, suffering from unspecified ill health, sailed to Jamaica to convalesce. She set off in a tearful farewell from the Glasgow docks, her father and siblings not at all sure they would see her again. That winter cholera struck Glasgow once more, and its victims this time included the 62-year-old and visibly aging Professor James Thomson. His end came quickly. He appeared weak but not overly unwell, then lapsed suddenly into a delirium, calling out urgently for his daughters and becoming calmer when he thought they were beside him. William described events in a letter to David King, Elizabeth's husband. "He burst out rather faintly into a very incoherent set of expressions of numbers in all varieties of arithmetical denominations, hurrying rapidly from one to another, and giving the answer or saying 'That's right! Now, what is seven hundred and eighty-six inches equal to?' and so on for several minutes." Aunt Agnes wrote to Elizabeth of her father's last moments. "Elizabeth! Elizabeth Thomson! Oh it is a dear name," he called out. He died on January 12, 1849. Anna Bottomley came over from Belfast as soon as she could but too late to see her father alive.

Further departures followed. Robert Thomson, the youngest child, had attained good health after surviving two surgeries to remove stones. In 1846 he joined the Scottish Amicable Insurance Office, starting on £20 a year—a tenth of the value of William's Cambridge fellowship. William bought stock apparently on Robert's advice, but was soon writing to his father asking for help on unloading it. In 1850, a year after his father died, Robert emigrated to Australia, where he married and had children. A letter from him to William survives, written in April 1885 on notepaper of the Colonial Mutual Life Assurance Society in Melbourne. It is a brief letter of introduction to William—Sir William Thomson by then—on behalf of a Melbourne colleague of Robert's who was coming to England for some months. He never returned to Britain and died in 1905.

Recovering from her illness, Elizabeth returned to Glasgow. Meanwhile James, perhaps feeling overshadowed by his brother's increasing reputation, moved to Belfast in 1851 and became a temporary professor of engineering at Queen's College. He won permanent appointment in 1854. He and Anna were close, but in 1857 she died, at the age of 37, leaving a son, James Thomson Bottomley.

***

Carnot's essay on motive power was not the only forgotten treatise that came to influence Thomson's early career. In one of his undergraduate publications, Thomson had found a mathematical equivalence between the flow lines of heat, described by Fourier's theory, and the geometry of electric lines of force, as proposed by Faraday. In this equivalence, contours of constant temperature corresponded to electrically charged surfaces. Thomson soon discovered he had not been as original as he first thought. In the *Journal de Mathématique* a few years earlier the Frenchman Michel Chasles had published related geometrical theorems, although he had not made the physical connection between heat and electricity. Thomson added a note to his paper in the *Cambridge Mathematical Journal* mentioning Chasles. But then he discovered a still earlier precedent. A brief citation in another paper suggested that both his and Chasles's results had been anticipated in a work titled *An Essay on the Application of Mathematical Analysis to the Theories of Electricity and Magnetism*, privately published in 1828 by George Green. A former Cambridge man, Green had died in 1841, and Thomson could find no trace of his obscure treatise.

According to another perhaps retrospectively enhanced anecdote, Thomson mentioned to Hopkins on the evening before he left Cambridge for Paris that he was intrigued by references to Green but hadn't been able to find the *Essay*, whereupon Hopkins said he thought he had a copy. Going to Hopkins's rooms, they found three copies, of which Thomson left with two, one for himself and one for Liouville. (An oddity of this tale is that Hopkins was surely familiar with Thomson's published papers, apparently knew of Green's *Essay*, but didn't make the connection until Thomson brought it up. Either Hopkins, like everyone else, hadn't read Green or he told Thomson about it earlier. But that would have spoiled the story.)

Green, Thomson now discovered, had established a whole range of mathematical theorems concerning the geometry of electric and magnetic forces and the distribution of charges and magnets. In Paris, word of Green got to another French mathematician, Charles Sturm, who had also published similar ideas. One evening an excited Sturm had come to Thomson's lodgings on the Rue M. Le Prince, eager to see the fabled *Essay*. Riffling through the pages, he exclaimed *"Ah! Voilà mon affaire!"* when he caught sight of Green's prior proof of his own theorem. Some years later Thomson arranged for the republication of Green's work in a continental journal, along with his own explanatory essay, and certain theorems first associated with other names are now correctly known as Green's.

At Liouville's prompting, Thomson wrote a short proof of the equivalence of Faraday's lines of force and the inverse square, action-at-a-distance picture preferred by the French. Now equipped with Green's resurrected mathematics, he developed to a high degree of sophistication a new geometrical account of electric forces and charges. This work owed something to formal French rigor, to his own Cambridge training, as well as to Faraday's vision. Most notable was his introduction of "images" in solving electrical problems. A conducting body, such as a metal sphere, carries electricity all over its surface when charged. The force between one such body and another, especially when their shapes are more complex, can be calculated in principle from the inverse square law, but only with difficulty. Between each point of one surface and each point of the other a force exists. The total force between the two extended bodies is the sum of all these increments. Such a problem, an archetypal exercise in integral calculus, was a specialty of the French mathematicians, but for complex geometries the solution quickly becomes intractable.

Thomson proved that in terms of their electrical effects a charged body of some given geometry must be equivalent to a set of suitably placed points of electrical charge. Yet again Fourier's treatment of heat flow provided the germ of the idea. A source of heat, or several, placed within some medium, will after a time lead to contours of temperature throughout the medium. Those contours bear a specific relationship to the heat sources that produced them. Likewise, Thomson showed, the conducting surface of an electrically charged body can be related to a set of charges with the appropriate arrangement—and calculating from a finite number of points is easier than dealing with an extended body of arbitrary shape.

Back in Cambridge, Thomson talked of his ideas at the British Association meeting there in June. The import, he explained, was that by taking Coulomb's inverse square law of electrical attraction and repulsion, and applying the mathematical methods devised by Green, which he had partly rediscovered for himself, a number of Faraday's assertions about the nature of electrical phenomena could be demonstrated. Faraday was at the meeting and spoke with Thomson, gratified that mathematical argument bore out his beliefs.

Despite this promising start, Faraday and Thomson corresponded only occasionally over the years. The two men's mental powers, both acute, worked in utterly different ways. Thomson could never fully understand or even contemplate a proposition until he had given it precise mathematical form. Faraday, by contrast, constructed his physics entirely without the aid of mathematics, for the simple reason that he knew no mathematics.

No scientist, I believe, not even Newton or Einstein, had a greater power of pure imagination than Michael Faraday. The son of a blacksmith who had moved down from Yorkshire to London during economically troubled times, Faraday was born in 1791, the third of three children. His father had difficulty finding work and was often in poor health. The family lived in cramped conditions above a coach house in an area that today is on the fringe of London's affluent West End. "My education was of the most ordinary description, consisting of little more than the rudiments of reading, writing, and arithmetic at a common day-school. My hours out of school were passed at home and in the streets," Faraday recalled. He left school at 13 and apprenticed to a bookseller and binder, George Riebau, who deserves recognition as one of the unsung heroes of scientific history. Apprenticeships were often little more than indentured servitude, but Riebau was a generous and large-spirited man. Faraday at first worked as an errand boy but soon began to learn book-binding. He began to read the works he bound, and Riebau encouraged young Faraday to stay after hours and study whatever interested him. He read about electricity and chemistry in the *Encyclopedia Britannica* and with a few spare pennies bought old glass jars from a rag-and-bone shop to do his first experiments.

Industrialization and urbanization in the 19th century brought hordes of poor and uneducated young men into the growing cities. Philanthropists and social progressives, in their paternalistic but sincere Victorian way, founded evening schools and discussion societies to bring education and intellectual discourse to the working classes. The City Philosophical Society was one such institution. Faraday, joining it in early 1810, when he was 18 years old, participated nervously at first in discussions of history, philosophy, and science. Not unlike William Thomson's father, Michael Faraday was single-minded in the task of self-improvement. Forming friendships with other young men, he sought to acquire good English and learn some French, and he put together a little chemistry laboratory to try out what he read.

Faraday was fanatical and orderly in taking notes and bound up his autodidactic writings in volumes that George Riebau showed off to some of his customers. One such regular, a Mr. Dance, was sufficiently impressed by the apprentice's avid work that he gave Faraday tickets to hear lectures by the celebrated chemist Sir Humphrey Davy at the Royal Institution, near Piccadilly Circus and barely more than half a mile from the Faradays' meager lodgings. Sir Humphrey was a dashing man and a thrilling speaker, apt to make the young ladies in his audience swoon. Faraday merely took notes and tried to perform at home the experiments Davy recounted.

When he was 21 Faraday's apprenticeship came to an end. He was by then too rapt by science to settle for the reliable but dull life of a book-binder. He wrote to Davy for a job at the Royal Institution and got sympathy but no immediate help. A few weeks later, as luck would have it, Davy injured an eye in an experimental mishap and called on Faraday to assist him. Soon after that an assistant at the institution was thrown out for unruly behavior, and Faraday, in 1813, began working there, with accommodation provided and use of laboratory equipment in his spare time. In his long life he never worked anywhere else.

Six months later Faraday embarked on an 18-month grand tour of Europe with Davy and his new wife, the wealthy widow Mrs. Apreece, on the understanding he was to be Davy's technical assistant. Lady Davy regarded him as a manservant. In the salons of the great cities of Europe, Davy parlayed his scientific talents into a kind of showmanship. In Paris he brought out his traveling chemistry kit and showed that a strange purple vapor was a new element, which he called iodine. In Florence he experimented on small diamonds that the Grand Duke of Tuscany sacri-ficed for science and proved that diamond was a form of pure carbon. Faraday met some of the great men of Europe, in between resisting Lady Davy's instructions to haul luggage or shine shoes.

By the time they returned to England in 1815, Faraday had learned some chemistry and other science, but above all he had learned that salon life was not for him. He did not so much despise society as wish to live apart from it. Faraday's family belonged to an exclusive and self-con-tained Protestant sect, the Sandemanians. They lived according to a strict and simple interpretation of biblical guidance written down in the middle 18th century by Robert Sandeman, who died in 1771 in Connecticut having failed to establish an American branch of his religion. The Sandemanians believed in salvation through faith and thus rejected as coarsely utilitarian the more usual Protestant idea of redemption through good works. They married among themselves, as Faraday did, marrying Sarah Bernard in 1821. Their social life was almost wholly among the Sandemanians. Faraday avoided as far as possible civic events and func-tions, even if he was the object of the honor, and in later years almost his only concession to the social graces was his annual attendance of the anniversary dinner of the Royal Institution. He had no students and rarely collaborated with others. He explained once: "I do not think I could work in company, or think aloud, or explain my thoughts at the time. Sometimes I and my assistant have been in the laboratory for hours and days together, he preparing some lecture apparatus or cleaning up, and scarcely a word has passed between us."

Central to the Sandemanians was a pious humility, a calm acceptance of the fallibility and imperfection of humanity. This attitude colored Faraday's scientific work. "In all kinds of knowledge I perceive that my views are insufficient, and my judgement imperfect. In experiments I come to conclusions which, if partly right, are sure to be in part wrong; if I correct by other experiments, I advance a step, my old error is in part diminished, but is always left with a tinge of humanity, evidenced by its imperfection," he wrote to his brother-in-law. Above all he turned away from worldly vanities, the false allure of reputation and public acclaim. Work was its own justification. The purpose of scientific investigation was to shed light, however feebly, on God's design, and thus praise Him.

Even in the innocent days of the 19th century, such an attitude was hardly conducive to the promotion of a scientific career. Faraday resisted occasional attempts to draw him into professorial positions elsewhere and only intermittently attended meetings and conferences where he might explain his findings and opinions. After spending his early research years mainly on chemical work (notably he succeeded in liquefying chlorine), he moved into electrochemistry (reactions stimulated by the passage of electric currents through solutions) and thence into his pioneering and utterly original studies of electricity and magnetism.

Faraday was to an extent influenced, via Davy, by German philosophical views. Davy was close to the poet Coleridge, who had become a great proselytizer for *Naturphilosophie* after he spent 1798 in Germany. Kant, for reasons best left to philosophy, believed that the idea of points acting on each other through empty space was inadmissible. Instead he argued that forces pervading space were fundamental and that matter was in essence the manifestation of a resistance to force, rather than an entity in its own right.

However dubious these propositions, they stimulated Faraday to think of electric and magnetic effects as influences spreading throughout space, rather than as the summation of discrete forces between isolated objects. He objected vehemently to the idea that a force could act instantaneously across empty space. Instead, he thought electric and magnetic influences must propagate from one place to another, conveyed by some presumed medium—hence his lines of force, which he conceived almost as elastic links, carrying tension and perhaps inertia. Thus, Faraday viewed electricity and magnetism as live, conjoined creatures inhabiting space.

Qualitative though this picture was—as it had to be, since Faraday lacked the means to translate it into mathematical propositions—it enabled him to design and perform quantitative experiments. His most celebrated discovery was probably his demonstration of electromagnetic induction. It had been known since 1820 that a current passing along a wire would make an adjacent compass needle deflect. If a current could create a magnetic force, it seemed to Faraday and many others that the complementary effect—a magnet creating a current—ought to occur too. However, a permanent magnet placed beside a wire will do nothing.

In 1831, Faraday took an iron ring six inches across and wound it with coils of fine wire on opposite sides. Connecting a so-called galvanometer to one coil to detect any current, he touched the other coil to a battery and quickly disconnected it again. It was the *pulse* of current, not a steady flow, that made the galvanometer twitch one way when he connected the circuit, and the other way when he broke it again. Faraday imagined that when he passed a current through one of the coils, magnetic lines of force sprang away from it and cut through the other, generating a current. Once the magnetism was steady, the lines of force remained static, and no current appeared. But when he disconnected the circuit, the magnetic lines of force collapsed back again, generating an opposite current as they retreated through the secondary coil.[3]

Thus did Faraday's conception of electromagnetism as a dynamic and extended phenomenon lead him to find a long-sought effect, where the old picture of static forces had borne no fruit. Later that year he showed that moving a plain bar magnet near a coil could also create a

---

[3]The American scientist Joseph Henry, who lent his name to the publisher of this book, made the same discovery of electromagnetic induction at the same time. Much of Henry's work on electricity and magnetism parallels Faraday's, though it is not as well known mainly because American science itself was so little known then.

current. The greater the number of turns in the coil, the more wires each magnetic line of force cut through as it moved and the greater the current produced. His conception of magnetism, qualitative though it was, yielded quantitative predictions.

William Thomson first encountered Faraday's science in the early 1840s, when he was taking classes in Glasgow from David Thomson, substituting for the ailing Meikleham. It was then, Thomson claimed in his old age, that he was "inoculated with Faraday fire," but in his Cambridge notebook from March 1843 we find him recording a long conversation with Gregory "in w^h Faraday and Daniell [another electrical scientist] got (abused)^2."

Faraday devised his scientific theories in pictures, almost in cartoons, and though his experimental demonstrations were admirable, his thought processes must have seemed to young Thomson quaint at best. Where was the analytical proof? Where was the reduction of physical phenomena to quantities amenable to mathematical manipulation? Even now there is a tendency to praise Faraday with a tinge of condescension as a great but uneducated experimenter, as if he were some sort of idiot savant with an inexplicable knack for putting wires and magnets and batteries together in clever ways. Those who can think of physical theories only in mathematical terms evidently have trouble understanding a theoretician who did not work the same way. Faraday was no mathematician, but it was he more than anyone who originated the modern view of the electromagnetic field. He was a magnificent experimenter, but guiding his experiments was a powerful vision of electromagnetism. He had one of the great theoretical minds in physics.

Thomson slowly came to appreciate Faraday's insight. Following their initial meeting and correspondence in 1845, Thomson began a series of papers under the title "Mathematical Theory of Electricity in Equilibrium," in which he devised a general system for dealing with distributions of electric charges and the forces they produced. This was mathematics of a high order; it was also practical physics. For the small but growing number of practitioners who wished to make electrical calculations for the purpose of building machines and devices, Thomson's methods were a boon.

Because this series of papers established relationships between charges, considered as points in space, and the influences they produced, considered as effects spread throughout volumes or on surfaces, they made a start on capturing in mathematics Faraday's picture of electricity as an extended and pervasive "tension" maintained somehow across space. But the point was arguable. Thomson's mathematics could equally be seen as a comprehensive elaboration of the consequences of inverse square laws and action at a distance. In the first paper of the series, Thomson hedged his bets as to what his results meant. He preferred to think of them "merely as actual truths, without adopting any physical hypothesis, although the idea they naturally suggest is that of the propagation of some effect."

The exchange of views between Faraday and Thomson, though slight, proved crucial. Thomson began to think, as Faraday did, of space as a medium supporting both electric and magnetic phenomena. How rapidly his thinking evolved is evident from a paper he wrote in 1847 and sent to Faraday. In it he showed, in a very preliminary way, how the properties of a physical medium could be connected to electric and magnetic influences.

Thomson imagined, in general terms, some kind of elastic solid, such as a lump of gelatin, which had both resilience and flexibility. Unlike a rigid solid, it would yield in response to an applied force; unlike a liquid, it would rebound to its original state on removal of the force. Thomson showed that from a purely mathematical standpoint, forces of electric attraction or repulsion had the same form as compression or tension in the medium. Magnetism was trickier. A compass needle deflected by a current passing along a wire, for example, might flick to the left above the wire but to the right below. The geometry of magnetic forces, Thomson showed, paralleled a mathematical description of a localized twist or rotation of the elastic medium (as if, loosely speaking, one held a lump of gelatin in one hand, stuck a fork in it with the other, and twisted the fork a little).

"What I have written is merely a sketch of the mathematical analogy," he explained to Faraday. "I did not venture even to hint at the possibility of making it the foundation of a physical theory of the propagation of electric and magnetic forces." Nonetheless a physical theory was what he had in mind, at least as a distant dream. He had abandoned the old action-at-a-distance philosophy in which one simply posited forces between particles in truly empty space. The idea of space as an electromagnetic medium was pressing on him. Since the same medium, he hoped, would eventually be seen to carry both kinds of effects, there was the ultimate prospect of connecting electricity and magnetism by means of a single fundamental theory. This was to be the preoccupation of a lifetime, but for the moment Thomson contented himself with working out a couple of long accounts of the geometry and mathematics of magnetic forces, as he had done for electricity. Further study of the presumed medium supporting these phenomena would come later.

***

Ten days after the death of Professor James Thomson, Hugh Blackburn wrote to Thomson to say he intended to put his name up for the vacant position. Like many Cambridge-trained mathematicians (at least those who didn't want to enter the church), Blackburn had gone down to London to take up a legal career, but unlike Archibald Smith, he was finding no satisfaction in it. Thomson, while offering encouragement to his old undergraduate friend, tried to interest his colleague George Gabriel Stokes in coming to Glasgow. Stokes, five years older than Thomson, had been senior wrangler in 1841 and then became a fellow at Pembroke College. Like Thomson, he applied his mathematical knowledge to questions of physics and achieved important results in optics and fluid mechanics. His Cambridge career was not yet secure, however, and the prospect of a lifetime appointment alongside his friend Thomson held many attractions. But he ran up against Glasgow rules. All professors had to sign the Westminster confession, by which they declared their allegiance to the Presbyterian Church of Scotland. This was meant to guarantee that faculty members would take no part in the kinds of religious turbulence that had disrupted Scottish life and politics since the days of the Covenanters.

With his detestation of religious prejudice and sectarianism, Thomson's father had campaigned against the religious test as a needless holdover from unenlightened times. It survived, largely because signing the confession was seen by younger men as a piece of meaningless bureaucracy. Stokes, however, was the son of an Anglican minister in County

Sligo, Ireland, and his three brothers all became churchmen. He was moreover a punctilious man. He arranged to have testimonials sent to the Glasgow faculty but immediately regretted doing so. A month after James Thomson's death he wrote to William to explain that after consulting his older brother he decided he could not go through with the application. The "straightforward course is, to decline to take [the religious test] unless I am prepared to become a thorough Presbyterian, which certainly I do not mean to become. . . . It was all along a very doubtful question with me whether I could sign the test in a lax sense."

Laxity was fine with Thomson, however. Whereas his father had argued for the abolition of the test, as a matter of principle, Thomson's solution was to let the thing slide. He himself, he explained to Stokes, regularly attended the Episcopal Church in Glasgow (the Scottish counterpart of the Church of England) and went to the Established Church (of Scotland) no more than "once or twice or three times in the course of a session." Neither he nor his colleagues found anything objectionable in this. He told Stokes that "the *amount* of conformity to the Established Church which a conscientious observance by one in your position of the obligations imposed by the tests, would really be in no way inconvenient, or repugnant to your feelings."

He added: "It will be a very serious blow to the interest of this University if an honest member of the Church of England should never be able to be a candidate for any *situation or office* connected with it, however valuable an acquisition he might be; on account of an act of Parliament framed at a period of great political & ecclesiastical excitement; and allowed to continue unmodified in these settled times, merely because the modifications that those who have the interests of the University most at heart would be inclined to have made, are such that only those parts of the Act which are at present practically inoperative, would be abolished." No doubt this was an accurate assessment. The rule was still on the books, but everyone agreed to look the other way, except perhaps for some of the older and more conservative men, whose opinions hardly counted anyway.

Thomson was a conventionally religious man all his life and believed firmly that the rational working of the universe and the ability of science to describe it were signs of God's immanent power. But for niceties of doctrine and points of observance he had no patience. Long ago, in the summer of 1834 or 1835, the vacationing Thomson family had gone to services at a local parish church where there happened to be a revival meeting. Their father was not there, and Elizabeth had charge of the boisterous youngsters. As the service proceeded and the preacher became more animated, members of the congregation began to moan and sway and throw their bibles in the air. This set William snickering, which set off the other children. The preacher, hearing the disturbance, interrupted his sermon, glared at the Thomson children, and pointed his finger at William. "Ye'll no lach when ye're in hell!" he thundered, at which William and the rest collapsed into a helpless fit of the giggles, leaving Elizabeth, red faced, to hustle the young heathens from the church.

As an adult William learned to maintain his decorum, but he seems to have regarded church going as a necessary formality, bearing little relationship to his thoughts on God and the nature of the physical world. Stokes, by contrast, took these things seriously and would not go along with Thomson's plan to sign the Glasgow religious test with his eyes closed, so to speak. In April 1849 the open mathematics chair went to Hugh Blackburn. Later that year Stokes became Lucasian Professor (Isaac Newton's old position) and remained at Cambridge the rest of his life. Blackburn performed adequately as a teacher of mathematics but made no original contributions to his discipline. Stokes wrote voluminously on mathematical physics, ended up with a theorem,[4] an equation in fluid mechanics, and some optical phenomena named after him, served for a long time as secretary of the Royal Society, oversaw in minute detail the production of the society's *Proceedings*, and acted, through his indefatigable correspondence, as a guide and mentor to numerous mathematical physicists in Britain, Thomson included. That this career was lost to Glasgow University because of an antiquated rule might have been, in Professor James Thomson's hand, an additional spur to long overdue reform. To William Thomson it was a matter of keen but passing regret.

---

[4]Stokes's theorem actually came from Thomson in a letter in 1842. In the tripos a few years later Stokes asked candidates to prove this still unpublished result, and his name became attached to it. Thomson had no complaint about this, so far as I know.

He wrote to Stokes that "no case can prove the noxiousness of the [Glasgow] law . . . than the present one," but he took the question no further.

By contrast, Thomson threw himself with great energy into his new duties as Glasgow professor. He composed an opening lecture for the incoming class, which he delivered with minor variations each year for many decades. Science, he explained, began with natural history, which was the close observation and classification of material phenomena, and rose to the level of natural philosophy, which was the attempt to understand and connect those phenomena by rational means, expressed ultimately in the language of mathematics. Mechanics was the most mature of sciences, while electricity and magnetism were approaching that pinnacle. He threw in snippets from Francis Bacon and talked of the practicality and applications of science. He added a dash of religion, to say how science aimed to illuminate God's handiwork: "When I consider thy heavens, the work of thy fingers, the moon and the stars which thou hast ordained; What is man that thou art mindful of him, and the son of man that thou visitest him?" Science was to be undertaken in a spirit of humility and with a sense of wonder and beauty; nor should the ability of science to improve the lot of mankind be ignored.

This was the nearest Thomson ever came to a philosophical statement of purpose. He struggled to compose the lecture and was not pleased with his first delivery. "According to his own account, it was a total failure," Elizabeth wrote to her husband. "I think he had been very nervous, and he read much too fast. . . . He is very much disheartened, poor fellow." As a lecturer, Thomson tended to be more enthusiastic than orderly. He tried to keep to a plan but could never resist the digressions that rained in on his mind. At his best, to an appreciative and sophisticated audience, he could be thrilling, inventing profound and provocative science as he spoke. When he gave his introductory lecture he generally tried harder to stick to his script, and the struggle detracted from his fluency and intensity. James Clerk Maxwell, the young Scots physicist who began to make inroads into electromagnetism in the middle 1850s, said that Thomson's annual introductory lecture never managed to fill the hour and that "the lecturer was greatly downhearted at its conclusion."

With this weak essay at a grand purpose out of the way, however, Thomson moved into his regular scientific lectures with eagerness and delight. There he was never downhearted, always ready to share the joys of discovery and enlightenment with his pupils. But in his success at imparting information into lesser minds than his own, he got mixed reviews. Bright students liked his style. One recalled him as "an enthusiastic and inspiring teacher; he aroused and sustained the intelligent interest of his students. . . . No one could listen to him without being imbued with his spirit and being borne along the path he was travelling. . . . He was always in earnest, and when dealing with great problems spoke with the fervour of a missionary charged with a weighty message." But another student, less overawed, more overwhelmed, said: "Explanation, it has to be confessed, was never his forte. He would say, 'Look, see it and believe it.'" And from another: "Even in his introductory lectures Thomson soared to heights which made many of his class feel giddy and helpless."

Though he was patient with slower students, he seemed to think they were being obtuse, not that they had genuine difficulty understanding. He would prompt a struggling pupil more and more minutely, in smaller, easier steps, and finally say with bafflement rather than exasperation: "Now, Mr. Macintosh, why could you not have said so at first; why will you have me drag the information from you sentence by sentence, clause by clause, nay word by word?"

Particularly distressing to Thomson was what he took to calling aphasia—the inexplicable but frequent phenomenon by which capable students were reduced to helpless, struggling silence by the most elementary of mathematical questions. These students, he said, "will not answer when questioned, even when the very words of the answer are put in their mouths, or when the answer is simply 'yes' or 'no.'" Thomson read mathematics as easily as he read words and could not understand why others did not have the same facility.

As a professor of natural philosophy he deserves credit for one fundamental innovation, which was the teaching of practical science through student experimentation. With his brother James he had made mechanical toys, but not until his visit to Paris did he attempt any measurement or manipulation in a scientific laboratory. Neither at Glasgow nor at

Cambridge nor anywhere else in Great Britain was experimental science taught; instead, men such as Joule figured it out for themselves, with some advice from their elders, while in Cambridge Stokes and others began to take up laboratory work on their own initiative and using their own resources.

William Meikleham had not taught any experimental science and undertook no significant research either. When Thomson assessed his new professorial domain, he "found apparatus of a very old-fashioned kind. Much of it was more than a hundred years old, little of less than fifty years old, and most of it was of worm-eaten mahogany. . . . There was absolutely no provision of any kind for experimental investigation, still less idea, even, for anything like students' practical work." He credited Thomas Thomson, his former teacher and now faculty colleague at Glasgow, with having founded a laboratory for his chemistry students to work in. He wanted to work on problems in electromagnetism, and after setting up a laboratory he engaged his students to assist him. This became an essential part of his course in natural philosophy. He did not merely teach mathematical methods and explain what crucial information had emerged from experiments by others. He got the students to do the experiments themselves and to explore the subject in a practical way. It is hard to imagine now how physics could have been taught in a purely abstract way. But before 1846 it was, and Thomson was the first instructor in experimental physics.

To further both his teaching and his research, he embarked on a battle with the Glasgow faculty reminiscent of his purchase of the "funny" only a few years earlier. He applied for money to buy equipment, spent it, spent more, and argued with his colleagues until they paid up. He took over vacant rooms near his lecture hall, turning an old wine cellar into his first laboratory, and when a larger room became vacant after some administrative change, Thomson "instantly annexed it (it was very convenient, adjoining the old wine-cellar and below the apparatus room); and, as soon as it could conveniently be done, obtained the sanction of the Faculty for the annexation." There were protests and exchanges of letters, as there had been with his father, but Thomson committed the fait accompli and got his way.

In both lecture room and laboratory, Thomson invariably slid from an exposition of textbook material into wide-ranging discussions of un-solved problems currently on his mind. This might confuse his students; it could also enthrall them. Sometimes they tried out standard exercises in the use of laboratory instruments; sometimes they helped Thomson with his latest research project. In overseeing such enterprises the young professor "had none of the air or manner of a superior." As a teacher he may have flown often above the heads of his students but "he was never dull, never trivial, never commonplace." "What I liked best," said an-other student, "was when he left us to follow or not as we could, and went on thinking aloud, as he sometimes did. His mind was full of fan-cies, brimming over with metaphors."

* * *

With his father and brother John dead, both sisters married and, after 1851, brothers James in Belfast and Robert gone to Australia, Will-iam Thomson was alone in the old family home except for his aunt and housekeeper, Agnes Gall. He went to dances with Jemima Blackburn, Hugh's wife, as chaperone. Flirtations of an indeterminate nature came and went. His old friend Ludwig Fischer, at St. Andrews, wrote to him early in 1850: "I suppose it is out of the question your coming next week. Else we mean to have a Bachelor's ball on Thurs the 28th, and I might have to offer you a faint resemblance of what you enjoyed at Edinburgh. Of course the Ladies of Scotch Craig, I spoke to you of have been invited. But I doubt whether they would come; nor have I heard whether the beautiful Fanny will be there."

Early in 1851 Thomson was elected fellow of the Royal Society, a few months before his 27th birthday. In that same year he twice pro-posed marriage to Sabina Smith, sister of Archibald.[5] Although Archibald Smith had encouraged Thomson's academic career, congratu-lated him warmly on his successes, and in the end made no serious move to compete for the Glasgow chair, he advised his sister that "I really do not think you would be suited to each other." She duly turned William

---

[5]This romance was uncovered by Crosbie Smith and M. Norton Wise and re-counted in their book *Energy and Empire* (pp. 141-142), from which I have condensed these details.

down. A year later he tried again, for a third time. Again, at her brother's urging though apparently against her own inclination, Sabina said no. William confessed to Sabina's sister his "bitter bitter grief" and many years later Sabina wrote of her regret at not resisting Archibald's influence: "It was the extremity of folly to think I c^d go on refusing a man, & yet have him at my disposal whenever I choose!" By this time, however, she had seen young William Thomson evolve, over the decades, into the wealthy and celebrated Lord Kelvin, a great figure in the land. This may have amplified Sabina's remorse.

Thomson was on his own now. His father was no longer around to advise him or to plan his campaigns. In any case, James Thomson had detested old Mr. Smith and was no great admirer of Archibald, so perhaps would have advised William against this entanglement in the first place.

In July 1852, only three months after his final refusal by Sabina Smith, William became engaged to Margaret Crum, whom he had known since childhood. She was the daughter of Walter Crum, a Glasgow cotton merchant who was first cousin to William's father. Margaret knew his sisters, and their letters to him while at Cambridge mention a number of visits by her but say little about her activities, interests, or personality. The betrothal was sudden. To Elizabeth William wrote of news which "I think will please you as much as it will surprise you." He told Stokes that "sometime, probably early in September, I am going to be married, to a Miss Crum. I cannot describe her exactly to you, but I am sure that is unnecessary to ensure your good wishes at present, and when you come down to see us in Scotland, I am sure you will be glad to make her acquaintance." Unfortunately no one else seems to have exactly described Margaret Crum either. She and William appear to have embarked on some kind of alliance rather than a romance. As Margaret explained to Elizabeth, "We have one interest in common that can never fail, and as I told Mrs. Gall, I feel that in William's love for his sisters and her, lies my best security for the continuation of those feelings on which the happiness of my life must now depend." William apparently set aside passion and deep feelings of the soul after he finished reading *Evelina* and *Wilhelm Meister* in Cambridge, and his disappointment over Sabina Smith stifled any resurgence.

William Thomson and Margaret Crum were married in Glasgow on September 15, 1852. He was 28 years old; she was 22. From a brief honeymoon in north Wales the following week, William sent this account of an afternoon with his new wife: "The day is somewhat dark and cold, and some people might say dreary, but it does not seem so at all to me. I scarcely think it does so to Margaret either, although she has just been saying to me that it is, and what is more, laying particular emphasis on the most dismal parts. Perhaps she is only joking, but whether or not, she looks cheerful, and has quite got rid of her cold. In fact, I do not think either of us are going to apply to Dr. Brown to undo what he did on Tuesday."

One may best interpret this as a piece of dour Scots humor or as the effort of a young man trying to impress an older married sister with his newly acquired sophistication. Few other impressions of Margaret are to be found. Jemima Blackburn recounts how William was a great friend and frequent visitor to the Blackburn household, both before and after his marriage, but his wife makes no appearance in her recollections. Hermann von Helmholtz, encountering her for the first time, wrote of "a rather pretty woman, very charming and intellectual" and the novelist William Thackeray, having met the Thomsons in Glasgow through mutual friends, asked to "give my best regards to . . . Thomson please with his nice wife." On the other hand, she was an amateur poet. The first lines of her poems (published privately after her death) display a grim consistency, and a selection of them can be arranged almost to form a poem themselves:

> *They have sung to thee, O grave!*
> *Ours is a short and evil day*
> *Wounded, bleeding*
> *I long to die*
> *I saw a shadow in the night*
> *When thou dost come for me*

Margaret's verses are doleful, monotonous, and self-absorbed. This was a woman at least half in love with easeful death. To be fair, she had grounds for misery. In May 1853, eight months after their marriage, the

Thomsons went on a Mediterranean tour, taking in Gibraltar, Malta, and Sicily. To William, long accustomed to hiking in the highlands, this was a mere jaunt, but Margaret wore herself out, came home weak and ill, and though the nature of ailment remains unclear she was an invalid for the rest of her life. Back in Scotland she stayed several weeks in Edinburgh for "surgical nursing." Helmholtz, after meeting her, offered this explanation: Margaret was "in a wretched state. A year after her wedding she suffered an abdominal inflammation,[6] and for two years has been in such a state that she can't walk, stand or sit upright without pain, and can only lie on her back." Whatever the cause, Margaret's health was such that she composed a poem, "On Pain," from which the following selection is ample:

> There's many a wight sings of delight,
> Who courteth her in vain;
> But I, more true, will tell to you
> Of what I know—'Tis pain. . . .

> My quick young feet him soon did meet,
> When they life's race began
> Said he, "As friend, thy steps I'll tend"
> Ah, me! no more I ran. . . .

> Joy did I meet and haste to greet,
> He seized my hand instead;
> Love did I find and seek to bind,
> Before his face it fled. . . .

> Now where is rest, when such a guest
> Me ever followeth,
> Nor lets me clasp with desperate grasp
> The outstretched hand of Death.

---

[6]A miscarriage possibly? Pure speculation on my part.

Thomson became nursemaid to his wife. By next summer her health had not significantly improved, although she had occasional better days that William seized on as grounds for hope. He reported to Elizabeth that "she looks much better . . . but she has not at all advanced in walking power. I always carry her up and down stairs, and often from one room to another. A walk half round Miss Graham's garden lately knocked her up for several days. But by avoiding all such exertion she keeps tolerably free from pain, and has much the appearance of good health. I take her a drive nearly every day and sometimes twice in a little pony carriage."

The appearance of good health is the most Margaret Thomson ever subsequently attained and that only intermittently. It is easy for the modern reader to infer some variety of malingering on Margaret's part, perhaps amplifying an underlying problem. William was clearly of so accommodating a nature that his wife had no great incentive to improve, especially if, as her poems suggest, she had developed a fond intimacy with chronic pain and morbid thoughts. But we should not forget that in the middle of the 19th century all kinds of internal disorders were undiagnosable and untreatable.

In the years thereafter Thomson applied himself diligently and uncomplainingly to the care of his invalid wife. She became a duty rather than a passion. Perhaps that suited him.

***

After three years of thinking mostly about electricity and magnetism, Thomson returned to his Paris discovery of the science of steam engines. Still he had not seen Carnot's essay and knew it only through Clapeyron. But that was enough for him to turn one aspect of Carnot's theory into an important realization, both theoretical and practical, about temperature. Hot and cold, of course, are direct physiological sensations, but temperature is a difficult concept to make quantitative. Around 1590, Galileo invented his "thermoscope," in which the expansion of air with rising temperature provided a crude numerical scale. Early in the 18th century Gabriel Fahrenheit came up with recognizably modern thermometers that relied on the expansion of colored alcohol or mercury. There was no theory to speak of behind these instruments, only the empirical fact that gases and liquids tend to expand when they get hot. But these early thermometers at least allowed measurements by different investigators to be recorded, compared, and calibrated against each other.

During the 18th century many experimenters studied the relationship between pressure and temperature for a given volume of gas. Their results came together in a simple law: pressure times volume rises in proportion to temperature. In particular, for a gas maintained at constant pressure, volume changes linearly with temperature, which meant that the expansion of any suitable gas would serve to make a thermometer and that all gas thermometers ought to yield the same temperature scale.

It was readily apparent, though, that this wasn't quite true. The simple rule relating pressure, temperature, and volume became known as the ideal gas law, on the understanding that all actual gases departed from this ideal in ways small or large. Gas thermometers, therefore, gave slightly different temperature scales depending on which gas was used. One scale could always be calibrated against another, but since there was no independent way of measuring temperature, there was no way to say which temperature was most nearly correct—that is, which one corresponded most closely to the temperature implicit in the ideal gas law.

Also implicit in the ideal gas law, it appears, is that temperature cannot fall without limit. On cooling, the volume of any gas decreases, and since no physical object can have zero volume, let alone a negative volume, there would seem to be an absolute zero of temperature, below which it cannot fall further. As early as 1699, the Frenchman Guillaume Amontons estimated that this endpoint corresponded to a temperature of about −248° Celsius, and in 1847 Thomson's friend Regnault came up with −272.75°C, very close to the modern value of −273.15°C.

But this is highly misleading. A volume of steam, for example, cooled below 100°C at normal atmospheric pressure, turns into water. Below that temperature the gas law no longer applies. Likewise, although air could be cooled to much lower temperatures without apparent change, no one in the late 18th or early 19th century imagined that the temperature could really be reduced down to −270°C or thereabouts without some substantial physical change intervening. Regnault and his contemporaries therefore regarded a numerical value such as −272.75°C as a calibration point for a gas-based temperature scale, and nothing more. Absolute zero as a physical concept did not yet exist.

In Regnault's lab and elsewhere the air thermometer had become the de facto standard, and it could be calibrated with some consistency against mercury or alcohol thermometers. Still, no rational temperature scale—meaning a temperature that was defined quantitatively in terms of known physical laws and standards—had been devised. Thomson concluded that the vagaries of individual gases and their failure to live up to the ideal gas law prevented any gas-based thermometer from yielding an absolute temperature.

In 1848 Thomson wrote a short paper explaining how Carnot's theory could solve the problem. Carnot had established that the maximum work any engine can produce from a known quantity of heat can depend only on the upper and lower temperatures between which the engine cycles, and that this efficiency is the same no matter what the working substance, whether steam, air, or some other gas. Thomson defined a temperature scale by asserting that a Carnot cycle operating through a one-degree interval always produced the same amount of work from a given quantity of heat. That is, a certain cycle would yield the same amount of work operating between 100° and 99° as it would operating between 99° and 98°, and so on.

Thomson's noteworthy conceptual innovation here was to define a temperature in purely mechanical terms, independent of the properties of this or that gas or liquid. On the other hand, since building an ideal Carnot cycle was no more possible than finding an ideal gas, his temperature definition was at this point theoretical rather than practical.

By asserting that an ideal engine had the same efficiency at all temperatures, Thomson was able to say that "all degrees have the same [mechanical] value." But this was no more than an assumption. Thomson could cite no evidence or argument for it.

A corollary of this assumption was that Thomson's temperature scale had no zero. Drop a degree, get some work; drop another degree, get the same work again; and so on without limit. His temperature scale therefore went down to minus infinity. He knew that the temperature scale defined by an air thermometer came to a halt at −273°C, when volume went to zero, and so concluded that "infinite cold [on his scale] must correspond to a finite number of degrees of the air-thermometer below zero."

Histories of science sometimes claim that Thomson's 1848 paper established the existence of absolute zero as a physical concept. This is not true. He clearly regarded it as a virtue of his temperature scale that all degrees had "equal value" and that it went down to "infinite cold." By contrast, he said, "the value of a degree . . . of the air-thermometer depends on the part of the scale in which it is taken." In other words, the fact that gas temperature had a zero in the vicinity of $-273°C$ he did not regard as physically meaningful but as a misleading consequence of the way it was defined.

In setting out these conclusions, however, Thomson admitted to one nagging anxiety. He still had not resolved the apparent contradiction between Carnot and Joule. He persisted with Carnot's view that the production of mechanical work during a cycle came from the transmission of heat from a higher to a lower temperature. But in a footnote he referred to "Mr. Joule of Manchester," who had unarguably converted work into a proportional amount of heat. Joule believed that the reverse must also happen, but Thomson declared that "the conversion of heat (or caloric) into mechanical effect is probably impossible, certainly undiscovered." Rather helplessly he could only conclude that "much is involved in mystery with reference to these fundamental questions of Natural Philosophy."

Just weeks after this paper appeared, Thomson acquired a copy of Carnot's essay on motive power from Lewis Gordon, who had been appointed in 1840 the first professor of engineering at Glasgow. Thrilled to see at last the source of ideas he had been pondering so long, he immediately set to turn Carnot's wordy discussion into a logical and mathematical exposition. He talked of his discoveries to colleagues, and in April 1848, J. D. Forbes, a professor at Edinburgh, wrote to Thomson urging him to write up his analysis for the Royal Society of Edinburgh. "As you have taken so much trouble about this Theory of Carnot's," Forbes wrote, "I think it would be reasonable to expect you to print a little notice of it for the benefit of people in general." Excited though Thomson may have been, Forbes had to prod him again in November: "I write to remind you of your promise to give us an Abstract of the *Motive Power of Heat* for the R.S. When can we have it?"

Delivered in January 1849, Thomson's "Account of Carnot's Theory of the Motive Power of Heat, with Numerical Results deduced from Regnault's Experiments on Steam" played a pivotal historical role. He brought Carnot, ignored by the French scientists and unknown to the English, before a new audience and did it in a way his colleagues could grasp. A direct translation of Carnot into English might have had as little effect as his original publication in French. Thomson's interpretation and amplification of Carnot not only rescued the Frenchman's seminal work from a quarter century of obscurity, but showed how these new ideas could be expressed in the modern language of mathematical reasoning. Thomson here bestowed a new name on this area of study. "Thermodynamics" he called it—the dynamics of heat, the connection between heat and work.

Even as he was preparing his account of Carnot, Thomson's grasp of this new science evolved in fits and starts. Writing to Forbes, he teasingly mentioned that he had thought up a trick for producing ice "*ad libitum* without the expenditure of mechanical effect." A Carnot engine running in reverse moved heat from a lower to a higher temperature. It struck Thomson that if both temperatures were the same, then an ideal engine (with frictionless moving parts, no heat leakage, and so on) would still move heat from one side to the other but would consume no physical effort because there was no change of temperature. So he imagined an engine operating between two reservoirs of water, each at precisely 32°F. Extracting heat from one side would turn the water into ice but with no mechanical cost. Hence his note to Forbes.

But when William tried this out on James Thomson, his cautious and thoughtful brother immediately saw a difficulty. Water expands when it freezes, and the expansion could be made to push a piston and do work. If that were the case, William's ice machine would apparently create mechanical effort out of nothing, which they both deemed unacceptable. James found the answer. He concluded that the melting point of ice must fall slightly when pressure is applied. Then any attempt to make the ice do work would unfreeze it, and it could not push a piston. Some months later William did experiments to check this prediction and found that James was exactly right, and had accurately calculated the magnitude of the effect. The result thrilled William. In his Glasgow lectures he always described this finding with delight, explaining that it was the first quantitatively precise prediction to be derived from Carnot's theory of engines.

The failed ice machine didn't appear in his presentation to the Edinburgh Royal Society, but plenty of other puzzles remained. The disagreement between Carnot and Joule worried him still. He fully accepted by now Joule's numerous demonstrations of the conversion of heat into work, but he continued to insist on Carnot's principle that, as he put it, "the thermal agency by which mechanical effect may be obtained, is the transference of heat from one body to another at lower temperature." Heat is not consumed. Somewhat desperately, he claimed that this principle had "never been questioned by practical engineers," although he would have been hard pressed to find a practical engineer who even understood the question, let alone had an answer.

In a footnote Thomson illustrated his perplexity. A bar of metal, he noted, will conduct heat from a hot body to a cooler one without producing any mechanical work, whereas passage of the same amount of heat through a Carnot engine *will* produce work. Carnot himself had at least indirectly made the same point, but either was not troubled by it or left it as a matter for later consideration. Thomson, however, saw a problem: "When 'thermal agency' is thus spent in conducting heat through a solid," he asked, "what becomes of the mechanical effect which it might produce? Nothing can be lost in the operations of nature—no energy can be destroyed. What effect then is produced in place of the mechanical effect which is lost?"

Here, remarkably, Thomson was tiptoeing around a universal law of conservation of energy (using a word, indeed, which had hardly any currency and no precise meaning at the time), yet he didn't seem to fully grasp the significance of what he was saying. Joule, a couple of years earlier, had argued that all forms of energy can transform, in suitable circumstances, one into another, but that energy as a whole cannot be created or destroyed. In his footnote Thomson seemed to concur—except that in the body of the paper he held fast to the rule that heat could not be transformed into mechanical work. Thus he was left clinging to two contradictory propositions.

Another inconsistency showed up. Thomson tried to harmonize the thermodynamic implications of Carnot's theory with new measurements by Regnault on the heat absorption capacity of steam at different temperatures and with Joule's evidence for the conversion of work into heat at a constant rate. As before, he assumed that the efficiency of a Carnot cycle was the same at all temperatures. He found it impossible to establish consistency. In particular, his calculations told him that work should turn into heat with a conversion factor that was not constant but varied with temperature.

To a reader with some knowledge of physics and the benefit of hindsight, the most inexplicable flaw of Thomson's 1849 paper is that he had already seen the answer to this last puzzle but had failed to absorb it. A few weeks earlier Thomson had written to Joule describing some of his calculations and expressing consternation that the results didn't seem to match up. In a letter dated December 9, 1848, Joule, no mathematician, had casually thrown out the resolution. Casting his eye down Thomson's lists of numbers, he observed that if one assumed the efficiency of the Carnot cycle to be inversely proportional to temperature, rather than constant, then everything fell into place. Work would convert into heat at a constant rate, as Joule had long argued. But Thomson would not let go of his interpretation of Carnot and therefore did not give Joule's proposal the consideration it merited.

To avoid more confusion than we have unwisely waded into already, it is helpful at this point to jump forward a year, to 1850, when the German physicist Rudolf Clausius published his answer to these difficulties. Clausius, incidentally, noted that he not yet seen a copy of Carnot's original paper and was relying on the expositions by Clapeyron and Thomson. His solution seems ludicrously simple, not to say obvious. In a Carnot cycle, he argued, *some* of the heat passes from hot to cold unchanged, but *some* is converted into work. Moreover, the relative proportions are such as to reproduce Joule's suggestion that the efficiency of the cycle is inversely proportional to temperature.

This was staring Thomson in the face when he read Joule's letter of December 1848. So impressed was Thomson by Carnot's conclusions, it would seem, that he feared scrutinizing the assumptions too closely in case the whole elegant piece of reasoning should fall apart. Clausius understood perfectly well the problem, but took a very different attitude. "I believe we should not be daunted by these difficulties," he wrote in his

1850 paper. "Then too I do not think the difficulties are so serious as Thomson does, since even though we must make some changes in the usual form of presentation, yet I can find no contradiction with any proved facts. It is not at all necessary to discard Carnot's theory entirely."

As Clausius went on to explain, Carnot's general conclusions still held true even when his assumption about the nonconvertibility of heat was amended. Thomson was fully capable of seeing this. Apparently he never looked.

This failure, moreover, stands in contrast to the flexibility Thomson had shown in other cases, for example in his reconciliation of Faraday's portrayal of electricity and magnetism with the apparently very different picture of action at a distance. In that case he sifted what was important and necessary from what was extraneous and incidental, and reconciled the two views. In the case of Carnot and Joule, he could not see beyond the apparent contradiction to the underlying consistency. It emerged some years later that Carnot himself, before his premature death, probably from cholera, in 1832, had seen that his assumption was wrong. In notes discovered only much later, he had written that "wherever motive power is destroyed, there is a simultaneous production of an amount of heat exactly proportional to the motive power that is destroyed. Conversely, wherever there is destruction of heat, motive power is produced." This is precisely Joule's position, reached a decade before Joule began his justly celebrated experiments.

As far as the science of thermodynamics is concerned, these questions are of no great consequence. Laws were established, it hardly matters by whom. Thomson is one of several people associated with the birth of this new science. He could easily have been, after Carnot, the most influential. Hindsight is dangerous, of course. Especially in science, everything is obvious once someone has figured it out. Still, Thomson's stubbornness in sticking with Carnot's false principle and doubting Joule—perhaps for no greater reason than that Carnot's theory became lodged in his brain first—is an indication of a certain lack of flexibility, or an inability to take a leap in the dark, that inhibited Thomson's scientific imagination.

***

Clausius was not alone in suggesting how to amend Carnot's argument. William John MacQuorn Rankine was a Glasgow engineer who had, like Thomson's brother James, learned his science through a mixture of schooling and practical work. In 1850 and 1851 he wrote a couple of long, tortuous, occasionally confused yet remarkably inventive papers that approached thermodynamics from a different perspective. Rankine had long been impressed with the idea of heat as a form of motion, and he took up an essentially atomic or molecular view. He conceived of a gas as a collection of atoms and supposed that heat was nothing but the motion of these atoms. This would be precisely the modern view were it not that Rankine settled on a model of atoms as tiny vortices, so that heat was essentially the rotational energy of these little whirlpools. Nevertheless, having made heat explicitly a kind of atomic motion, Rankine found it obvious that heat could turn into mechanical work, since both were merely different kinds of motion.

Rankine reached the same conclusions that Clausius did about the proportion of heat converting into work in a Carnot cycle. But because his reasoning sprang from a highly debatable atomic model, the generality of his conclusions was far from evident. Clausius, in fact, suffered from a lesser version of the same problem: He had implicitly assumed an ideal gas as the working substance of the engine he analyzed, and in places he relied, or appeared to rely, on his own model of a gas as a collection of particles in motion.

With the essentials of thermodynamics now in place, though marred by dubious assumptions and gaps of reasoning, Thomson's particular intellectual powers came into their own. As he had done with electricity and magnetism, Thomson wrote over the next couple of years a series of long papers, "On the Dynamical Theory of Heat." Like much of Thomson's work, these papers represent an inextricable mix of originality and exposition. To some extent he simply took the principles established by Carnot, Joule, and Clausius, and set them down in a systematic way. On the other hand, there are many places where he showed a profound sense of logical and mathematical rigor and employed it to derive thermodynamic relations that depend as little as possible on unwarranted assumptions about the nature of heat or the constitution of gases. This was Thomson's standard way of constructing a theory, going back once again to his reading of Fourier: Apply sound reasoning to empirical knowledge and thereby create a theory that was sweeping and general but at the same time founded on fact.

Thomson showed how thermodynamics rested on just two basic principles. The first, which he credited to Joule, was that in any interconversion of heat and work, complete or partial, the sum of both quantities remained the same. This is conservation of energy, also known as the first law of thermodynamics. His second principle, credited to Carnot with the essential modification by Clausius, was that an ideal engine, capable of working forward or backward, extracted the maximum possible amount of work from a given amount of heat.

It was still not altogether clear, however, how best to formulate this second principle. Carnot had originally argued from the fact that one could not create work out of nothing, which makes the principle of maximum efficiency appear to be a consequence of the first principle, that energy cannot be created or destroyed. But this is a spurious association, coming about in essence because Carnot stuck with the idea that heat and work were distinct. With the recognition that an engine converted heat into work, discussion of the efficiency of that conversion became a separate issue.

Thomson stated the second principle thus: "It is impossible, by means of inanimate material agency, to derive mechanical effect from any portion of matter by cooling it below the temperature of the coldest of the surrounding objects." That is, a machine can derive work when temperature flows from hot to cold; it can't produce work by making some object colder than everything else. This was not very different from what Clausius had said, though Thomson seemed to think his statement a little more profound. Mainly, it appears, he wanted to distinguish his own words from those of Clausius, to whom he gave generous credit— "the merit of first establishing the proposition upon correct principles is entirely due to Clausius"—which he instantly took back: "I may be allowed to add, that I have given the demonstration exactly as it occurred to me before I knew that Clausius had either enunciated or demonstrated the proposition."

In his series of papers Thomson also picked up the question he had posed in a footnote the year before about the passage of heat by conduction from a hot body to a cold one without any concomitant production of work: This heat, he said, was "irrevocably lost to man, and therefore 'wasted,' although not *annihilated.*" The meaning of "waste" here was left hanging, but Thomson came back to it in 1852 in a short note with the striking title "On a Universal Tendency in Nature to the Dissipation of Mechanical Energy." He wrote this after he had come across a famous essay, "On the Conservation of Force,"[7] published five years earlier by Hermann von Helmholtz. Although this essay is sometimes credited with establishing the law of conservation of energy, Helmholtz mainly rounded up various arguments and bits of evidence from other authors to show that a universal principle indeed exists. He discussed energy of all forms— energy of motion, gravitational attraction, heat, even chemical—and argued that transformation among all these forms is possible, but never creation or destruction.

Thomson's reading of Helmholtz crystallized a qualitative notion into an absolute rule. In his note "On a Universal Tendency," Thomson built on his new understanding to argue that when heat is "wasted" but not lost, it distributes itself in such a way that no further work can be obtained from it. The ultimate fate of any system is for the temperature to become the same everywhere. Thomson introduced here a remarkable number of new ideas and definitions. First he made a distinction between "statical" energy, such as is possessed by a weight suspended at some height, and "dynamical" energy, which is the energy of motion when the weight falls. Rankine later introduced the terms "potential" and "actual," and Thomson then substituted "kinetic" for actual. These are the modern terms.

He also sharpened the distinction between "reversible" and "irreversible" processes. Carnot had recognized that a reversible engine gives maximum efficiency, but he had apparently not stopped to wonder about the fate of the heat lost, through conduction or escaping steam, in a less efficient and therefore irreversible engine. In an irreversible process,

---

[7] *Über die Erhaltung der Kraft.* In the middle of the 19th century force had not been clearly distinguished from what we now call energy. Conservation of energy was nonetheless Helmholtz's topic.

Thomson now explained, heat that moved from high to low temperature without creating work did not signify any overall loss of heat energy, but the possibility of obtaining work from that heat was gone forever.

Because reversibility was the ideal, never realized in practice, Thomson argued that all natural processes, probably including biological and animate ones as well as purely physical and chemical changes, represented a loss of potential work from heat. From this he jumped to a "cosmic" conclusion: "Within a finite period of time past the earth must have been, and within a finite period of time to come the earth must again be, unfit for the habitation of man as at present constituted, unless operations have been, or are to be performed, which are impossible under the laws to which the known operations going on at present in the material world are subject."

The idea of the universe running down to a state of enervated uniformity has become known as the "heat death," a name and idea often attributed to Helmholtz or sometimes Clausius. But Thomson clearly made the point in 1852, although he didn't come up with the catchy phrase.

Finally, and a little sneakily, Thomson managed in the course of his several papers on heat and the new thermodynamics to work in a revision of his absolute temperature scale. He stuck to the essential idea of defining a temperature difference according to the amount of work produced by a Carnot engine, but having finally understood that the efficiency of a Carnot engine itself varied with temperature, he had to adjust his original argument. It was a straightforward thing to do, and his old temperature emerged as simply the logarithm of his new temperature.

In a note written in 1881 for the publication of the first volume of his collected papers, he made light of this adjustment, saying only that the new scale was "practically more convenient" than the old one. But there were two significant differences. First, the revised temperature corresponded to the temperature defined by an ideal gas thermometer and was therefore closely related to the practical temperature scales that laboratory scientists had long used. Second, where the old scale descended to minus infinity, the new scale had a zero. With his revised understanding of Carnot, Thomson could see that the efficiency of an ideal engine would approach 100 percent as the temperature dropped to absolute zero. Temperatures below that would make no sense for a variety of reasons, among them that it would then seem possible to get more work out of an engine that was put in as heat.

It is thus in his 1852 paper that Thomson truly established the existence of an absolute zero of temperature. In the modern picture of heat as the motion of atoms, it is obvious that when all motion has ceased, heat is absent, and temperature can go no lower. But Thomson could not and would not avail himself of any such unwarranted assumption. It is a tribute to the power and thoroughness of his reasoning that he could deduce the existence of an absolute zero without any reference to the physical nature of heat itself. He used only what was known empirically of heat's properties. Fourier would have been proud.

By this time the first law of thermodynamics was understood by all to be the conservation of energy. No one person can truly be credited with its discovery or enunciation. Many people made qualitative proposals for such a principle, on more or less philosophical grounds, whereas James Joule came to the idea through careful measurement and experimental test. Thomson, adamant that theoretical proposals without experimental support were next to worthless, became a great advocate of Joule, to an extent that generated controversy and unseemly attacks that swirled about for decades to come.

The origin of the second law of thermodynamics is murkier still. In modern textbooks the second law is a statement about a quantity called entropy. In reversible changes, entropy stays the same; in irreversible ones it increases. Entropy can never decrease. Therefore, the entropy of the universe as a whole is always rising. This is a more precise statement of Thomson's cosmic conclusion of 1852.

One of the few undeniable truths about the second law is that the word "entropy" (from a Greek construction meaning "transformation") was proposed by Clausius, in 1865, when he bestowed the name on a quantity he had first formulated in 1854. Rankine in 1850, however, had defined a "thermodynamic function" that looks very much like entropy, except that it was tied to his vortex model of atomic motions, and Thomson in his 1852 account of "Universal Dissipation" had written down an expression that, with a bit of adjustment and translation, can readily be identified with entropy. At the time, though, he evidently didn't feel the urge to isolate this new concept and give it a name. In any case, Rankine, Thomson, and Clausius, in these earlier papers, observed only that in reversible processes a certain quantity stayed the same. They didn't yet delve into the significance of this quantity, if any, for irreversible processes.

If the second law of thermodynamics is simply the principle that reversibility implies maximum efficiency, then credit goes to Carnot. But Carnot was thinking only of engines, not universal processes; he was working with a false assumption about caloric that precludes definition of either a first or a second law of thermodynamics; and his own words suggest that he did not see a separate principle here, only an instance of the general prohibition of perpetual motion. Subsequently Rankine, Clausius, and Thomson all made contributions to the statement of a second law, both in its physical conception and in the mathematical demonstration of its universality. Clausius in the end christened the child and most often gets the credit.[8]

Scholars continue to debate, rather fruitlessly, who said what and when, and what their fumbling words meant. One lesson is that science is harder than it looks. Of Thomson's participation, though, it is difficult to avoid the judgment that he didn't do as much as he might have done. He had an exceptional ability to sort and clarify, to resolve confusion and contradiction, and many of the standard elements of classical thermodynamics trace back to his definitions and arguments. On the other hand, at crucial points he needed a prompt from someone else. He developed the full theory of Carnot engines only after Clausius had supplied the essential idea that heat was consumed, not just transferred from one place to another. He made use of a quantity that eventually became entropy but did so apparently without seeing the general utility of it, as if he found it convenient for a specific purpose but failed to look beyond.

In the 1850s thermodynamics was imperfectly understood even by its creators. Nevertheless it was abundantly clear that scientific under-

---

[8]Writers of physics textbooks almost always stick to a principle of "one idea, one inventor" so as to not to distract readers with the messy complications of history. A couple of generations of this and history really does become simple!

standing of heat and work and energy and their interrelationship was no longer a cause of qualitative mystery, but could be captured in a handful of precise mathematical expressions. William Thomson, both as originator and expositor, was unquestionably one of a handful of people who had turned ill-defined notions into a new and fundamental discipline of physical science. His work in electricity and magnetism, though not developed to the same degree, nevertheless gave to those subjects a range and coherence they had not previously possessed.

***

In the summer, Thomson often traveled to Bad Kreuznach in the Rhine valley, where he could hike and think, and his ailing wife could take the waters. She was not at all well. One year Thomson told his brother James that "she suffers much after the driving and walking and is quite unable to sit up without much pain in her own room. . . . Dr Johnson . . . says it will do good notwithstanding the pain & fatigue, to a limited degree, but he says she is not in a fit state for almost any exercise." They went to Kreuznach a number of times, and Thomson took a little comfort in the fact that his wife was even allowed to try the iron waters there.

During his 1855 visit to Germany, Thomson arranged to meet Hermann von Helmholtz, whom he had admired since reading his influential 1847 essay on the conservation of energy. Like Thomson, Helmholtz had wide-ranging knowledge across all of science (he had begun in medicine, moved into physiology, learned physics and mathematics in order to understand the science of perception, then began a career in physics proper) as well as an ability to synthesize arguments and evidence into a coherent whole. Helmholtz had come to Britain in 1853 and made a trip to Scotland, before attending the British Association meeting in Hull, in order to search out Thomson. He recounted his fruitless journey to his wife: "From Edinburgh I traveled for a couple of hours in the afternoon through a heavily built-up hilly area, with a variety of ruins, to Glasgow. This is a very big (pop. 300000) industrial city, horribly noisy and busy, swarming with poor, red-haired, dirty, unhealthy-looking workers. It did not make a pleasant impression. I was looking for a physicist, Prof. Thomson, who has worked a great deal in matters concerning the conservation of energy, but he had gone away to the seaside, so I strolled about in the streets until I'd had enough, then came back."

Two years later they succeeded in meeting. Helmholtz again recorded his impressions in a letter to his wife: "As he is one of the leading mathematical physicists in Europe, I expected to find a man somewhat older than myself, and was not a little astonished when a very youthful, exceedingly blonde young man, almost girlish, appeared before me. . . . He exceeds, I might add, all the scientific greats I know personally, in sharpness, clarity, and quickness of mind, so that at times I felt dull-witted beside him."

Helmholtz's surprise is understandable. Thomson had been publishing important papers for almost 15 years. He had established theorems in applied mathematics, extended Fourier's studies of the flow of heat, clarified the relation of electric charges and magnets to the forces they produced, as well as the interaction of magnetism with currents, and done as much as anyone to establish the foundations of classical thermodynamics. No other scientist in Europe at the time could lay claim to such range and depth of achievement. He had celebrated his 31st birthday just a few weeks earlier.

3

CABLE

On August 23, 1850, a small boat trailing an unwieldy black rope from its stern sailed clumsily across the English Channel from Dover to Calais. With lead weights attached every 100 yards, the rope sank the 30 feet or so to the seabed. When the boat reached the French coast, the end of the rope was attached to a wooden box equipped with brass knobs and a dial bearing the letters of the alphabet. At Dover the man running this curious operation, John Brett, attached similar equipment to the other end. After fiddling with the device for a while, he reported that signals were now traveling back and forth across the channel, and he produced slips of paper on which printed letters could be seen. Locals who had gathered on the beach were skeptical. Some "expressed their great astonishment, and inquired if paper and print had all made its transit by the wire." Others scoffed, imagining the rope as a sort of immense underwater bell-pull. "What a mad scheme!" someone said. "Why a sailor, or anyone who knew anything about seafaring matters, would declare it was impossible to pull such a line 25 yards, let alone that number of miles, over such a rough and uneven surface as the bottom of the channel."

Brett proclaimed, as only a true Englishman could, that this communication across 21 miles of shallow water represented the first telegraphic

114

link "from one continent to another." The rope was, of course, an electrical cable, of exceedingly amateur design. Hundred-yard lengths of copper had been wrapped in strips of gutta percha, a gummy tree sap discovered a few years earlier in Malaysia. Michael Faraday, given some samples of gutta percha in 1848, had sent a short note to the *Philosophical Magazine* recounting the material's excellent qualities for the laboratory electrician. It was soft and moldable when warm, resilient yet still flexible when cold, and made a good electrical insulator. Faraday used it for plugs and supports and insulating sheets in his various experiments. It also resisted water, which is why Brett had chosen it to insulate his underwater telegraph cable. To assemble the hundred-yard lengths into 25 miles of cable, the exposed ends of the copper wires were twisted together, and globs of warm gutta percha were applied to the joints and squeezed crudely into shape with a wooden press.

This cable was not robust, to say the least. After only a few hours, it broke somewhere, or else rocks chafing at it wore through the gutta percha and allowed seawater to reach the copper wire inside.[1] Once that happened, any electrical current traveling down the wire would conduct away into the watery deep. Whether Brett's first cross-channel cable ever worked at all is debatable. Charles Bright, in his 1898 history of submarine telegraphy, allowed that "some few, more or less incoherent, letters appeared here and there . . . but intelligible words were conspicuous by their absence." Willoughby Smith, another telegraph engineer, suggested in his memoir that the letters allegedly received were no more than random firings of the equipment. According to him, the operators at Dover wondered at the time if their colleagues on the French side had overindulged in celebratory champagne. Still, Brett's transient claim seemed momentous enough to the London *Times*, which remarked that "the jest or scheme of yesterday has become the fact of to-day," while the *Spectator* presciently observed that in the future "a man in London might sign a

---

[1]Gordon (2002) repeats an anecdote from the journalist W. H. Russell (1866, p. 4) that the 1850 cable failed because a fisherman pulled it up and cut out a section, thinking he had found a new kind of gold-bearing seaweed. However, Brett himself offers no such tale. It may be that pieces of the discarded cable were hooked up later, giving rise to fishy stories.

bill in Calcutta, transmit it for endorsement to St. Petersburg, and receive cash for it on authority in Cairo, in the space of an hour or so."

Brett's cable was not the first underwater telegraph, but it was the first of any length. In 1838 a Colonel Pasley had experimented at the Chatham docks in London with a wire wrapped in tarred rope. At the behest of the East India Company, Dr. William O'Shaughnessy trailed wires across the Hooghly River the following year. In the United States in 1842 Samuel Morse, whose 1837 electromagnetic telegraph and the code that he also devised became standards of the new technology, transmitted signals across New York Harbor. Charles Wheatstone, an English scientist who also made crucial contributions to telegraphy, signaled in 1844 from a boat in Swansea Bay, Wales, to a lighthouse on the shore. His cable had several copper strands twisted together, wrapped in hemp, and then soaked in boiled tar. The following year Ezra Cornell, founder of the university, laid a line 12 miles across the Hudson, from Fort Lee to New York City. His cable had two cotton-covered copper wires wrapped in rubber, further protected by a lead sheath. It worked for a few months, until a chunk of ice in the river severed it. Then gutta percha came on the scene, and in 1848 Werner Siemens ran a cable across the harbor at Kiel, in northern Germany. Siemens made the important invention of a press that molded warm gutta percha around a metal wire in a continuous run, producing seamless insulation.

In 1845, as news of these ventures and their mixed success was getting about, John Brett sent a telegram to Sir Robert Peel, the British prime minister, with a proposal both outlandish and grandiose. Fired by a vision of technical innovation allied to patriotism, he described a combination of "oceanic and subterranean inland electric telegraphs" by means of which "any communication may be instantly transmitted from London, or any other place, and delivered in a printed form, almost at the same instant of time, at the most distant parts of the United Kingdom or of the Colonies." The advantages to a great and growing colonial power would be immense. The global might of the British Empire would surely come to depend on this technology and profit from it. Brett offered his technical abilities to his country, in return for financial support and some guarantees of exclusivity. For the sake of imperial might, the government must surely support his endeavors.

Her Majesty's government did no such thing. The classically educated mandarins inhabiting Whitehall's upper reaches were singularly uninterested in feats of untested technological novelty. As civil servants they were extremely adept, however, and repulsed Brett's enquires by means of a practiced game of departmental handoff. Peel told Brett to try the Admiralty, which had authority over transoceanic communication. The Admiralty demurred on the grounds that they were not in a position to make business or financial arrangements; that was something the Treasury would have to deal with. Policy initiatives were not the Treasury's bailiwick, however. Might this be a question for the Foreign Office to decide? But the FO's remit was diplomacy and statesmanship, not technical questions of sending and receiving messages.

Rebuffed by one department after another, Brett made overtures to the French government with the more modest suggestion of a telegraphic link across the English Channel. The French showed a hint of interest, and Brett then returned to the Admiralty for permission to land a French-sponsored cable on British soil. The Admiralty said it had no particular objection, provided there was no expense to Her Majesty's government and provided too that Brett would agree to cut the cable at once if their Lordships, for whatever reason might in the future occur to their wise heads, should so command. In a final stroke of bureaucratic genius, they advised Brett that if he wished to attach an electrical cable somewhere along the lovely southern coast of England, he would surely need to obtain permission from the Commissioners of Woods and Forests.

Brett's correspondence with various offices of the British government appears in his book *Brett's Submarine Telegraph*, which mainly purports to show how many important ideas were originally his and how unfairly history had already treated him by 1858, when he published his memoir in London. Nevertheless, reading the various irresistible explanations offered by civil servants and politicians of why they could not possibly take action on Brett's proposal, one begins to wonder how the British Empire ever spread much beyond the Chatham docks.

In the end Brett gave up on London and made a deal with the French government. For his channel connection, let alone the global network he imagined, John Brett and his brother Jacob were woefully underprepared. Jacob had some technical experience, while John, a former antiques dealer, supplied entrepreneurial talents. Their first contract, signed in 1847, lapsed after a year because they hadn't done anything. They renewed in 1849 with the same stipulation that they must demonstrate the feasibility of their plan within a year. The 1850 cable barely beat the deadline.

With renewed financial backing, Brett laid a second cross-channel cable in September 1851. Where the first wire had been ridiculously flimsy, the second went to the opposite extreme. It consisted of four copper wires individually insulated with strips of gutta percha, applied cold (evidently they hadn't heard of the Siemens machine for applying insulation). These conductors were then twisted together with strands of tarred Russian hemp, that assembly being wrapped and wound with tarred yarn, the whole thing then being wound about with 10 galvanized iron wires for strength and protection. The cable was almost two inches across and weighed a massive seven tons per mile. Laying the cable proved fairly easy, except that their boat drifted off course, so that Brett's crew ran out of cable a mile from the French coast, off Cap Gris Nez. They improvised by filling in the gap with a length of the old wire but a month later managed to pick up the cable at the join, splice in an additional length of the new strengthened line, and complete the job.

Building on this success, the brothers Brett formed the European and American Telegraph Company and made a creditable attempt to corner the market on this new business. The Irish Sea at first proved too rough and deep for cabling, but they laid a 70-mile link from Dover to Ostend in 1853 and then began to attack the Mediterranean—an adventurous project, since no one knew how deep the water was in its middle sections.

Although the Bretts had remarkable entrepreneurial energy, they had inevitably to deal with the uncertainties and pure ignorance implicit in any new technology. John Brett, in his book, remarks a little testily that their endeavors were held back by his brother Jacob's infatuation with a telegraphic letter-printing machine devised by Mr. Royal E. House of Vermont. That instrument, like the teletype machines of a few decades ago, printed messages on strips of paper in capital letters. Charles Wheatstone invented a similar device, similarly unwieldy. These machines, equipped with a rotating circular dial inscribed with the letters of the alphabet, had to receive a string of pulses down the telegraph wire in order to know at which letter to stop and stamp out a mark. They were both slow and unreliable.

Samuel Morse's famous code stands as an innovation of middling genius. He saw how to get two kinds of intelligible signal—dots and dashes—down a telegraph, and he devised a system for turning bursts of dots and dashes into the letters of the alphabet in an economical way. We take this for granted nowadays, it seems so elementary, but at the time it arrived like a godsend. As Werner Siemens put it: "The simplicity of Morse's apparatus, the relative facility of acquiring the alphabet, and the pride which fills everyone who has learned to use it, and which causes him to become an apostle of the system, have in a short time ousted all dial and older letter-printing apparatus." Jacob Brett's devotion to the letter-printing device of Mr. House represented years of wasted time.

More mundanely, John Brett displayed a habit of packing not quite enough cable on his early expeditions. That was not such a silly mistake as it might seem. Keeping a vessel on course while it was dropping cable off the stern was not easy, as Brett had found on the 1852 cross-channel project. A more serious difficulty showed up four years later when he tried to lay a connection from Sardinia to the north coast of Africa. This was a distance of some 150 miles, across water up to 1,600 fathoms deep. In August 1856 Brett and his crew loaded up with cable and sailed south in perfect weather, with clear sky, little wind, and a calm sea. But as they moved into waters deeper than anything they had experienced thus far, their cable unreeled over the stern with alarming and increasing speed. The explanation is simple. The deeper the water, the greater the length of unsupported cable that dangles from the ship; this greater hanging weight pulls the cable from the ship uncontrollably fast. In earlier and easier Mediterranean expeditions, Brett had added a brake to the drum that paid out the cable, for precisely this reason. But in these deep waters they could barely restrain the cable.

Sailing to their utmost, they found themselves 13 miles from the African shore with 12 miles of cable on board. Desperately, they lashed and buoyed the cable to keep the remainder from spilling into the sea and sent a request for additional cable back along the cable they had already laid. Then the weather turned bad. Squalls blew up and the ship, pitching and rolling, hung on desperately to the heavy cable end. After five days their ordeal came to an abrupt end. Chafing against the stern of the ship, the cable finally broke on August 19, "not ten minutes after receiving a telegraphic reply through it from London, to inform us that the extra cable was in progress, and would speedily be forwarded to us," as Brett ruefully recorded.

Despite such setbacks, John Brett established himself for a time as the leading figure in submarine cabling. When Frederick Gisborne, a globe-trotting Englishman then resident in Canada, began to think seriously of an underwater telegraph across the Atlantic Ocean, connecting the New World to the Old, it was Brett whom he first contacted. Gisborne was struggling to overcome enormous difficulties of geography and climate in order to lay a mix of overland and underwater cables connecting the east coast of the United States to the tip of Newfoundland. From there it was less than 2,000 miles to the west coast of Ireland. The idea of a transatlantic cable was not unique to Gisborne. Morse had written in August 1843 to the secretary of the U.S. Treasury making just that proposal, with casual and quite unfounded confidence that the project would present no new difficulties beyond the obvious logistical ones.

But Morse simply made a suggestion. Gisborne spent years hacking through the wilderness of Newfoundland in preparation for a practical attempt. He first proposed a route that included a way station in Iceland, but his Canadian ventures proved far more costly than he had imagined, and by the mid-1850s he was on the verge of bankruptcy. His savior was not Brett but a newcomer to the cabling business, a New York entrepreneur and financier by the name of Cyrus West Field. In the end both Gisborne and Brett faded from their pioneering roles, and it was Field's stamina, imagination, and financial resources that saw the Atlantic telegraph project through to a successful conclusion. But money was not the only necessity. Scientific problems also stood in the way, and to resolve those Field needed the best scientific advice he could find.

***

Though William Thomson was never as keen a student of current affairs as his father had been, he surely knew of the blossoming of telegraphy, both overland and undersea, into a new industry that rapidly altered the pace of ordinary life. It was, moreover, the first commercial technology that depended on electricity, one of the many subjects in which Thomson held acknowledged expertise. Still, it was some time before he first began to think critically about telegraphy as an exercise in the theory and application of electricity. Overland telegraphy would not have fascinated him. A battery applied at one end of a wire produced a detectable signal (instantaneously, it seemed) at the other. The fundamental nature of that signal, how it moved, why it would pass through copper or iron but not through tar or cotton or gutta percha—these scientific arcana mattered not in the least to inventors such as Morse, still less to the money men Brett and Field. Telegraphy was a simple means of communication making use of an utterly mysterious physical phenomenon.

That was the case, at least, until the advent of underwater cables. Signals transmitted through the 1851 Dover-Calais cable, and more obviously in the 1853 Dover-Ostend and later Irish cables, suffered from a troubling degree of fuzziness. What should have been clear and unambiguous blips came through distorted and blurred, sometimes to the point that operators couldn't be sure whether they had registered a real signal or not. These difficulties alarmed George Airy, the astronomer royal, who had conceived a plan to link the London and Paris observatories by telegraph so as to allow simultaneous observations from both places. By this time there were enough telegraph lines around Britain and the continent that it was possible to set up test circuits in which signals traveled along hundreds or even thousands of miles of wire in the air, underground, and underwater. Experiments showed that underwater cables, and to a lesser extent underground ones, suffered a small but detectable delay in transmission. Instead of an instantaneous sharp pulse, operators would see a signal both delayed and smeared out.

Airy asked a young telegraph engineer, Josiah Latimer Clark, to look into the problem. One day in early 1854 Clark invited the renowned Michael Faraday to visit his cable works and observe some experiments. He had coiled 100 miles of cable in a tank full of water and demonstrated to Faraday that it transmitted signals more slowly and less clearly than a 1,500-mile circuit of overland cable looping around the country. Faraday immediately supplied a qualitative explanation. Any signal passing down a wire creates an electrical disturbance in its vicinity. Water, unlike dry air, has significant electrical conductivity, and an electrical disturbance passing through it creates local electric currents that act as a kind of inertia or brake on the primary signal. In essence, Faraday told Clark, a signal passing through a submerged wire has to work harder to get from one place to another—hence the delay and degradation in the signal.

Faraday published his analysis of the problem in the *Philosophical Magazine*, where he also compared a long insulated conductor, immersed in water, to the familiar laboratory device known as a Leyden jar. A Leyden jar (named after the Dutch city where it was invented) was a glass vessel lined inside and out with separate layers of metal foil. With the external layer grounded, the inner layer could be charged with static electricity, which the jar would then retain. The two metal layers, separated by glass insulation, acted as a storage device for electric charge—in modern parlance, a capacitor.

An underwater cable, Faraday observed, had a conducting core surrounded by a layer of insulation, which was surrounded in turn by an earthed conducting body, the ocean. Such a cable did not simply conduct electricity but stored it too. Its characteristics were therefore quite different from those of a plain wire, but Faraday, in his usual way, perceived the essential physics of the matter without being able to calculate anything. Nevertheless, if poorly understood electrical phenomena were already causing trouble on the 70-mile cable from England to Holland, the prospects for a link of 2,000 miles or more across the Atlantic Ocean must be questionable.

The problem finally came to Thomson's attention in a roundabout way. At the close of the 1854 British Association meeting in Liverpool, a young man had introduced himself to Thomson as the son of the Dublin mathematician William Rowan Hamilton. He wanted to ask an electrical question. Thomson had to rush away to catch a steamer to Glasgow and handed the young man off to his friend George Stokes. The question concerned Faraday's analysis of undersea cables. Stokes, no electrical expert, couldn't help and so passed the problem back to Thomson in a letter dated October 16, 1854. Thomson was by then at Largs, on the Ayrshire coast. He had a couple of weeks remaining before the Glasgow session began and spent the time catching up with correspondence but also, as he told Stokes, "devoting myself as much as possible to the open air & the sea." He did not have access to the *Philosophical Magazine*, so could only infer Faraday's arguments from Stokes's brief account of them. But that was all he needed. "In taking up your letter this morning to answer it," he wrote, "I find that the whole may be worked out definitely as follows." In several pages of calculations Thomson worked out, as no one had done before, the theory of the transmission of a pulse of electricity down an insulated underwater cable. A second letter, two days later, added further details, notably some calculations of the feasibility of a telegraphic connection to America. This was an exercise in applied science, carried out by Thomson with his customary speed and brilliance, and done simply to satisfy his curiosity about a physical phenomenon that was new to him. He did not immediately feel any great urge to polish his analysis into a scientific paper, nor did it occur to him that his findings might have practical not to say commercial importance. Through November he exchanged further letters with Stokes, working out some additional wrinkles. Stokes helped by coming up with a simpler way to obtain solutions to the fundamental equation of telegraphy that Thomson had worked out.

Thomson's innocence ended abruptly. On December 1, 1854, he wrote asking Stokes to keep quiet about the contents of his previous letters because he had applied for a patent on "the remedy for the anticipated difficulty in telegraphic communication to America." Joining in this application were William Rankine (whom Thomson knew from his work in thermodynamics) and John Thomson (not William's deceased brother John, obviously, but a son of the other William Thomson, the medical professor). Writing to his brother James the following January, Thomson explained that it was Rankine, the experienced professional engineer, who had "suggested the plan of taking a patent, w^h I had no idea of at first. In a few days I expect it will be secured to us: in the meantime don't say even as much as I have said to you, on the subject. I am not very hopeful of making anything of it, but it is possible it may be profitable."

Before Thomson's theoretical analysis, no one had designed an underwater cable except in a crude way. There had to be a copper conductor down the middle, surrounded with gutta percha for insulation, made watertight with tar and pitch and hemp and rope or whatever else came to hand, and finished off with some sort of iron binding for strength and protection. Thomson modeled such a cable as a combination of resistance and capacitance, with the magnitude of these factors depending on the construction of the cable. The thicker the wire, the less the resistance, but the thicker the insulation, the greater the cable's capacitance. In effect, an electrical pulse traveling down an insulated underwater wire had to charge up the cable as it went. The consequence, Thomson showed, was that a sharp pulse applied at one end spread out, as it moved along, into a rolling wave of increasing length.

Thomson obtained the curious result that the arrival time of this changing signal, if measured by the moment the crest of the wave reached the far end, increased with the *square* of the distance traveled. In other words, the signal had no fixed speed. Although the front of the pulse moved at a constant rate, the crest of the following wave lagged farther behind, the farther it went. Alternatively if the diameter of both the conductor and the insulation of a cable were increased in proportion to its total length, then the signal delay and what Thomson, groping for technical language to describe the clarity of the signal, quaintly called the "distinctness of the utterance," would remain the same. Collectively, these assertions became known as the law of squares in telegraph theory.

This seemed at first a discouraging discovery. If a cable 100 miles long was an inch or two in diameter and weighed a ton or two per mile, one could hardly countenance 2,000 miles of cable measuring a foot and a half across. Thomson argued, though, that with a strong enough signal and sufficient patience and understanding on the part of the operators, signals could be sent and received across the Atlantic, though at a limited rate compared to what had been achieved over the modest subocean distances traversed thus far.

Thomson published his paper "On the Theory of the Electric Telegraph" in the *Proceedings* of the Royal Society for May 1855. He reproduced with little modification the reasoning he had worked out within a few hours of reading the letter from Stokes. His solution once again owed a good deal to his youthful reading of Fourier. An electric pulse moving down a wire against both resistance and capacitance, he argued, was directly analogous to heat migrating along a metal bar. In later papers he found an alternative analogy: He likened the pulse to a surge of water passing down a rubbery pipe that expanded in response to increased pressure. This he called his "peristaltic" model of signal transmission. Thomson was never happier than when he found analogies between one problem and another. It indicated the universality of his reasoning. It maintained his strategy of modeling phenomena from empirical and observational laws, rather than striving for some fundamental a priori theory that would yield results as mathematical theorems. What electricity was, in some essential way that would satisfy continental adherents of *la physique* or devotees of German *Naturphilosophie*, was of no consequence. What mattered, in Thomson's view, was to find a solution to the problem at hand, not to worry about questions of "metaphysics."

***

Even in the middle of the 19th century, the application of science to technology had barely begun. Thomson, with others, had worked out the fundamentals of thermodynamics, but the builders of steam engines mostly worried about cracked cylinders and poor insulation. The pioneers of telegraphy were less scientifically aware still and even those who pretended to a little knowledge of electricity found Thomson's broad-ranging science and powerful mathematics beyond them. One who failed to understand his reasoning but disputed his findings anyway was the splendidly named Edward Orange Wildman Whitehouse, a successful physician in Brighton who had caught the telegraphy bug and begun experimenting with cables and electricity not long before Thomson came across the subject. Attending the British Association meeting in 1855, in Glasgow, Whitehouse heard Thomson announce the law of squares. At the BA the following year he recounted his own tests of signal transmission through cables of various lengths, which he claimed contradicted this supposed law.

Thomson, in Germany with his invalid wife, did not hear this rebuttal but read about it soon after in the *Athenaeum*, a London magazine, which reported that Whitehouse "has been able to show most convincingly that the law of the squares is not the law which governs the transmission of signals in submarine circuits." Whitehouse's account from the BA meeting itself was so confused, both as to what he did and whether he understood what he was doing, that Thomson had difficulty responding.

Whitehouse had tested three cables, each 83 miles long, "coiled in a large tank in full contact with moist earth, but not submerged," which he could join to make a cable of 166 or 249 miles in total length. He also had access to a longer cable, presumably an underground one, that gave him a total of 1,020 miles. Whitehouse may or may not have understood that an underground cable, surrounded by damp earth, represented an intermediate case between a cable in dry air and one immersed in water. Later he remarked without explanation that he thought a cable wrapped in iron could be regarded as identical to one underwater. Without describing in any detail what exactly he measured, he claimed the transmission time was proportional to the length of the cables he tested, not the square of the length. This boded well for the Atlantic project, he said, and he concluded with an airy dismissal of Thomson's so-called theory of the telegraph, implying that ivory-tower academics shouldn't meddle in the affairs of practical men: "And what, I may be asked, is the general conclusion to be drawn as the result of this investigation of the law of squares applied to submarine circuits? In all honesty, I am bound to answer, that I believe nature knows no such application of that law; and I can only regard it as a fiction of the schools, a forced and violent adaptation of a principle in Physics, good and true under other circumstances, but misapplied here."

Thomson replied briefly at first, saying without elaboration that he thought Whitehouse's results were consistent with the law of squares, despite any appearance to the contrary. Whitehouse then sent Thomson a more detailed account of his tests, to which Thomson wrote a thorough rebuttal. He explained that the law of squares applied to uniform tests, wherein precisely the same signal was applied to a cable, and the time of maximum response at the other end was recorded. Whitehouse had not arranged for a constant input and timed his detection at the other end as soon as he saw something. Responding again in the *Athenaeum*, Whitehouse seized on Thomson's admission that the applicability of the law of squares "depends on the nature of the electric operation performed at one end of the wire, and on the nature of the test applied at the other extremity" and argued that the practical issue was to get a useful signal down the wire, not to operate according to some theoretical ideal.

This was a fair point. Although Whitehouse clearly didn't understand Thomson's theorizing, it was also true that Thomson had not fully thought through the implications of his theory for practical telegraphy. His analysis of the telegraph illuminated both the strengths and the weaknesses of Thomson's intellectual style. He began with a handful of basic empirical propositions about electricity, used them to formulate a simple model of the properties of an insulated submarine cable, and proceeded to write down a differential equation that captured the desired solution. In his first reply to Whitehouse he had expressed his confidence in this approach by saying that his theory, "like every *theory*, is merely a combination of established truths." One does not have to be a deep philosopher to perceive the narrowness of this view. There must be more to theorizing than simply combining old knowledge in new ways, else where would new ideas come from?

There was also the problem that the "established truths" of electricity known to science at that time were far from complete. Thomson's telegraph theory, as it turned out, had serious flaws. It was not for another decade that a full theory of electricity and magnetism came into being, which would eventually allow a comprehensive treatment of signal transmission. Thomson's blithe certainty in his analysis seems at best like overconfidence, at worst an indication of a blindness to or incuriosity about the evolving nature of scientific understanding.

On the other hand, Thomson's venture into telegraphy gave at least a preliminary explanation for the unexpected behavior of submarine cables and showed that engineers would ignore the arcane lessons of natural philosophy at their peril. For the time being, the exchange between Thomson and Whitehouse concluded with protestations of good will on both sides and acknowledgment by both that anyone proposing to build an Atlantic cable would be wise to test and investigate thoroughly before proceeding with so ambitious and expensive a project.

Cyrus Field was just the man not to do this. That same year, 1856, he came to England on one of what would eventually total 56 transatlantic voyages, each costing almost two weeks of his life. Born in Stockbridge, Massachusetts, in 1819, Field had worked his way up from junior clerk in a New York dry goods store to become the preeminent paper merchant in the city before he was 30 years old. He had that power of spontaneous adaptability essential to business success. Anticipating the modern cliché, he saw opportunity in every problem. He specialized in high-quality papers for an upmarket clientele, avoiding the low-margin trade in newsprint. He saw an interest in colored paper and urged his suppliers to see what they could come up with. Dyeing was an uncertain process and batches came out in unpredictable hues. Field rose to the challenge. When he took delivery of a parcel of red paper that was a little paler than it should have been, he called it "salmon" and marketed it at a premium. When the blue came out a little darker than usual, he wrote to his privileged customers to tell them of their unexpected opportunity to obtain a quantity of "extra blue" paper that had come his way.

By the late 1840s Field was selling up to $500,000 worth of paper a year, but he began to tire of business and took off with his wife on a tour of Europe. There, especially in London, he encountered a level of industrialization and technological development he hadn't seen before and saw the energy and affluence that both produced it and derived from it. By 1852 he was one of the 30 richest men in New York, worth more than $250,000. Quixotically he then left his paper business in the hands of colleagues, set off on an unhappy expedition to South America with the painter Frederic Church, and returned to New York in 1854 with enormous wealth and ambition but no settled purpose.

Meanwhile, his brother Matthew had teamed up a couple of years earlier with Frederick Gisborne, the Newfoundland telegraph engineer. So far they had spent huge amounts of money tackling the intractable and dangerous Canadian wilderness, had failed to complete their planned telegraph line from New York City to St. John's (a distance of more than 1,000 miles), and were piling up debt. Now here was a project momentous enough for Cyrus Field. Field contacted Morse, who assured him blithely that no serious technical problems stood in the way of an Atlantic cable. While Matthew Field and Gisborne toiled away in the distant wastes of eastern Canada, Cyrus Field took over the project, formed a consortium, raised money, and in 1855 sailed for England to meet John Brett, who at that time could claim the greatest success and expertise in the laying of submarine cables.

The only commercial manufacturers of undersea cable were in Britain. Field, with Brett's assistance, ordered a quantity of cable for the marine segments of the Newfoundland cable (across the St. Lawrence and from Cape Breton to Newfoundland itself). By the following summer the American end of the Atlantic cable project was close to completion, at a cost exceeding $1 million. The transatlantic link itself would cost considerably more than that. Field tried but largely failed to raise money in New York and in 1856 sailed for England again. In a hectic trip lasting several months, he consulted Brett and his assistant Charles Bright, who at only 23 had already overseen the laying of a cable from England to Ireland. Field ordered 2,500 miles of insulated copper wire from the London Gutta-Percha Company, with spiral-wound iron sheathing to be supplied by two other companies, R. S. Newall and Glass, Elliott. He obtained conflicting advice from Whitehouse, Thomson, and Faraday about the delay and distortion inherent in undersea transmission. In October he formed the Atlantic Telegraph Company, with Brett as president, himself as vice-president, Bright as chief engineer, and Whitehouse (who now gave up his Brighton medical practice altogether) as chief electrician.

The appointment of Whitehouse was fateful but by no means foolish. Thomson had only just begun his foray into applied science, and though he had made the general point the cable design ought to be guided by scientific principles, it was by no means clear that electrical science was thus far well enough advanced to be useful. Samuel Morse, moreover, had visited in England in 1856, where he tested long cables in collaboration with Whitehouse and pronounced both the man and the results satisfactory. Thomson was 32, five years younger than Field; Whitehouse was 40, had practical experience in cabling and electrical testing, and in his career as a physician had acquired business sense.

The Atlantic Telegraph Company issued 350 shares at £1,000 each, which Field, racing around the country giving inspirational speeches in all the big industrial cities, sold in less than two weeks. This represented $1.75 million in capitalization. Among the subscribers was William Thackeray, who had met William Thomson and his "nice wife" some years earlier. With the selling of the shares Field also established an unpaid board of directors, which included Thomson.

If Whitehouse and Thomson agreed on one thing, it was that cable design ought to be thoroughly tested before the great adventure began. But with Field in charge, there was no time. He wanted to order cable now, for a voyage the following summer. Thomson didn't like the design adopted (he thought the copper core too thin), but Whitehouse, with his more optimistic view of signal transmission, saw no problem. On his own initiative, Thomson embarked on a study of the quality of copper supplied by several British foundries and to his alarm found that the electrical resistance of copper wire of the same alleged gauge and purity varied in some cases by more than a factor of two. But it was the middle of 1857 when he discovered this. Cable for the first attempt had already been made, and Field had persuaded the U.S. and British governments to lend him two large ships, the U.S.S. *Niagara* and H.M.S. *Agamemnon*. With armaments removed, interior structures torn out to create vast holding tanks, and with systems of drums and brakes and pulleys mounted on the stern, the two vessels became the world's first ocean-going cabling ships. Field needed both, because no single ship was large enough to carry 2,500 miles of cable.

Early in August 1857, *Niagara* and *Agamemnon* lay at anchor a mile or two out from Valentia Bay in the southwest corner of Ireland. Bright had argued that the ships should meet in mid-Atlantic, splice their cable ends together, then sail for their respective home shores. The cable might then be laid in only a week, if all went well. But others, notably Whitehouse, insisted on starting from Ireland, so that progress reports could travel down the cable as it was laid. The departure of the ships was a gala occasion, with speeches and toasts and festivities. Field delivered a message from President Buchanan, inviting Queen Victoria to send the cable's first message to him. On August 5 a small ship brought one cable end ashore, where it was hooked up to the telegrapher's office. Whitehouse had planned to sail on the *Niagara* to oversee communications to shore, but he either fell ill or suffered an attack of the nerves, and Thomson went in his place.

It was a brief trip. The *Niagara* steamed about four miles out, when the cable got tangled in the paying-out machinery and broke. The ship returned and tried again. By noon on Sunday, August 9, signals were coming to shore from almost 100 miles away. The crew struggled constantly with the clumsy system for letting the cable go from the stern of the ship. The crude device for maintaining even tension proved hopelessly inadequate, and as the *Niagara* rose and fell the threat of losing the

wire was ever present. As the ship sailed into deeper water, the weight of
cable hanging from the stern became increasingly unmanageable, and the
crew had endless difficulty braking the drum enough to stop the cable
from reeling out but not so much as to snap it. On Tuesday the signal
through the cable ceased. Later in the day (at a time, Bright helpfully
noted in his memoir, when he was away from the machinery), an ill-
timed application of the brake as the ship rose on a swell put the strain on
the cable past breaking point—and break it did. About 300 miles of
cable dropped uselessly to the seafloor.

Back at Valentia the engineers tallied the remaining length of cable.
Just over 1,800 miles—about 10 percent more than the distance from
Ireland to Newfoundland, they reckoned, but that was not a sufficient
margin of error. With the enthralled crowds of a few days earlier now
vanished, the directors quickly decided to abandon the attempt but not
before agreeing to try again next year. *Niagara* and *Agamemnon* sailed
back to Plymouth, where the cable was off-loaded into covered tanks of
water for storage through the winter. (Gutta percha dried out and became brittle under prolonged exposure to light.) Another 600 miles of
cable was ordered. This first attempt at the great project, Field was quick
to assert, had been far from ignominious. They needed better paying-
out machinery. The cable itself proved adequate. Next year would be
different.

***

Field, concurring with Bright and the other engineers, believed that
improvements in the paying-out system would solve all their problems.
Thomson contributed some thoughts to the design of tensioning and
braking equipment, but Field put such matters in the hands of William
Everett, chief engineer of the *Niagara*, who adapted existing ideas and
designs to the task of cabling. In any case Thomson's interest lay mainly
in electrical questions, in which he did not share at all Whitehouse's complacency. In his paper on the quality of commercial copper, he said he
"was surprised to find differences between different specimens so great as
most materially to affect their value in the electrical operations for which
they are designed" and argued "how important it is to shareholders in
submarine telegraph companies that only the best copper wire should be

admitted for their use." He had evidently grasped by this time that the way to convince businessmen of the gravity of a scientific problem was to show that it would cost money if not solved. Still, at least in Thomson's own account, it took much stubbornness and persistence on his part to bring the directors around to his point of view.

At Thomson's insistence, the board added a clause to its contract with the Gutta-Percha Company demanding an insulated wire of verified high conductivity. No can do, was the first response. The board then asked what price the company would charge to conform to the new terms: £42 per mile instead of £40, came the answer, which the board agreed to. Thomson then helped set up a testing station at the factory so that the quality of the wire could be constantly monitored. Thomson's determination on this point thus led to the first scientifically informed quality control system for the manufacture of a commercial product. As he later commented, "It was not until practical testing to secure high conductivity had been commenced in the factory, that practical men came thoroughly to believe in the reality of the differences of conductivity in the different specimens of copper wire, all supposed good and supplied for use in submarine cables."

A second matter on which Whitehouse was complacent and Thomson nervous was that signaling across the Atlantic placed new demands on the sensitivity of the receiver. In standard telegraphy equipment, from the letter-printing machines of Wheatstone and House to the superior Morse receiver, a current ran through a coil, creating a magnetic field, which attracted or repelled an adjacent permanent magnet. The energy to move the magnet ultimately came from the current—a point that derived ultimately from Joule's experiments on the energy carried by electricity, although such thinking was not yet familiar to practical engineers. Thomson began to think of detecting the signal with a galvanometer, a laboratory instrument for detecting small currents. At the end of 1856 he wrote to Helmholtz asking for details of an instrument he had designed. The principle of a galvanometer was the same—a coil produced a force on a magnet attached to a pointer—but a good one was carefully made and well balanced, with lightweight components, and was far more sensitive than the heavy devices found in telegraphy offices.

Thomson at first imagined he would simply take apart one of

Helmholtz's galvanometers and see what he could do to reduce the mass of the moving parts. But in a stroke of inventive brilliance he saw how he could reduce the mass of one moving part to nothing at all. Inspired, so he liked to claim, by light reflecting off a monocle dangling around his neck, he substituted for the moving magnet-and-pointer arrangement a tiny piece of magnetized steel that he glued to the back of a piece of mirrored glass and suspended by a short fiber. (In the first attempt he used a hair plucked from his dog; later he substituted a silk thread from one of his niece Agnes's dresses.) A current passing through the nearby coil created a field that twisted the magnet one way or another, and by directing a light beam onto the mirror in such a way that the reflected spot swung back and forth across a graduated scale, he created a weight-less pointer for his galvanometer.

The mirror galvanometer, as he dubbed it, was the subject of Thomson's second patent, taken out in 1858. Having made a prototype, he requested the substantial sum of £2,000 from the Atlantic Telegraph Company to build a number of instruments for use with the cable to be laid later that year. The directors, yielding to Whitehouse's opinion, turned him down, but later he managed to get £500, along with permission to test the mirror galvanometer during the voyage. (His professorial salary was not much more than £200 per year.) In April he went to Plymouth to test the cable stored there and got three letters per second through the entire length—some 2,700 miles. At the end of May the *Agamemnon* set course for the Bay of Biscay to conduct deep-water tests of the new paying-out machinery. Whitehouse was supposed to go along to oversee electrical tests, but again he backed out at the last minute, leaving the field to Thomson and his new galvanometer. All went well, both mechanically and electrically, but by the time the *Agamemnon* and *Niagara* returned to Plymouth to prepare for the transatlantic voyage, Whitehouse was firmly ensconced in the electrician's office and doing his utmost to resist Thomson's appeals for better equipment and more testing.

Dissension simmered among the officers and directors of the Atlantic Telegraph Company. Thomson's initiative, inventiveness, and obvious enthusiasm for the project contrasted with the increasing recalcitrance of Whitehouse, who complained openly about "the frantic fooleries of the

Americans in the person of Mr. Cyrus Field." Morse, who had also clashed
with Field over technical choices, dropped out of active participation.
But Whitehouse was still chief electrician and Thomson an unpaid adviser. Both appeared eager to travel with the 1858 cabling voyage,
Thomson because he wanted to demonstrate the virtues of his mirror
galvanometer, Whitehouse to prove he was in charge. Field tried delicately to make sure they went on different ships. But at the last minute,
as he had done previously, Whitehouse announced he couldn't or wouldn't
go. Thomson boarded the *Agamemnon* while Whitehouse arranged to go
to Ireland to await signals coming down the wire.

Also working to Thomson's advantage was the fact that the directors
had now agreed to Bright's preference of having the ships meet in
midocean and lay the cable from there out to both shores simultaneously.
Whitehouse would have nothing to do unless or until a cable end reached
Valentia, while Thomson, though still acting in what was formally described as an advisory role under engineer C. W. de Sauty, became the de
facto electrical authority on the *Agamemnon*. On June 10, 1858, the two
ships, so weighed down with cable that "experienced mariners gazed in
apprehension at their depth in water as they left the shore" departed for
the mid-Atlantic rendezvous, accompanied by a fleet of smaller vessels.
The project almost ended in catastrophe before it began. Ten days out a
monstrous storm blew up. The *Niagara*, the larger and stouter ship,
steered clear of the worst. The *Agamemnon* came close to sinking. There
was not enough room below deck for all the cable she carried, and some
250 tons was lashed on the upper deck, making the ship dangerously top
heavy. She became unsteerable and sat helplessly in seas that heaved over
the decks, rolling her over at 45 degrees to one side then as far to the
other. Deck planks, already strained by the weight of cargo, separated
and let water flood below. The electrical cabin, with Thomson striving to
save his equipment, was washed out. Coils of cable broke loose and flailed
about; below, coal burst out of the holds and crashed back and forth as
the ship lurched from side to side.

After a perilous night, the storm began to abate. Ten sailors had been
injured, but the ship remained seaworthy and no cable had been lost.
The *Agamemnon* steamed on to the rendezvous, joining the *Niagara* on
June 25. The next day the crews attempted to splice together the cable
ends from the two ships but encountered an absurd difficulty. Half of the

cable sheathing had been made by R. S. Newall, the rest by Glass, Elliott. Because of the haste of manufacture and lack of planning, it turned out that one company had wound the protective iron sheathing clockwise, while the other had done the opposite. Had the two ends been spliced directly together, tension on the cable would have caused both windings to unravel. The engineers had to improvise an ungainly wooden bracket through which the cable ends were wound and secured, allowing them to be joined.

Finally, on June 26, the ships began to sail apart, connected by a cable through which they maintained electrical contact. After only a few miles, the cable snagged in the *Niagara*'s paying-out machinery and broke. By prior agreement, if contact was lost, both ships were to return to the rendezvous and try again. On the second attempt they managed about 40 miles before the cable broke again. A third time they tried. The *Niagara* had sailed a little over 100 miles, the *Agamemnon* almost 150, when the cable parted as it was disappearing over the latter's stern. Now fog had descended, and the ships failed to find each other. Both returned to Ireland.

In two voyages Field and his colleagues had succeeded only in scattering several hundred miles of costly cable at various places on the floor of the Atlantic. Field, along with most of the engineers and electricians, wanted to try again. They still had plenty of cable and plenty of time before winter weather would begin to threaten. But many of the financiers, who had by now seen hundred of thousands of dollars slip to the bottom of the sea, were ready to wrap up the Atlantic Telegraph Company and label the entire enterprise a noble failure. Field, the consummate salesman, prevailed again, and by the end of July the fleet, recoaled and reprovisioned, was back in the middle of the Atlantic.

Despite the catalog of mishaps and errors thus far, the third attempt was a charm. Around midnight on July 28 the splice was made. The *Niagara* sailed west and arrived on August 5 in Trinity Bay, near the optimistically named hamlet of Heart's Content, Newfoundland, trailing behind it a cable that was still receiving signals from the other ship. The *Agamemnon* had a slightly harder journey, against difficult weather. On the first day Thomson and his colleagues suffered through an hour and a half of anxiety, after the mirror galvanometer abruptly ceased to register the periodic signal sent from the other ship. Thomson emerged from the

electrical cabin "in a fearful state of excitement. The very thought of
disaster seemed to overpower him. His hand shook so much that he could
scarcely adjust his eyeglass. The veins on his forehead were swollen. His
face was deadly pale," wrote the London *Times* reporter sailing with the
expedition. Thomson told Bright he thought the conductor somewhere
in the cable was broken but that the insulation was intact. He waited
anxiously "in a perfect fever of nervous excitement, shaking like an aspen
leaf, yet in mind clear and collected, testing and waiting, with a half-
despairing look for the result." So he and Bright and the rest waited, in
dread of another failure. At one point someone saw the light spot from
the mirror galvanometer twitch through an unmistakable 40 degrees, but
Thomson, dashing into the operations room, saw nothing. Then just a
few minutes later signals from the *Niagara* began to come through again.
The engineers convinced themselves that the cable had suffered a minor
fault as it was sinking to the seabed but that once laying there securely, in
frigid temperatures and under enormous pressure, the gutta percha had
healed and all was well again. It did not pay to think too much about
what might have gone wrong.

By August 5 the *Agamemnon* had reached Valentia, where Thomson
was obliged to hand over the cable to Whitehouse's care. Field, in Newfoundland, telegraphed an announcement of the success to New York.
"The electrical signals sent and received through the whole are perfect,"
he declared. "By the blessing of Divine Providence it has succeeded." The
unexpected news, after such lamentable beginnings, set off hysteria in the
press and in the streets. From Bangor, Maine, to Washington, D.C., and
inland to Cincinnati and Chicago, church bells pealed out, bonfires
blazed, cannons roared. Mayors pontificated, ministers offered up grateful sermons. "The Great Event of the Age . . . Triumph of Science . . .
London within a Flash of New York . . . This news will send an electric
thrill throughout the world" blared the New York *Herald* on the morning
of August 6.[2] In succeeding days newspapers carried more tidbits of news,

---

[2]In its digest of joyful reports from across the nation, the *Herald* also included this:
"Skepticism of the Vermonters: The news of the successful laying of the Atlantic cable is
received here with feelings of suspicion. The Rutland *Courier* is out with the despatch in
an extra, but very few believe a word of it."

delivered in triumphal style. But the tone of the reports gradually changed. Messages from Newfoundland said that all was well, that adjustments were in hand, that signals were coming through. But where, the press began to ask, was the inaugural message from Queen Victoria to President Buchanan?

On August 7 the New York *Post* felt obliged to assure its readers that "the rumors of deception and trickery, &c., &c., have not the least foundation, so far as we know or believe," but just a week later the paper published a letter from a knowledgeable correspondent saying that the emanations from Whitehouse and his aides to the effect that they needed another five or six weeks were "enough to awaken in the sanguine unpleasant apprehensions, and to strengthen the doubters of the enterprise. The question is continually asked, Why should six weeks, or even one week, or even one day, be required for making the 'experiments', when everybody knows, who knows a little of practical telegraphy, that if the connection is good, one hour is sufficient for putting up the batteries and adjustment?"

In London the board of the Atlantic Telegraph Company was growing similarly restless. Thomson had left Ireland a few days after landing, and Whitehouse, still insisting on the need for unspecified adjustments, refused to say what he was doing in the telegrapher's hut at Valentia. The New York newspapers reprinted confused comments from the London *Times* and added their own scraps of intelligence from Newfoundland, such as this item of noninformation issued on August 13. In response to numerous inquiries, said the telegraph operators, "we are unable to return any other answer than that the cable remains all right—the electrical signals passing through its whole length satisfactorily—but that the electricians have not yet concluded their arrangements for putting their recording instruments into operation."

"Where's the Queen's message? Is the insulation perfect? Will the Atlantic telegraph work? Why don't they give us the information?" inquired the exasperated editors of the *Herald* on August 16, who complained further about the secrecy surrounding Whitehouse's "experiments."

Just as skepticism erupted openly, however, jubilation squelched it. The following morning the *Herald* was back with stacked triumphal headlines: "The Queen's Message to the President of the United States . . . The President's Reply . . . Another Great Problem Solved . . . Tremendous Sensation Throughout the City . . . Everybody Crazy With Joy . . . Now's the Time for a Universal Jubilee . . . &c., &c., &c." The *Post* remained more skeptical, editorializing thus: "True, the Queen's message bears no date, neither do we have any intimation of the time it has taken to transmit it—whether an hour, day, or week—nevertheless, we are assured, upon the faith of the Atlantic Telegraph Company, that it was actually transmitted from Ireland to Newfoundland by a submarine electric telegraph."

Then in succeeding days came actual news. England and France had concluded a treaty with China, ending hostilities there. The Indian Mutiny was coming under the control of imperial forces, and the British government sent an order through the telegraph countermanding the dispatch of a regiment of troops from Canada to India. This action alone, boosters of the cable were fond of pointing out, saved the government £50,000. For a couple of weeks, hundreds of messages went back and forth: news, political communications, commercial transactions. New York City threw an enormous gala for its heroic son, Cyrus Field, on September 1, with half a million people thronging the streets, a parade that took hours to pass down Broadway, and a great banquet that went on past midnight. "Glorious Recognition of the Most Glorious Work of the Age . . . Reunion of all the Nationalities . . . Art, Science, Commerce, Agriculture, Literature and the Mechanic Forces Joined Hand in Hand," trumpeted the *Herald*.

Among its many virtues, the cable would bring peace to the world, or so said the *Post* the following morning: "It is the harbinger of an age when international difficulties will not have time to ripen into bloody results, and when, in spite of the fatuity and perverseness of rulers, war will be impossible." But just four days later there was a sharp change of tone: "It is rather unfortunate that, during the whole week that was spent by our City Fathers in celebrating the electrical union of the Old World with the New, we have not been favored with a single evidence of its usefulness. Not a single public despatch has traversed the wire for some ten days or more." As September wore on there were only enigmatic reports of further difficulties and reluctant admissions by the Atlantic

Telegraph Company that no signals were at present being received, though experiments and tests continued. It seemed there was a difficulty at the Irish end, "near the shore, and remediable." Shares of the company, sold at £1,000 apiece, were down to £500 or less. By the end of the month the hard news could no longer be concealed: The cable had fallen silent.

The exact cause of death could never be established. The cable had worked, but it had never worked well. Signals, often fragmentary or unreadable, often had to be repeated over and over until a message successfully got through. It had taken more than 16 hours, it emerged later, for the Queen's brief communication to be clearly received in Newfoundland, though mysteriously the operators there managed to send the same message back the other way for verification in only 67 minutes. Throughout September communication was slow, error ridden, and untrustworthy. Days went by when nothing came through.

Ordering an investigation, the board of the Atlantic Telegraph Company managed to pry Whitehouse from his station in Valentia. Thomson and others took over, to try to reconstruct events and see if the project was salvageable. When he had handed over cable operation at the beginning of August, Thomson had been receiving clear signals on his sensitive mirror galvanometer. Whitehouse immediately connected his own equipment—heavy electromechanical receivers of standard design for overland telegraphy, requiring large currents. To supply those currents, he hooked up a gigantic induction coil (a kind of transformer) five feet long, supplied by a series of powerful battery cells, and yielding up to 2,000 volts. This, Whitehouse believed, would be more than enough to blast signals from Ireland to Newfoundland, and eventually, by brute force, he got the Queen's message through. But he could detect no reply.

Then, at least in some accounts, he substituted Thomson's sensitive galvanometer, began to receive signals from across the ocean, but had an assistant manually feed the messages into one of his own devices, so he could sent printed strips to London that appeared to come from his receiver. This was why the Queen's message traveling back from Canada came through so quickly—it was received by a mirror galvanometer, though Whitehouse pretended otherwise.

For Thomson and the others, sifting through the wreckage of

Whitehouse's miscalculations and deceptions was dismal work. Writing from Ireland to his friend James Joule at the end of September, Thomson contrasted his initial enthusiasm with the subsequent disappointment. "Instead of telegraphic work, which, when it has to be done through 2,400 miles of submarine wire, and when its effects are instantaneous exchange of ideas between the old and new worlds, possesses a combination of physical and (in the original sense of the word) *metaphysical* interest, which I have never found in any other scientific pursuit—instead of this, to which I looked forward with so much pleasure, I have had, almost ever since I accepted a temporary charge of this station, only the dull and heartless business of investigating the pathology of faults in submerged conductors."

Learning what had really happened at Valentia during August and September, the directors fired Whitehouse. But the damage was done. Probably there had been a partial fault in the cable, a flaw in the insulation hundreds of miles from the Irish end. This was the old cable, hastily manufactured to a poor design for the 1857 expedition, then stored through the following winter in tanks of water at Plymouth. The following summer Field wrote to Thomson to say that on examining some cut-up sections of the cable that he had sold to Tiffany's in New York as mementos, he found that in places the copper wire was distinctly off center, in some cases almost piercing through the gutta percha to the surrounding layer of tarred hemp. "I should like much to know to what cause you attribute these imperfections. What is in your opinion the cause of the Cable ceasing working?" he asked. Thomson speculated that the gutta percha had been applied too hot or that the cable had been bent before it had properly cooled. Whether winter storage of the cable had caused additional problems he could not say.[3]

A pair of electricians tested samples of the cable with the huge voltages that Whitehouse had applied. A section with perfect insulation, submerged in seawater, suffered no harm when they applied thousands of

---

[3]A piece of 1857 or 1858 cable found in Ireland in the 1980s suffered the same problems that Field described. D. de Cogan (1985) speculates that gutta percha, an impure organic material, may have suffered a kind of bacterial fermentation while stored at Plymouth.

volts to it. But when they made a pinprick hole in the gutta percha and repeated the test, "the interior of the jar lit up as if it were a lantern" and the hole in the insulation burned out big enough to put a thumb in.

In all likelihood the 1858 cable had too many imperfections to have lasted long. But Whitehouse's unauthorized experiments and desperate application of larger and larger voltages undoubtedly brought it to a pre-mature end. Naively, Thomson at first tried to defend Whitehouse, tell-ing the board he had acted unwisely, as it turned out, but not maliciously. But the directors, who had now seen close to £2 million drowned and lost forever, were beyond magnanimity. One director wrote sternly to Thomson: "I must not hide from you that the course you took in relation to our recent difficulties with Mr Whitehouse added greatly to our troubles . . . & I am therefore much pleased to find that you are at length convinced that we acted wisely in dismissing Mr Whitehouse. . . . This great undertaking has been jeopardized & perhaps ruined by placing the electrical department in the hands of a man so inefficient, selfish & un-scrupulous."

Thomson learned his lesson. Perhaps, despite all his misgivings about Whitehouse, he clung to the belief that a man of science must necessarily be honest and sincere. Even after their first dispute, Thomson had thought about proposing Whitehouse as a fellow of the Royal Society. Honest disagreement was how science made progress. Thomson could believe that Whitehouse genuinely thought his telegraph system supe-rior; he could not grasp that Whitehouse resorted to trickery because he could not bear to be upstaged by some young, unworldly academic. Even-tually, faced with direct evidence of Whitehouse's dishonesty, Thomson blinked a couple of times before he could believe it. But believe it he did, in the end.

The dispute burst into the correspondence pages of the London *Times*, Whitehouse attacking the board and Thomson, and the directors responding in kind. Thomson wrote privately to all parties, making clear that Whitehouse was now telling falsehoods—in particular, he claimed that the president's reply to Queen Victoria was received on one of his devices, whereas in fact it came through Thomson's mirror galvanometer. Official statements from the Atlantic Telegraph Company made plain their confidence in Thomson and utter distrust of Whitehouse.

Even so, Whitehouse did not entirely lose his reputation. He acted as consultant to Glass, Elliott in the construction of a Mediterranean line from Malta to Alexandria in 1861. But that was his last involvement with telegraphy. He returned to Brighton and died there in 1890, at the age of 73.

After the jubilation of 1858 turned sour, rumors began to fly that the whole thing had been a hoax from the outset, a scheme by which Field could unload his expensive shares on innocent investors. Even so, prospects for another cable attempt did not immediately fade. But on returning to the United States, Field found the economy in a downturn and politics uncertain as the country headed toward civil war. In 1859 he was in England again, trying to win government support for another venture. But even his powers of persuasion were now inadequate. In New York a disastrous warehouse fire put his old paper business on the road to bankruptcy. Then came secession and war.

In Glasgow in early 1859, at a city banquet celebrating his contribution to the cable, Thomson sounded a heartening message of Victorian optimism and the inevitability of progress. "The foundation of a real and lasting success is securely laid upon the ruins which alone are apparent as the result of the work hitherto accomplished. . . . What has been done will be done again. The loss of position gained is an event unknown in the history of man's struggle with the forces of inanimate Nature." Thomson may have firmly believed that the obduracy of nature could be overcome, but Field had to contend with money and politics. It was some years before an Atlantic telegraph again engaged anyone's attention.

* * *

Thomson's urgent effort to introduce quality control into the manufacture of commercial copper wire came up against numerous obstacles, not the least of which was the absence of any standardized procedure for measuring electrical conductivity or its inverse, electrical resistance. There was at that time no scientific unit of resistance, nor indeed of voltage or current. Galvanometers, including Thomson's ultrasensitive mirror galvanometer, did not strictly speaking measure electric currents. Rather, a current passing through the device made a needle or a light beam swing, but how much it would swing in response to a given current varied from one instrument to another. So, for example, one sample of wire could be said to have twice the resistance of another when, if both were connected in circuit with the same battery and galvanometer, the needle swung to half the amplitude for the first sample as for the second. (Though the underlying science was still fuzzy, it had been established that a given type of battery, say a zinc and a copper electrode immersed in an acidic solution, always produced the same electric potential, or voltage. The standard household battery produces 1.5 volts for this reason.)

The inability to perform accurate electric measurements mattered little for overland telegraphs covering modest distances. Either they worked or they didn't. Engineers most often tested for a signal by touching a tongue to the bare wire: An ordinary battery produces a titillating tingle. But submarine telegraphs, as Thomson more than anyone knew, displayed a spectrum of intermediate conditions between working clearly and not working at all. Sporadic failures of the insulation could let some of the current trickle into the ocean. Variations in temperature or pressure might alter the capacitance of the cable, influencing both the strength and the timing of emerging signals.

Telegraph engineers learned a number of tricks for locating faults in an underwater cable. The simplest case was an outright failure such that the sea came into contact with bare copper, effectively earthing the wire at some unknown position. A known voltage applied at the shore end would pass some current down the wire as the electricity ran to ground at the fault. The greater the current so produced, the smaller must be the resistance of the wire it was passing through, therefore the closer the fault must be to the shore. At first, engineers used the method of comparison. They kept beside them miles of cable, coiled up, so as to compare the resistance of the faulty cable to some known length of wire. This was hardly convenient, especially when dealing with thousands of miles of underwater cable. Some absolute standard of resistance, and equally important some way of measuring resistances against the absolute scale, became increasingly necessary.

This problem fell naturally into Thomson's range of interests. He had already proposed an absolute way of measuring temperature, based on Carnot's theory of engines. In 1851 he had brought before the English-speaking scientific world his expanded and revised version of a system of electrical units proposed on theoretical grounds by Wilhelm Weber in Germany. Still, it took his involvement with telegraphy to fully convince him of the need for practical measurement systems based on sound scientific principles.

Weber showed how to connect electrical phenomena with the familiar system of mechanical measurements by using Coulomb's inverse square force law. Electric charge can be measured according to the force produced between two equal charges at a known separation. Current is the rate at which charge flows down a wire. According to Ohm's law, enunciated by Georg Simon Ohm in 1827 though previously hinted at by many others, the current flowing down a wire is equal to the voltage applied divided by the wire's resistance. But if you only know the current, there are two unknowns: you would know the voltage if you knew the resistance, and vice versa, but if you don't know either, where do you start?

Thomson, expounding Weber's ideas, filled in this gap by using one of his friend Joule's early results. Joule had shown that the heating produced when electricity flows down a resistive wire is proportional to the product of the voltage and the current—what we now call the power of the electric flow. This gives an independent relationship between current and voltage, and allows resistance to be defined in an absolute, mechanical way—that is, using only measurements of force and energy.

Theoretically neat though it may have been, this so-called electrostatic system of units did not lend itself to practical application. There was no way to manufacture electric charge in reproducible amounts, and in any case the unit of charge implied by the metric unit of force over a separation of one centimeter was enormous, orders of magnitude bigger than anything encountered in the laboratory or the telegraph room.

Weber had also set out an alternative system, based on the force between magnets rather than charges. Permanent magnets were no more standardized or controllable than static electric charges, but Weber observed that a current passing through a coil of known dimensions would create an electromagnet that would feel a measurable force from the earth's magnetic field. Here was the prospect of a more practical system: A coil could be made with some possibility of sameness from one laboratory to the next, and the earth's magnetic field was at least approximately the same everywhere, once allowance for the laboratory's latitude had been made. The force produced on an electromagnetic coil therefore offered the chance of creating a standardized electric current, by which any scientist anywhere could in principle calibrate a galvanometer.

So elaborate a procedure, difficult enough for laboratory scientists, was far beyond the expertise of the telegraph engineers and technical men who actually needed standard measurements. At the 1861 British Association meeting in Manchester, the veteran telegraph engineers Charles Bright and Latimer Clark made a plea for the adoption of standardized measures that telegraphers had devised. Their voltage standards took the form of known electrochemical battery cells, which always produced the same potential, while their resistance standards were approximately reproducible pieces of metal. The German scientist M. H. Jacobi had in 1848 made in his laboratory a number of lengths of copper wire whose resistances, so far as he could measure with a cell and galvanometer, were identical. These he distributed to his colleagues throughout Europe, though they never found widespread use. Charles Wheatstone's favored unit of resistance was a one-foot length of copper wire weighing 100 grains which, if well made, would have a fixed and uniform cross section. Werner Siemens, on the other hand, argued for the use of a column of mercury contained in a glass tube one meter tall and one square millimeter in cross section. For none of these units was there any scientific or rational justification. They were just convenient, or equally inconvenient, as far as telegraphers were concerned.

Bright and Clark wanted the BA to bestow an official imprimatur on one or more of these standards. But the scientists of the BA, aware of the scientific as well as practical importance of choosing units, assembled a committee to look into ways of devising a system that was generally applicable but also had a sound theoretical foundation. Clark disliked the way his and Bright's initiative had been taken out of their hands. In the *Electrician*, the world's first journal of electrical engineering, Clark voiced his concern that "the gentlemen who constitute the Committee . . . are but little connected with practical telegraphy, and there is a fear that while bringing the highest electrical knowledge to the subject, and acting with the best motives, they may be induced simply to recommend the adoption of Weber's absolute units, or some other units of a magnitude ill adapted to the peculiar and various requirements of the electric telegraph."

This was unfair. Thomson, who more than anyone combined theoretical understanding with direct experience of telegraphy, took a leading role, and the committee included practical men such as Wheatstone and Joule, as well as more refined theorists such as the young James Clerk Maxwell. Nevertheless, Bright and Clark refused at first to serve on the committee, though they joined after a year or two. Their eagerness to take part had been deflected when the nascent committee, at Thomson's urging especially, agreed to use Weber's magnetic system as a theoretical foundation and refer any practical measurements, such as the telegraphers preferred, to these absolute standards. Even when applied science and engineering had hardly moved out of infancy, distrust and wariness already existed between the academics and the practical men.

Tension developed at the 1861 BA meeting in part because Thomson, a friend to both sides, was not there in person. Just before Christmas the previous year he had been amusing himself on the ice at Largs with the Scottish game of curling, when he had fallen badly and broken his left leg. The local doctor diagnosed a fracture, but the supposedly more expert physician summoned from Glasgow claimed it was only a sprain of some sort and recommended bed rest with frequent application of hot bandages. A week of this treatment produced no improvement, and when a third physician came from Edinburgh and pronounced Thomson to have broken his leg after all, near the top of the thigh bone, irreversible damage was already done. The leg was set as best it could be, with Thomson repeatedly under chloroform for the pain. He was on his back for many weeks, and only by Easter of the following year was he able to hobble about on crutches. Eventually he recovered, but his left leg remained an inch and a half shorter than the right, a lameness somewhat concealed by the way he would dart about at great speed, his left hand pressed to his hip.

Unable to come to Manchester for the BA meeting, Thomson communicated his views on units in letters to a young engineering colleague, Fleming (pronounced Flemming) Jenkin. Thomson would perhaps have been able to soothe and charm the telegraphers Bright and Clark, but Jenkin had a tendency to lecture. We would know little of Fleeming Jenkin except that an account of his life came to be written by none other than Robert Louis Stevenson. In the late 1860s Stevenson, son and grandson of the Stevensons who made a name for themselves building light-houses, attended Edinburgh University ostensibly to become an engineer. For this he had no interest or aptitude and went to classes only to idle about and make jokes in the back row. Jenkin, then professor of engineering, brooked no such unseemliness in his lecture room. "At the least sign of unrest his eye would fall on me and I was quelled," Stevenson recalled. "Such a feat is comparatively easy in a small class; but I have misbehaved in smaller classes and under eyes more Olympian than Fleeming Jenkin's. He was simply a man from whose reproof one shrank."

Stevenson cut the class altogether but struck up a friendship with Jenkin through a common interest in amateur dramatics. At the end of the session he had to obtain certificates for his classes, which he generally was able to seduce from his professors whether he had attended their lectures or not. But Jenkin resisted. "You see, Mr. Stevenson, these are the laws and I am here to apply them," said Jenkin. "I could not say but that this view was tenable," Stevenson observed, "though it was new to me."

Eventually Stevenson wangled his certificate even out of the obdurate professor, and he came to admire the man for his rectitude, though he could be forbidding on first acquaintance. "He seemed in talk aggressive, petulant, full of a singular energy; as vain, you would have said, as a peacock," Stevenson wrote of Jenkin. But on closer acquaintance he proved honest and rational, always ready to engage in serious discussion. He also turned less severe and judgmental as he got older, but when he attended the 1861 BA meeting as Thomson's unofficial deputy, he was only 28 years old and full of the righteousness of a new convert to the world of scientific engineering. He had come to Thomson's attention a few years previously, when he was working at R. S. Newall in Birkenhead, near Liverpool, overseeing the manufacture of the Atlantic cable. He had earlier sailed with John Brett on the cabling voyage from Sardinia to Africa.

Jenkin, careful and assiduous, strove to instill the notion of quality control in technical manufacturing as insistently as Thomson had done. They were natural allies. Before the 1861 BA, Jenkin had already made an effort to measure the insulating properties of samples of gutta percha systematically, instead of throwing lengths of cable into a tank of water, applying a current, and trying to detect electrical leaks, as had been the usual practice. A failure of insulation too small to show up in such crude tests might nevertheless cripple a 2,000-mile undersea cable.

Thomson impressed upon Jenkin the importance of understanding electrical tests in a sound theoretical way as well as through experience. To men such as Bright and Clark, these niceties seemed like needless fussiness. No doubt they were eager to adopt practical guidelines and move on, but no doubt too they rather feared the intrusion into their livelihoods of scientific principles they could not follow. Telegraphy was the foundation of electrical engineering as a profession before it became an academic subject. (The modern British Institution of Electrical Engineers began life in 1871 as the Society of Telegraph Engineers.) Neither Bright nor Clark nor Jenkin had any formal university education in the technical applications of electrical science; no such course was available to them. Instead, they learned some mathematics and physics and picked up engineering principles on the job, as apprentices, just as William Thomson's older brother James had done. They had, often, the difficult pride of the autodidact. Scientific rationalization of electrical units, minor matter though it may seem now, threatened to take away from the pioneers of telegraphy control of the subject they had invented.

On the other hand, Werner Siemens's column of mercury was making headway on the continent as a resistance standard, and if the engineers and scientists could agree on one thing, it was that British units should rule the world. Thomson came up with an ingenious extension of Weber's method that made the magnetic system into a feasible basis for practical definitions. He mounted a circular wire coil so that it could rotate around a vertical axis. At the center he suspended a small permanent magnet, hanging horizontally like a compass needle. With the coil stationary, the magnet lined up with the earth's field. When the coil rotated, its wires cutting through the lines of the terrestrial magnetic field, a current began to flow, creating a secondary or induced magnetic field that acted to twist the small magnet. With the coil rotating at constant speed, the magnet shifted to a new stable position, in the modified magnetic field it now experienced. The clever and elegant result, Thomson proved, was that the deflection of the magnet depended only on the dimensions of the coil, its resistance, and its rate of rotation. Because both the direct and induced forces on the central magnet depended on the earth's magnetic field, the position at which these forces cancelled didn't depend at all on the strength of the field, which was only approximately known. The need to know the earth's field was the great defect of Weber's original proposal. Thomson's solution got around that problem.

Maxwell, by this time professor at King's College in London, oversaw experiments to establish a British Association unit of resistance using Thomson's method. It was important to get the length of the wire accurately, without stretching, which Maxwell and his collaborators did by unwinding the coil and laying the wire into a convenient groove between long floorboards at the laboratory. In Weber's magnetic system, resistance turns out to be measured in the same units as a velocity,[4] and the BA settled on 10,000 kilometers per second as its unit, this being a convenient magnitude for measuring resistances encountered in day-to-day work. The pedantically correct unit would have been one meter per second or one centimeter per second, depending on which of two competing metric systems one chose, but either one would have been an impossibly tiny amount of resistance. This was the drawback to Weber's theoretically elegant structure.

At the 1863 BA meeting, committee members announced that they had produced a single physical sample, the so-called June 4 standard, with a resistance measured at 107,620 kilometers per second—in other words, a little over 10 BA units. Over the next few years they produced half a dozen such standards, made of platinum-silver alloy, all with slightly different resistances but measured, so it was claimed, to good accuracy. In fact, discrepancies among these standards, as well as in comparison to Jacobi's old standard, to resistances that Weber himself had made, and to the mercury column favored by Siemens, existed at the level of five percent or more for many years.

By virtue of the BA's scientific influence as well as the leading role that British manufacturers played in the telegraph industry, the BA unit

---

[4]This is best regarded as a purely algebraic equivalence, arising from the way electrical measurements are derived ultimately from a force measurement. For comparison, Weber's electrostatic system gives resistance the dimensions of the reciprocal of a velocity.

became by the late 1860s the de facto standard. In Paris in 1881, by force of intellectual power as well as personal charm, Thomson led a successful effort at the first International Conference for the Determination of Electrical Units to win official adoption of the BA definition as the universal standard.

Even with Thomson's innovation, however, calibrating resistances on the absolute scale proved troublesome. As long as different BA standards varied by a few percent, the practical utility of the system was questionable. Engineers did not have the means to calibrate resistances themselves; use of Thomson's method demanded high experimental expertise. Siemens, attending the Paris conference, insisted stubbornly and not without reason that the BA standard was all very well from an intellectual standpoint but of little help for engineers. He continued to push hard for his mercury column, on the grounds that it was easily defined and reproducible in simple laboratories. Adopting the BA unit on the admission that its precise value had yet to be determined was, Siemens insisted, a strategy bound to cause more confusion than it resolved.

It happened that Werner Siemens had a younger brother, Wilhelm, who had gone to London as a young man to market an electroplating method that the two of them had developed. There he fell under the spell of Britain's entrepreneurial culture, though not without a hard assessment of its detractions. He wrote to his older brother: "I have had the opportunity to hear much about the character of the Englishman and have arrived at the conclusion that it is composed of pure egoism; an Englishman, for example, does not feel any shame in deceiving another person and there is no greater triumph for him than to hoodwink a foreigner, especially a German. . . . Yet as a people they are great, because they are free; and the people in Germany cannot imagine what freedom is. When I have lived here for a full year, I will be spoilt for Germany for the rest of my life."

So it was. He stayed in England, became a British citizen, anglicized his name to William, took up telegraphic and electrical engineering in earnest, and became an acquaintance of William Thomson. He too was at the 1881 Paris meeting, as was Thomson's great friend Helmholtz. Debate over the resistance standard came to a stalemate, with Siemens mustering a good deal of support for his position. The chairman of the session, not wanting the effort to end in deadlock, adjourned the public discussion. A smaller group reconvened in the salon of a hotel, where deal making commenced. William Siemens persuaded his brother into a compromise by which he accepted the theoretical superiority of the BA definition, along with a firm commitment that his mercury standard would be calibrated and approved for practical use.

This was on a Saturday evening. Names for the units had still to be chosen, and national pride from many quarters demanded satisfaction. In their original proposal to the BA, Bright and Clark had suggested *galvat* (from Luigi Galvani, discover of "animal electricity" in frog's legs) for current; *ohma* (Ohm), for electromotive force or electric potential; *farad* (Faraday) for electric charge; and *volt* (Alessandro Giuseppe Volta, who invented the electrochemical battery) for resistance. Clark transformed these into galvad, ohmad, farad, and voltad, and suggested that a millionfold of these units should be named galvon, ohmon, faron, and volton. C. F. Varley, another veteran of the Atlantic cable voyages, wrote to Thomson suggesting *ampère* for the strength of a magnetic pole, in order to get a Frenchman into the picture, and added: "I object to Galvad because Galvani discovered next to nothing." Varley also disliked Clark's names for the multiples, on the grounds, among other things, that Fleeming Jenkin "writes so badly that . . . Ohmad and Ohmon will be confounded in indiscreet writing"—an objection, he noted, that also applied to himself and Thomson.

Issues of penmanship aside, the conferees at Paris succeeded in assuaging chauvinism while appropriately honoring certain scientists. In another late-night meeting over hot chocolate, Thomson and the rest settled on the modern system. Giving an official name to the colloquial BA unit they chose ohm, since it was Ohm's law that clearly defined electrical resistance. Ampere got the unit of current (with Thomson insisting that the accent be dropped for international usage), volt became the unit of electric potential, and Coulomb, who had established the force law between charges, was honored with the unit of electrostatic charge. Farad turned into the unit of capacitance, a sort of secondary honor and arguably less than the man deserved. Thomson may have been thinking of Faraday's early understanding of the role of capacitance in the retardation of undersea telegraph signals. But equally, Thomson never wholly grasped the character of Faraday's individual genius, so different from his own brilliance at mathematical problem solving, and therefore may have been disinclined to push for a greater recognition.

The compromise between the BA unit and Siemens's mercury standard, along with the names of the basic measures, came to the conference as faits accomplis when it reconvened on Monday morning. Thomson and Helmholtz hammered the deal through, each smothering discontent from their own countrymen. The French had no axes to grind and were presumably happy to get two of the four basic units, ampere and coulomb.[5]

The 1881 meeting left the resistance standard in an unhappy state. The BA definition was theoretically sound but hard to put into practice, and neither the BA wire standards nor Siemens' mercury column were good to more than a few percent. The BA program continued for some time to make better-quality standards. In 1884 a third international conference settled on a column of mercury 106 centimeters long and one square millimeter in cross section, at the temperature of melting ice, as equivalent to one ohm. (This was refined to 106.3 centimeters at a meeting in Chicago in 1893, by which time the numerous standards agreed to within 0.1 percent.)

Speaking to the Institution of Electrical Engineers in London in 1883, Thomson portrayed the saga of the BA unit as a victory in the long term, with the mercury standard an interim solution until the wrinkles were worked out. Thomson's effort in setting electrical measurements on a trustworthy theoretical foundation represents one of the most influential if little known achievements of his career. In no other person did experience of telegraphy combine with profound knowledge of elementary principles, still less in anyone as energetic, articulate, and forceful. Latimer Clark, who had at first doubted the necessity for the principles Thomson espoused, came eventually to see their importance. Writing the evening before his 1883 lecture to remind Thomson, in case

---

[5]Especially since the meeting was in France, the Americans might legitimately have pushed for franklin over coulomb for charge. But their only representative was Henry A. Rowland of Johns Hopkins, who was seriously outnumbered by Frenchmen, Germans, and the British.

it had slipped his mind, of his and Bright's original suggestions, he concluded: "I was not mathematical enough to see the enormous value of an absolute system, founded on mass, time, & space. It is this which has gained for the British system of Electrical Measurement its universal acceptance by mankind."

***

Jenkin revered Thomson so much that his young wife, Annie, feared meeting the great man. She imagined "Professor Thomson as an aged and severe philosopher and rather dreaded an introduction to him. One evening I was sitting reading by the lamplight, when I heard hurried steps coming up the stairs: the door opened and in came a tall, fair-haired young man, who, not waiting to be announced, said with a most radiant smile, 'Where is Fleeming? Are you his wife? I must see him. I am William Thomson.' I saw for the first time that benevolent bending of his eyes on the person to whom he spoke that always remained and increased, I think, with the years. But the splendid buoyancy and radiance, which made me say to my husband when he came in later, 'I have had a visit from Professor Apollo,' I never saw again. It was in the following winter that Professor Thomson met with the accident which lamed him for life." This was in 1859, when Thomson was only 35 years old but already a powerful figure in the British scientific community, an authority on every aspect of physics, and with the beginnings of a public reputation after his adventures with the Atlantic cable voyages and the noisy dispute with Whitehouse.

The business of telegraphy claimed an increasing part of his life. Always rushing hither and thither, Thomson had never been one for slow cogitation, and now had no time for it anyway. If he could solve a problem in a few hours, as he had done when learning from Stokes of the submarine cable difficulties, then solve it he would. If not, he would put the matter aside until he could spare an hour or two at some later date. While laid up for months with his broken leg he had overseen researches at Glasgow by sending letters, often several a day, to his technical assistant, Donald M'Farlane, demanding a detailed account of the results of yesterday's experiments and ordering the next series to be done at the instant. During this convalescence he kept beside him a green notebook, whose pages he rapidly filled with mathematical ideas, experiments to be attempted, drafts of papers, and any other technical thoughts that came to him. For the remainder of his life he never went anywhere without a green notebook and would pull one out on his numerous train journeys between Glasgow and London, at home during a lull in the conversation, in the middle of dinner, or when someone was speaking directly to him.

In the early 1860s the Jenkins lived in London. Thomson frequently went there on cable business and would squeeze in a visit to his friends. "I say we dined hurriedly," Annie Jenkin recalled, "because [Thomson] always did, or seemed to me always to do, everything at topmost speed. When he came, it was always in a hansom cab, in front of which he stood, urging the driver on and guiding him by pointing his stick to our house, the address of which he never could learn though he came thither constantly, and when he went he was whirled away just in time to catch some mysterious train which started for Glasgow at the earliest possible hour in the morning." As he became more busy and more famous, he would send a message to the stationmaster in Glasgow that he wanted to catch the last train to London, and the stationmaster would delay it until Thomson got there, clutching a green notebook as he hurried from cab to carriage.

His scientific publications proceeded apace, but their character changed. He wrote numerous short notes on problems of telegraphy, on the properties of copper and other conductors, on varied phenomena in electrical induction and transmission and the like, on the mechanical stresses on a cable dangling from the end of a ship, and so on. He had even, in 1852, presented to the Glasgow Philosophical Society his idea for a double-piston machine that could both heat and cool air for domestic purposes; this was a kind of heat pump, an antecedent of systems that have become popular in recent years for home heating and air conditioning. The great themes of his youth—the nature of electricity and magnetism, the foundations of thermodynamics—sank from view. In Germany, Clausius was polishing and refining his formulation of what would become known as the second law of thermodynamics, the law of increasing entropy, so that the significance of Thomson's fundamental but not fully resolved contributions began to fade. At home, James Clerk Maxwell, picking up on Thomson's mathematical analysis and geometrical depiction of Faraday's lines of force, began his long journey to a comprehensive theory of electromagnetism.

Everywhere in the natural philosophy of the mid-1800s, throughout the great systematization that became known as classical physics, lay the scattered evidence of Thomson's brilliance and originality. Yet he never quite finished things off in a way that would allow history to judge him the true creator of any of the subjects he tackled. Telegraphy distracted him from real science—or so it is easy to think. But the cable did not pull him away, so to speak, of its own accord. He went willingly. He began to enjoy the company of engineers and men of business. His patent on the mirror galvanometer and other innovations brought him money. He received fees for consulting and advising on other projects. He traveled about the country at breakneck pace, mixing scientific with business meetings, flourishing in the world of commerce and enterprise.

In the dismal days after the failure of the 1858 cable, Thomson had written to Joule from the little telegrapher's cabin at Valentia complaining of the drudgery of locating faults, but only after saying how, for him, transatlantic communication possessed "a combination of physical and (in the original sense of the word) *metaphysical* interest, which I have never found in any other scientific pursuit." Telegraphy didn't distract Thomson from science, in other words; it was for him what science was all about. He loved to solve problems, especially practical rather than philosophical ones. His contributions to electromagnetic theory and thermodynamics were in that vein. He saw how to reconcile opposing views and bring mathematical models in line with experimental and engineering reality. In devoting so much time and energy to the creation of a system of electrical units, he brought high principles to bear on empirical questions, and he helped engineer an international solution. Science for science's sake could never have been Thomson's motto. He was not, in that sense, an intellectual but rather an astonishingly clever and brilliant man. The point of science was to make things happen, to get things done, to resolve puzzles and difficulties. Above all, Thomson was good at that.

***

With the embers of the American Civil War barely cooling, Cyrus Field mustered support and money for a new Atlantic cable venture with remarkable alacrity. Although the Atlantic Ocean remained unbridged, submarine cables of increasing length had been laid with growing reliability in other parts of the world. By 1862 the Gutta Percha Company had manufactured some 9,000 miles of insulated wire. Glass, Elliott had put down dozens of underwater telegraph links, including a 1,500-mile section from Malta to Alexandria and a 1,400-mile connection across the Persian Gulf, part of a chain that gave London instant contact with India. A number of British financiers and entrepreneurs became interested in the Atlantic project, but as Thomson said later, "Cyrus Field, from the other side of the Atlantic, helped keep it alive; he gave help and impulse where they were required; worked with those who did not require revivification; and he, with his English colleagues, revived the undertaking in 1865."

In 1859 the British government had set up a formal parliamentary inquiry into the failure of the 1858 cable. During 22 hearings over a period of nine months, testimony came from scientists, engineers, oceanographers, manufacturers, and electricians. Latimer Clark provided a thorough account of the necessary properties of insulators and conductors and of the testing of cables, both during manufacture and when in use. Whitehouse returned to provide his own dissenting views of the operation of submarine cables, but Thomson and Wheatstone succeeded in portraying him as a man out of his depth in this new technology.

The parliamentary inquiry, in a massive and detailed report that stands even today as a model investigation of a technological enterprise, concluded, in short, that the Atlantic connection was unquestionably feasible but that the 1857 and 1858 attempts had been hasty and cavalier in their lack of attention to technical and engineering essentials. From a modern perspective, this is stating the obvious, but in those days the whole panoply of research and development, of feasibility studies and cost-benefit analyses, of prototypes and field tests, had hardly been thought of, let alone systematized. The first Atlantic cabling ventures had been driven by enthusiasm and a sense of adventure, even wonder, more than by hard-headed planning. As Werner Siemens commented after the success of Brett's English channel cable, "With the perseverance characteristic of the English in prosecuting their undertakings . . . the laying of a large number of other cables was at once planned and attempted, before the problem was ripe for a scientific and technical solution. Failures accordingly could not but occur."

By the mid-1860s, however, as Field rounded up his resources again, cable manufacture and laying had become practiced if not mature technologies. It was still an uncertain business, but it no longer seemed exotic. Even so, the Atlantic project had dissenters. Colonel Taliaferro P. Shaffner, formerly of the Union Army, had acquired some expertise in stringing telegraph lines around the interior of the United States and refused to believe that the 1858 cable had ever really worked. "A line of two thousand miles cannot be successfully operated for telegraphic purposes. . . . I express my opinion, that not ten consecutive words were ever sent through the cable in any one hour after it was laid," he declared in 1859. He won support from the governments of Denmark, Sweden, and Norway for a line that would run in sections from Newfoundland to Labrador to Greenland to Iceland to the Faroe Islands to Scotland and finally to Norway, the longest submerged section being about 600 miles between Greenland and Iceland. At each landfall messages would be received and sent on to the next, a reasonable strategy except that it required permanently manned stations in each of the desolate intermediate spots.

Shaffner didn't quite say that the direct link from Ireland to Newfoundland would fail, but speaking in 1859 to the merchants of Glasgow, Thomson's hometown, he suggested that "to operate a line of that distance would require men such as Faraday and your Thomson—men of the very highest science. But when they are gone, where will you find their equals to succeed them?" This missed the point, of course. Thomson's goal was always to enlist technology in support of systems that ordinary men could operate with confidence. In his history of the subject, Charles Bright credited Thomson's improved mirror galvanometer as an essential factor in the ultimate success of the telegraph to India. Such developments pleased as well as enriched Thomson. Science itself might be the domain of experts, but the products of science ought to make life easier for the everyday engineer. That was exactly why he had battled so hard to bring a rational system of electrical units into general use.

With thanks due in large part to Thomson, the electrical part of submarine telegraphy, even across the Atlantic, had ceased to be a major concern. The most likely cause of electrical failure was damage to the fragile gutta percha insulation, but improvements in the design and manufacture of iron outer coverings allowed Thomson, speaking at a meeting in London in 1861, to look forward to a time in the near future when "a submarine telegraph cable would be designed, constructed, and laid, with the same prospect of success and permanency as a bridge, or a railway."

It was getting the cable over the stern of the ship and safely down to the seabed that continued to pose the greatest difficulties. Enthusiastic amateurs suggested suspending an Atlantic cable from buoys so that it ran only 50 or 100 feet below the surface, or even dangling it from an array of hot-air balloons to avoid the water altogether. But by the time Field had organized a new Atlantic cabling voyage in 1865, engineers had developed impressive cabling machines, yards long, with drums and pulleys and tensioners, that allowed the crew to let the cable out at a controlled rate and, more important, pick it up again smoothly when a fault had been detected.

The most obvious change in the new expedition was that a single ship now carried the entire tonnage of cable. This was the *Great Eastern*, the vast, ill-starred creation of the renowned English engineer Isambard Kingdom Brunel, who had intended the vessel as a passenger and cargo ship that could travel from Britain to Australia on a single charge of coal. On one of his early trips to England, before the first cabling attempt, Field had met Brunel on the train from Bristol to London, Brunel being the builder of that track and the founder of the Great Western Railway Company. Learning of Field's project, Brunel had taken him out to the east of London, where the almost 700-foot-long hull of the unfinished *Great Eastern* loomed over the marshy Isle of Dogs. "There is your ship," he told Field, but not until September 1859 was the giant vessel floated, with difficulty, onto the shallow waters of the Thames estuary. Brunel only once, and briefly, saw his fondest creation moving under its own power. He suffered a stroke two days before the ship's launch and, partly recovered but feeble, saw it begin sailing into the English Channel. A few days later there was a disastrous explosion, killing a number of people and destroying the forward funnel. The great ship limped on. Brunel died on September 15, a few days after being told of the tragedy. He was only 53, a small, intense, combative man brought down, it was said, by a lifetime of financial struggles and commercial rivalries.

The *Great Eastern* plied fruitlessly back and forth across the Atlantic for a few years. The owners had difficulty finding enough passengers and cargo to make the voyages profitable, and on one of the first occasions when it appeared they might make some money, the ship ran aground off Ireland, incurring costly repairs. A few years later she was holed in Long Island sound, and the cost of repairs bankrupted its owners. By 1864 the *Great Eastern* was idle, in the hands of bondholders to the tune of just £100,000—a ship that had cost more than £1 million to build. Daniel Gooch, a railway engineer turned magnate and former colleague of Brunel, joined with a few colleagues and bought the ship at auction by buying out £25,000 of bonds for cash, with the holders of the other £75,000 agreeing to take shares in the new company.

Gooch had not long before this become a director of the newly formed Telegraph Construction and Maintenance Company, an amalgamation of Glass, Elliott with the Gutta Percha Company. He now struck a deal with Cyrus Field and the Atlantic Telegraph Company. In return for £50,000 in ATC shares, he agreed to use the *Great Eastern* to lay an Atlantic cable, with his company bearing all operating costs and handing the cable over to Field only after a successful voyage. The ornate ballrooms and luxurious cabins of Brunel's great ship were stripped out, leaving a cavernous space that was divided into three enormous tanks suspended within the hull on massive timbers. "It presents the appearance of a dead forest, all the trees of which have been roughly trimmed," wrote one young man who worked a cabling voyage. "Huge beams stretch in all directions, vertical, horizontal, and diagonal, tiring the eye by their similarity and numbers, and giving an idea of almost unnecessary strength."

The delegation of cable-laying operations to Gooch's company left Thomson and the other technical members of Field's team in an awkward position. The chief electrician for Field was C. F. Varley. On the 1865 voyage, in the words of a journalist, Varley "was ordered by his board not even to give his advice if he were asked for it, unless the demand were made in writing, and in that case he was only to answer in writing, and to insert in the written document a distinct declaration that the opinion given was not in any way to bind the company which he represented. Professor W. Thomson of Glasgow, whose name is known over Europe, and who is certainly one of the most distinguished and acute physicists in the world, was admitted on board as a sort of scientific aide-de-camp to Mr Varley, but he was not to depart from the course indicated by the board to his principal. So there were two gentlemen, full of suggestions and ideas and formulas, reduced to silence—two great guns, spiked as it were, but charged to the muzzle. . . . In the gravest discussion they held no part. The only way in which they could give utterance to their feelings was by asking questions."

The 1865 attempt almost succeeded. Soon after the expedition left Ireland on July 23, the electricians detected a fault in the cable that had just gone overboard, and the ship was brought to a halt to allow retrieval and repair. A splinter of sharp iron was found piercing the insulation, from core to exterior. Splicing out the damaged section, the crew resumed their tasks. Then a few days later exactly the same thing happened again. Now there was talk of sabotage. Under questioning, the crew all swore their devotion to the project. Watchmen were posted to oversee the uncoiling of the cable from the holding tanks up onto the deck, through the paying-out machinery, and into the sea. Then a little later an alert crewman saw, as the cable wound around one of the drums, a splinter of brittle iron separate from the outer covering and lodge in the machinery. He removed it before it could do any damage, and the mystery was solved. "What we had taken for assassination might have been suicide," as one commentator put it.

The *Great Eastern* plodded serenely on through heavy seas, "steady as a Thames steamer," the cable unreeling smoothly over the stern. About two-thirds of the way across, 600 miles from Newfoundland, detection of another fault brought the ships to a halt. As the crew prepared to reel the last few miles of cable back in again, the rolling sea caused it to chafe gently against the ship's side. Conditions were not bad, and work was proceeding smoothly, when a slight change of the wind or an unusual swell made the ship heave momentarily in a different direction. Without warning, the cable snapped and disappeared below the waves. The shock was abrupt and stunning. "I will never forget this hour or the effect it had upon all engaged. Had we been one family and just lost a dear father or mother, our faces could not have worn a more down cast expression," Daniel Gooch wrote later.

Cabling crews had by this time learned to retrieve lost wires by dragging a grapple across the seabed, but the lost cable of 1865 lay 2,500 fathoms down, in some of the deepest waters of the Atlantic Ocean. Nevertheless, they grappled four times and hooked up the cable on three occasions, only to find that their ropes and tackle were not strong enough to pull it all the way to the surface. They had to give up because every time the grappling line broke, they lost hundred of fathoms of it, and finally did not have enough to make another attempt.

After this "sad and dreadful discouragement . . . we were all dispirited in a sense, but not discouraged," Thomson said later. "I remember well a night in the cabin of the *Great Eastern*, when the enterprise of 1865 was finally seen to be a failure that the rest of us wished to go to bed and sleep in discouragement after the labors of a fortnight. But Field would not sleep until he had the prospectus elaborated which led to success." The cable had shown no electrical problems. The paying-out machinery had worked well. The attempt could have been successful had the planners thought to include stronger lines for picking up the lost cable, and more of it. This they would be sure to do next time.

Field only briefly returned to the United States before coming to Britain again at the beginning of 1866, to Thomson's great relief. "I am very pleased to learn that you are again in this country. You are not come too soon as the [Atlantic Telegraph Company] seems to require your impulse and I am sure will be much the better for your presence," he wrote. For legal and financial reasons, Field started up yet another company, Anglo-American Telegraph, which he quickly floated for £600,000. On Friday, July 13, a foggy as well as inauspicious day, the *Great Eastern* set off once again from Ireland. Apart from a stoppage to unravel a tangled section of cable, the voyage proceeded uneventfully. On the morning of July 28 the tiny fishing village of Heart's Content came into view, Thomson and the others on the *Great Eastern* having maintained unbroken contact with Valentia. The cable end was landed and hooked up. Announcements traveled down the wire to New York and London. Congratulations pulsed in from San Francisco and Alexandria, Egypt, and places in between, all now part of a seamless telegraph network.

There was little of the wild exuberance that had broken out in 1858. International telegraphy was not exactly routine, but it lacked the immediacy and novelty it had possessed eight years earlier. The British, this time, seemed more excited than the Americans. The Atlantic cable had become almost entirely a British project, in money and technology and ships, except for the essential presence of Cyrus Field—who, because of his close ties to Britain and Britain's unconcealed support for the south during the Civil War, attracted some criticism and dissent in American newspapers for an excess of anglophilia.

After just a few days the flotilla headed to sea again in an attempt to pick up and complete the 1865 cable. Thomson and some of the other engineers had devised a plan by which three ships would drag for the cable and pull it up part way to the surface, distributing the immense weight on three grappling lines. Still it took weeks for the plan to succeed. Dropping a line to the seabed took two hours; hauling it up again, with or without a cable at the end, took several more. After a number of excruciating near successes, the lost cable end was dragged aboard the *Great Eastern* early on the morning of September 2 and hooked up to the electrical room. Back in Valentia, operators at length noticed the flickering light of a mirror galvanometer on the long-dead cable, and cheers erupted in Ireland and in the middle of the Atlantic Ocean as messages of confirmation passed back and forth.

Now there were two transatlantic lines, and commercial telegraphy began in earnest. Old shares of the Atlantic Telegraph Company finally began to pay dividends. The cables of 1865 and 1866 lasted, in fact, only a few years. By late 1870, both had failed forever. But by then there was a cable from France to Newfoundland, and so profitable was this business that by the mid-1880s a dozen cables crossed the Atlantic Ocean.

Thomson, who had traveled on five cabling voyages to Newfoundland and back without direct payment, started earning money from the licensing of his patents for the mirror galvanometer, which all telegraph operators used. His expertise was rewarded with contracts to advise on other cable projects around the world. Letters from lawyers and patent agents in subsequent years reveal sums of hundreds or a few thousands of pounds coming to Thomson in license fees for patents he owned outright or shared with men such as Jenkin and Varley—this at a time when £200 a year was a substantial middle-class income.

As well as wealth, there came public recognition and official honor. At the age of 42, in honor of his extended efforts in bringing the transatlantic cable to reality, Thomson became Sir William Thomson. To his colleagues in the academic world, it might have seemed that he had abandoned his true calling, but so far as Sir William was concerned, the technology of the telegraph was science in action. There was nothing lowly or shameful about it. Speaking in 1874 as president of the Society of Telegraph Engineers he declared that "in no other branch of engineering, indeed, is high science more intelligently appreciated and ably applied than in the manufacture and use of telegraphic lines."

Thomson saw no fundamental distinction between his scientific analysis of undersea cables and his earlier analysis of Carnot's heat engine, which had led him to the foundations of thermodynamics. He could pass from one to the other and back again without feeling he had transgressed any intellectual boundaries. He had no interest in becoming a financial magnate, a pure entrepreneur. With the pioneer days of global telegraphy coming to an end, Thomson's interest naturally reverted to old concerns and brushed up against new ones.

4

CONTROVERSIES

After electing William Thomson to the Glasgow chair of natural philosophy in 1846, the faculty, as part of the formal admission procedure, directed him to prepare and read a dissertation in Latin on the subject "*De caloris distributione per terrae corpus*"—the distribution of heat within the earth. By marvelous coincidence, this was a subject close to Thomson's heart. No doubt he or his father put a word in someone's ear. The dissertation met with the faculty's entire satisfaction. The question it addressed haunted Thomson until the end of his life.

In one of his dozen undergraduate publications, Thomson had shown that Fourier's theory of heat flow implied a fundamental difference between past and future. From some initial distribution of temperature within a solid body, heat would flow inexorably as time marched on, ironing out temperature variations until a state of uniformity emerged. Going back in time, therefore, heat distributions necessarily became less uniform. Thomson proved in 1842 that extrapolating a plausible heat distribution backward in time would in general produce, at some point, an unphysical, impermissible pattern. Mathematically, the solution to the equations became discontinuous or double-valued or otherwise pathological. Physically, heat gradients would become infinite or the temperature indeterminate.

Following up this insight, Thomson applied his reasoning in 1844 specifically to the case of the earth. Regarding it simply as a cooling body with no internal source of heat, he showed that it could not have an unlimited past. In his 1846 dissertation to the Glasgow faculty, he developed this argument further. Among other things he explained how careful measurement of heat loss from the earth's surface could in principle yield estimates of the age of the earth. This was a novel and striking conclusion, but Thomson had no data to apply to the problem. Characteristically, he would not take the matter further until he could find something quantitative to say about the earth's age, and left the question as one of many that he would pick up again when the time was ripe.

By the mid-1850s he had begun to think about a closely related problem: the origin of the sun's heat. In this case, calculations could be attempted. From the known distance of the sun to the earth, and from estimates of the candle power of the sun's heat at the earth's surface, a rough idea could be obtained of the total amount of heat—now understood as a form of energy—produced by the sun every second. This energy must come from somewhere; it could not appear out of nothing. Either the sun had some initial reserve, which it was gradually using up, or else energy was coming into the sun by some other means so that it could continuously pump out heat at a prodigious rate with no net loss.

One or two scientists had suggested that a stream of meteors falling constantly on the sun's surface could provide a sufficient source of energy to account for the output of heat. This idea owed something to Joule's experimental demonstration that the energy of a falling body could turn into heat. Taking up this idea in 1854, Thomson calculated that about 10 pounds per hour (not quite 100 tons per year) of infalling material could supply enough energy to generate the sun's heat.[1] This mass, added to the sun every year, would not perceptibly increase its size on the sky even over millions of years, he asserted. Without further ado he concluded that the energy of the sun was "undoubtedly meteoric" in origin.

---

[1] This is a serious underestimate for at least two reasons. The sun produces considerable energy in forms other than direct heat (light, ultraviolet, radio waves, etc.) and heat received at the earth's surface is a poor measure of solar heat because a good deal is reflected. Thomson could not know either of these things at the time.

He had second thoughts, though, as soon as he had delivered the first version of this paper to the Royal Society of Edinburgh in April. A series of footnotes added in May brought out a number of difficulties. If the mass added to the sun in the form of meteors came originally from beyond earth's orbit, then the increased gravity would change the orbit and thus change the length of the year. He calculated it would have shortened by a month and a half since the beginning of the Christian era, which was surely not possible. As the meteors slowed down in the sun's outer atmosphere, friction (which generated heat) would significantly slow the sun's rotation in as little as 32,000 years. And if the meteors all spiraled in at the sun's equator, how would a uniform glow arise across the whole body?

Thomson was by this time in the habit of discussing his scientific proposals in his extensive correspondence with Stokes, who came up with a related difficulty. If meteors drifted toward the sun from beyond Mercury and Venus, the orbits of those planets would change as the mass of the sun increased. This proved to be a fatal problem. In 1859 the French astronomer Urbain Leverrier announced that he had detected a tiny drift in the orientation of Mercury's slightly elliptical orbit. Contrary to appearances, this was not good news for the meteor theory. Leverrier's careful observations set precise limits on the extent to which Mercury's orbit changed from year to year and forced the conclusion that, if a supply of meteors was to produce the sun's heat, those meteors would have to be contained within the orbit of Mercury from the outset. To have such a large mass of hidden material hanging about so close to the sun, and drifting inward in an orderly manner over perhaps millions of years, was unfeasible for both theoretical and observational reasons.

Not long after his first pronouncements on solar heat, Thomson became entangled with the Atlantic cable project and made no further progress for some years. He had satisfied himself that any conceivable chemical reactions would be too feeble to supply enough energy and that if the sun had been endowed at birth with some quantity of heat, which had been leaking passively away ever since, it would have cooled very rapidly early on and could not possibly maintain its current temperature even for centuries, let alone millennia or longer. And the meteor theory had run into all kinds of difficulties.

During the hiatus between the first and second series of transatlantic ventures, Thomson, in what was becoming his standard scavenging style, took up an idea originally due to Helmholtz, who had suggested that the sun was born through the coagulation of countless small meteors and other rocky bodies into a single gigantic sphere. As numerous bodies, originally scattered far and wide, came together under their own gravitational attraction, they would gather with increasing speed, and as the bodies finally coalesced that kinetic energy would turn into heat. Thus, Helmholtz argued, originally cold matter spread over a large volume of space could create a single condensed hot body.

Thomson at first put this idea aside because he thought the initial charge of heat would dissipate very quickly. But he came to see that continuing slow contraction of the sun under its own immense gravity would convert gravitational energy into heat in a gradual manner. This was the power of thermodynamic argument: Shrinking released gravitational energy, which had to emerge as heat in the end. The details of the transformation were unimportant. In an essay published in 1862 in *Macmillan's Magazine*, a journal of general interest, Thomson concluded that slow gravitational contraction could keep the sun in roughly its present state for a period of at least 20 million years, perhaps as much as 100 million, but certainly not as much as 500 million years.

At about the same time, during 1859 and 1860, Thomson got hold of data on heat loss from the earth, which allowed him to obtain numerical estimates of the planet's age from the methods he had proposed long ago. When the British Association met in Glasgow in 1855, Thomson had urged official endorsement of a program of measurements to establish the gradient of temperature with depth underground. It was well known that coal mines were hotter at greater depths, but Thomson wanted to know how fast the temperature increased with the descent. On his summer trips to Kreuznach, for his wife's health, he had noticed and pondered the warm water bubbling up from underground. With Thomson's encouragement, his Edinburgh colleague J. D. Forbes began measuring temperatures at a range of depths in local rock formations. Thomson applied his analytical powers—and, once again, his acute understanding of Fourier's methods—to estimate from these measurements the heat conductivity of the rocks and thus the rate at which heat from the earth's interior was flowing outward at the surface.

Finally, on April 28, 1862, Thomson read to the Royal Society of Edinburgh his account of heat flow from the earth, and his inferences about the planet's age. His opening sentence plainly declared a broader and fiercer intent: "For eighteen years it has pressed on my mind, that essential principles of Thermo-dynamics have been overlooked by those geologists who uncompromisingly oppose all paroxysmal hypotheses, and maintain not only that we have examples now before us, on the earth, of all the different actions by which its crust has been modified in geological history, but that these actions have never, or have not on the whole, been more violent in the past than they are at present."

Thomson began here with a characteristic revisionist flourish. "Eighteen years" refers to the 1844 publication of his first thoughts on heat loss from the earth. But to say that for all this time he had been distressed at geologists' ignorance of thermodynamics was a mite unfair, as the subject itself only came into being around 1850. This is mere rhetoric, however. Thomson's annoyance is against the geologists' embrace of unsound science. In the mid-1800s most geologists held to some version of a general philosophy by the name of uniformitarianism, according to which geological change on and within the earth was going on today at the same rate it had always gone on. Erosion of rocks by wind or water was understood as a slow processe that shaped the planet's topography, so geologists accepted that the earth had not always looked precisely the same as it happened to look in 1862. But broadly speaking they believed that change was slow, so that the planet had never looked qualitatively different. By tacit implication they also believed that the earth's past was infinite, or at any rate as long as it needed to be. In other words, geologists felt able to draw on an immeasurable account of time past in order to explain how slow processes had produced the modern world.

As a corollary, geologists also rejected the possibility that there had been eras of abrupt, violent, or catastrophic change in the earth's past. This explains Thomson's reference to those who opposed what he called "paroxysmal hypotheses." His argument was simple. Physics, particularly the new science of thermodynamics, dictated that a cooling earth must have a finite age. It simply could not have existed for the amount of time that geologists complacently assumed. Therefore, if the present appearance of the planet was to be explained as the result of natural processes, those processes must have at some time worked significantly faster than they did at present. "It is impossible," Thomson declared in his 1862 article, "that hypotheses assuming an equability of sun and storms for 1,000,000 years, can be wholly true."

Geologists had not altogether ignored this problem. Charles Lyell, whose 1830 *Principles of Geology* was a bible of uniformitarianism, had tried to explain the earth's heat by proposing a thermo-electric-chemical mechanism in the interior. Chemical reactions were supposed to generate heat, which would in turn generate electric currents (thermoelectricity, by which junctions of certain dissimilar metals generate electricity when heated, was discovered in 1821), and this electricity would then dissociate the compounds formed in the original reaction, so the cycle could start over again. This proposal Thomson contemptuously but accurately dismissed as a kind of perpetual motion machine, capable of generating heat yet returning to its starting conditions unchanged—further proof, were it needed, that even the greatest of geologists were innocent of the laws of thermodynamics.

Forbes found that the earth's temperature rose by about 1° Fahrenheit for every 50 feet of depth. Thomson assumed that the earth had long ago been a uniform sphere, at the same temperature throughout, sitting in empty and absolutely cold space. As heat flowed away from the surface, a temperature gradient would develop in the interior, and simple application of Fourier's method yielded a formula for the surface temperature gradient as a function of elapsed time. For the initial temperature, Thomson chose 7,000°F, which he got from estimates of the melting point of a variety of igneous rocks. If the earth was hotter in the past, then at some point it must have been molten. A molten earth certainly wouldn't look anything like the earth today, and life on it would have been impossible. There were numerous assumptions and uncertainties in this calculation, particularly concerning the physical state of rock in the earth's interior and its heat conductivity and capacity at high temperature and pressure. Nevertheless, allowing some reasonable latitude in such parameters, Thomson estimated, from Forbes's data, a planetary age of between 20 million and 400 million years.

A certain amount of educated guesswork went into these numbers. Even so, they came out compellingly similar to the numbers Thomson had produced for the age of the sun. He was onto something. The solar system—at least one containing a warm sun and a habitable earth—was around 100 million years old. This was all the geologists could have. It was all physics would give them.

Adding to Thomson's irritation was the incursion of biology into his subject, in the person of Charles Darwin, whose *Origin of Species* appeared in 1859. Darwin did not much discuss the amount of time he thought the process of evolution required, but he recognized in a qualitative way that it was a slow business and leaned toward Lyell's view of an essentially infinite past. In one of the lesser parts of his revolutionary treatise, Darwin undertook the ill-advised task of estimating, from geological analyses of rates of deposition and weathering, the age of the Weald, a sedimentary rock formation in the southeastern corner of England where Darwin had settled after his strenuous travels on the *Beagle*. The Weald was about 300 million years old, he reckoned, and if so modest a geological feature had so great an age, Darwin felt no unease at assuming much longer periods for the earth as a whole.

In his 1862 paper on the age of the sun, Thomson took a swipe at Darwin's presumptuousness. The Weald estimate rested on the uniformitarian assumption that present rates of erosion were unchanging throughout geological history, and he showed in any case that even within its limited scope it wasn't a very astute calculation. Five years later Thomson's friend Fleeming Jenkin wrote a long review of the *Origin of Species* and dismissed Darwin's attempt at quantitative geology as a calculation of the kind engineers refer to as "guess at the half and multiply by two." By that time, however, Darwin's book had gone through several editions in the course of which discussion of the age of the Weald had been quietly shelved. Darwin still wanted a long time for evolutionary history, but admitted defeat on this instance. This minor error of Darwin's forever after colored Thomson's view of the man and his theories.

Even by the mid-1860s, however, Thomson's arguments about the age of the earth had produced little change among geologists, who were not quantitative scientists in the modern way. It disturbed him that his rock-solid thermodynamic arguments were blatantly ignored on the grounds, it seemed, that physics was physics and geology was geology and never the twain shall meet. In 1867, when the British Association met at Dundee, Thomson accosted the geologist Andrew Ramsay, who professed he could happily contemplate 1 billion years, 10 billion if necessary, for the life of the sun. Thomson objected that the sun, being a finite body, could not possibly shine forever. Ramsay responded as if this point of physics had nothing to do with him: "I am as incapable of estimating and understanding the reasons which you physicists have for limiting geological time as you are incapable of understanding the geological reasons for our unlimited estimates." Thomson rejoined that "you can understand physicists' reasoning perfectly if you give your mind to it." It was another example of what he called aphasia, the habit of switching off one's mind as soon as mathematics was mentioned. So far as he was concerned, Thomson was not telling geologists how to conduct their science, only that their theorizing could not disregard the laws of thermodynamics. It took some time for this elementary point to be appreciated, no matter how hard Thomson pressed it.

***

On the east coast of Scotland in July and August, daylight comes early and leaves late. Every year from 1868 until the close of the century, if the weather was even halfway decent, a tall, rugged man was generally to be found at half past six on these summer mornings, marching determinedly around the windswept links of the venerable golf course at St. Andrews, whacking and walloping a golf ball as he went. He was frequently on his own, getting in a round before his colleagues were up and about, so that he was ready to join them for a second round when they were blearily beginning their first. On a good day he could squeeze in five rounds of 18 holes before twilight stopped him.

A close observer would discover that this man talked to himself a great deal as he went—didn't talk, rather, but sang or recited or chanted. He knew by heart considerable stretches of Greek and Latin verse and for reasons known only to himself found declamations from Horace or Homer the ideal accompaniment to solo golf. When the classics palled, he tried a song of his own devising, modeled on the popular ditty "Star of the Evening":[2]

> *Beautiful Round! Superbly played—*
> *Round where never mistake is made;*
> *Who with enchantment would not bound*
> *For the round of the morning, Beautiful Round?*

and so on for nine verses.

This man, six feet two and a half inches in his boots, with thick beard and high forehead, was Peter Guthrie Tait, professor of natural philosophy at the University of Edinburgh. He had taken up the position in 1860, when Forbes retired. J. M. Barrie, the author of *Peter Pan* and a former Edinburgh student, recalled Tait's fearsomeness in the lecture room: "The small twinkling eyes had a fascinating gleam in them; he could concentrate them until they held the object looked at; when they flashed round the room he seemed to have drawn a rapier. I have seen a man fall back in alarm under Tait's eyes, though there were a dozen benches between them." On the links at St. Andrews, Tait golfed with speed and determination and nary a second thought; he picked a shot, hit it firmly, and went on to the next one. He wrote mathematics and physics the same way. Seven years younger than William Thomson, he had enthusiastically embraced the new style of mathematical physics that was coming of age as he acquired an education. Like Thomson, he embraced the principle of energy conservation not merely as a law of physics but as a foundation stone to all of science.

Tait, born and raised in Edinburgh, had been senior wrangler at Cambridge in 1852. Never was the title more appropriately bestowed. Nothing gave him more joy than a fierce dispute, and he would turn even small points of scientific disagreement into full-blown wrangles if he could find a way. The argument over the age of the earth suited him perfectly, setting as it did the hard principles of mathematical science against the lax and ignorant speculations of geologists. Through the late

---

[2]Lewis Carroll mimicked the same song in the Lobster Quadrille: "Soup of the evening, beautiful soup."

1860s, a frustrated Thomson had repeated his charge against uniformitarianism, with little perceptible effect. At the end of 1865 he had read to the Royal Society of Edinburgh a short paper, "The 'Doctrine of Uniformity' in Geology Briefly Refuted." Brief refutation indeed: Geologists assumed an infinite past; physics dictated it must be finite. He presented once again his proof that the earth must have been molten around 20 million years ago. This remonstration drew little response.

In February 1868 he tried again with a lecture, "On Geological Time," delivered to the Geological Society of Glasgow. He began bluntly: "A great reform in geological speculation seems now to have become necessary." Once again he talked of the cooling of the earth and of the impossibility of the sun shining forever. He tried out a new, more complex argument, which he said he had first heard from his brother James, though it seemed to go back ultimately to the observation by Kant that ocean tides, through friction, would slow the earth's rotation. Making quantitative use of this undoubted fact was tricky. Thomson reasoned that if the earth was originally molten and spinning, it would have assumed a slight nonspherical shape, flattened at the poles and bulging at the equator. If it then solidified and remained rigid that original figure would stay the same. From measurements of the known departure of the planet from strict spherical form, Thomson hoped to deduce the original rate of rotation when it was born and use estimates of tidal friction to calculate how long it would take to slow down to the present rate of rotation, once every 24 hours. He claimed to get a limit on the age of the earth consistent with his other calculations.

This was further demonstration that the uniformitarian assumption must be mistaken. He returned in conclusion to his original arguments about energy loss from the sun and the earth, which he regarded as unanswerable. "Now, if the sun is not created a miraculous body, to shine on and give out heat for ever, we must suppose it to be a body subject to the laws of matter. . . . Imagine it as we please, we cannot estimate more on any probable hypothesis, than a few million years of heat. When I say a few millions, I must say at the same time that I consider one hundred millions as being a few." He would concede 100 million years but drew the line at 500 million.

Perhaps because he at last spoke directly to an audience of geologists, Thomson finally drew a response, though not for almost a year later and not from a scientist intimately associated with geological thinking. Thomas H. Huxley had earned the sobriquet "Darwin's Bulldog" for his tenacious debating on behalf of evolutionary theory, particularly in his contest with Bishop "Soapy Sam" Wilberforce at the 1860 British Association meeting in Oxford, when he famously declared that if he could choose his ancestors he would take an ape over the bishop. Huxley thought of himself as a generalist and an orator, but his first interest remained the biological sciences. It was Thomson's complaint about Darwin rather than uniformitarianism that mostly captured Huxley's attention. In letters to colleagues Darwin confided that "Thomson's views of the recent age of the world have been for some time one of my sorest troubles" and when thinking of the long periods of time evolution required, he noted ruefully, "then comes Sir W. Thomson, like an odious spectre."

In February 1869 Huxley used his presidential address to the London Geological Society to take up Thomson's challenge directly and defend the honor of the geologists and the biologists. He seized on Thomson's changing theories of the origin of the sun's heat and on admitted uncertainties in estimates of the age to suggest that the physical arguments were not nearly as secure as Thomson claimed. He explained that in any case Thomson was fighting a straw man: No modern geologist, Huxley asserted, hewed to the strict uniformitarian line anymore. He mentioned Thomson's limit of 100 million years and asked rhetorically whether any geologist had ever wanted more than this. This was a little slick, since Darwin himself had wanted 300 million years for the age of the Weald, though that number was no longer mentioned.

For all his eloquence, however, Huxley was out of his depth in dealing with mathematical physics and the laws of thermodynamics, and he resorted more than once to meaningless bluster. "The rotation of the earth *may* be diminishing, . . . the sun *may* be waxing dim . . . the earth itself *may* be cooling. Most of us, I suspect, are Gallios, 'who care for none of these things,' being of the opinion that, true or fictitious, they have made no practical difference to the earth, during the period of which a record is preserved in stratified deposits."[3] Not unlike Whitehouse when

[3]Gallio, a Roman deputy, refused to try Paul for alleged breaches of Jewish law "for I will be no judge of such matters. . . . Then all the Greeks took Sosthenes, the chief ruler of the synagogue, and beat him before the judgment seat. And Gallio cared for

attacking Thomson's theory of the submarine telegraph, Huxley hinted that niceties of academic theory were somehow inapplicable to the practical world of geology and biology. He concluded with a fine flourish: "I speak with more than the sincerity of a mere advocate when I express the belief that the case against has entirely broken down. The cry for reform which has been raised without, is superfluous, inasmuch as we have long been reforming from within with all needful speed. And the critical examination of the grounds upon which the very grave charge of opposition to the principles of Natural Philosophy has been brought against us rather shows that we have exercised a wise discrimination in declining to meddle with our foundations at the bidding of the first passer-by who fancies our house is not so well built as it might be."

At last Thomson had an opponent willing to speak out. Reading of Huxley's dismissal of the case, he responded in April at the Geological Society of Glasgow, lamenting once again that "so many geologists are contented to regard the general principles of natural philosophy, and their application to terrestrial physics, as matters quite foreign to their ordinary pursuits." He dredged up remarks from a number of recent geological writings to show that, in some quarters anyway, belief in the possible infiniteness of the past still existed. He had nothing new to say scientifically but bristled at Huxley's accusation of meddling: "I cannot pass from Professor Huxley's last sentence without asking, Who are the occupants of 'our house,' and who is the 'passer-by'? Is geology not a branch of physical science? Are investigations experimental and mathematical, of underground temperature, not to be regarded as an integral part of geology? . . . For myself, I am anxious to be regarded by geologists, not as a mere passer-by, but as one constantly interested in their grand subject, and anxious, in any way, however slight, to assist them in their search for truth."

Into this more or less gentlemanly contest the rumbustious Professor Tait now inserted himself. In a lengthy review of the addresses by Thomson and Huxley, along with some other contributions, Tait

---

none of these things" (Acts 18:15-17). Meaning, presumably, Huxley was content to see other parties duke it out over the application of physical laws to the earth, but he took no position.

marched in with all the unsubtlety he could muster. He faulted Thomson only for responding in "the mildest and meekest spirit" to Huxley's charges; this, Tait feared, "weakens, not his cause but, his chance of a hearing by not sufficiently showing his teeth." Tait had no such reservations. To Huxley's suggestion that application of mathematical methods to geology was somehow inappropriate, he responded scornfully: "Mathematics . . . cannot be usefully introduced until we have arrived at something a little beyond what may be called the mere 'beetle-hunting' or 'crab-catching' stage. . . . Let us then hear no more nonsense about the interference of mathematicians in matters with which they have no concern; rather let them be lauded for condescending from their proud preeminence to help out of a rut the too ponderous waggon of some scientific brother."

Huxley's address, according to Tait, was "clever, dashing, and plausible; but when perused with attention it is found to be seriously illogical." He noted that geologists varied widely in the length of the past they desired; Huxley even seemed to think that the 100 million years Thomson would allow was not so bad. But then, Tait wondered, why did Huxley side against Thomson with those geologists who persisted in thinking the past infinite? He had a nice answer: "As we have but too lately seen, when two Irish mobs are engaged in the sweet pastime of murdering one another, the interference of the police at once reconciles the hostile factions into one great brotherhood."

From time to time Tait paused to inject a compliment toward Huxley as one of the foremost men of his discipline, but these pleasantries only served to introduce further insults and charges of scientific ignorance. Having laid waste to the foolish and insupportable beliefs of geologists and biologists, as opposed to the clear-eyed facts that Thomson had set before them, Tait closed his review of the subject by tightening the screws further: "In truth, when we come to examine the question as a whole, giving its full weight to each of the separate details, we find that we may, with considerable probability, say that Natural Philosophy already points to a period of some *ten* or *fifteen* millions of years as all that can be allowed for the purposes of the geologist and paleontologist; and that it is not unlikely that, with better experimental data, this period may be still farther reduced."

"A most engaging boy, brimful of fun and mischief, a high intellectual forehead, with fair, curly hair and a beauty that was almost girlish."

William Thomson remembered here by a fellow Cambridge undergraduate. (Pencil portrait of William Thomson at 16 by Elizabeth Thomson. Courtesy of the National Portrait Gallery, London.)

Plate 2   *Degrees Kelvin*

"I have had a visit from Professor Apollo. . . ."

William Thomson in 1859, age 35, reading a letter from Fleeming Jenkin concerning the Atlantic cable project. Annie Jenkin, after their first meeting, recalled Thomson's "splendid buoyancy and radiance." (Photograph by David King, from King, 1925.)

"One of the most valuable of these truly scientific, or science-forming ideas. . . ."

For James Clerk Maxwell (pictured here), Thomson's ingenious mathematical analogy between the geometry of electric forces and the flow of heat was the first step in turning Faraday's acute insights into the modern theory of electromagnetism. (Photograph from Lewis and Garnett, 1884.)

Plate 4

"Telegraphic work ... possesses a combination of physical and ... metaphysical interest, which I have never found in any other scientific pursuit."

William Thomson overseeing the operation of his mirror galvanometer on the *Great Eastern* during the telegraph voyage of 1866, which finally succeeded in laying a cable from Ireland to Newfoundland. (*Electric room on the* Great Eastern. Painting by Robert Dudley, courtesy of Metropolitan Museum of Art, New York. Gift of Cyrus W. Field, 1892 [92.10.43].)

"We never agreed to differ, always fought it out. But it was almost as great a pleasure to fight with Tait as to agree with him."

Peter Guthrie Tait, remembered in an obituary notice by Kelvin. (1870 photograph, from Knott, 1911.)

Plate 6   *Degrees Kelvin*

"A friendly and unconstrained party. . . ."

William Thomson (left), Hugh Blackburn, James Thomson, Hermann von Helmholtz, and an unknown child and man discussing the physics of bird flight at Blackburn's house on the Moidart Peninsula. (Watercolor by Jemima Blackburn, September 1871. Courtesy of Robert Fairley.)

"Lord Kelvin is a gentleman of exceedingly pleasant manners [and] amiability of disposition . . . a remarkable example of a great man whose native character has remained unchanged despite . . . the elevation to a lofty social position."

A reporter's impression from Kelvin's visit to the British Association meeting in Toronto, 1897. (Cartoon from *Toronto Globe*, August 21, 1897.)

"What I liked best was when he left us to follow or not as we could, and went on thinking aloud, as he sometimes did. His mind was full of fancies, brimming over with metaphors. . . ."

A student recollects Kelvin as a teacher, photographed here at his last lecture, in 1899. (Photograph from Thompson, 1910.)

Plate 8 *Degrees Kelvin*

"Our discussions did not always end in agreement, and I remember his admitting that a certain amount of opposition was good for him."

Lord Rayleigh (left) remembering Lord Kelvin. In 1900 they were photographed in Rayleigh's laboratory at Terling, Essex. (Photograph from Strutt, 1968. Courtesy of the University of Wisconsin Press.)

Now this was remarkable. Having thoroughly castigated Huxley and the rest for offering hopeful opinions instead of mathematically precise arguments, Tait finished with an embellishment of Thomson's reasoning drawn from thin air. He had no reason to say 10 million years rather than 100 million, apart from his eagerness to make life uncomfortable for the geologists. And his judgment that new data, as yet unknown, would most likely reduce the number further is as pure a piece of illogic as anything he criticized in Huxley. If Huxley was Darwin's bulldog, Tait was aiming to be Thomson's terrier. A few years later, in lectures intended for a general audience, Tait reiterated his severe views: "We cannot give more scope for [geologists'] speculation than about ten or (say at most) fifteen millions of years." If geologists found this irksome "so much the worse for geology. . . . [P]hysical considerations from various independent points of view render it utterly impossible that more than ten or fifteen millions of years can be granted."

Tait even managed to suggest that 10 million years might be generous, using an exaggerated version of the tidal friction argument in which he assumed that the earth solidified instantly when it had cooled sufficiently and remained absolutely rigid thereafter. For neither of these extreme assumptions did he have any good grounds, but they produced a small age and that was recommendation enough.

Thomson himself never insisted on an age unequivocally less than 100 million years, yet he never clearly dissociated himself from his friend Tait's opinion that anything more than 15 million years was "utterly impossible." Plenty of geologists could see that Tait's vehemence was mostly hot air—Darwin thought his views "monstrous"—but they lacked the mathematical and physical skill to take him on. Tait succeeded in making the battle fierce and uncompromising, which was greatly to his liking, but his strategy in the end backfired. Compared to Tait's shrill prosecutions, Thomson's admonitions began to seem reasonable, and his allowance of 100 million years generous. That, perhaps, was a period of time the geologists could work with.

***

This was not the first time Tait had taken up the cudgels on behalf of his excessively polite friend Thomson. In 1862 the two had written an article entitled "Energy" for the magazine *Good Words*, which was edited by Thomson's friend the Reverend Norman Macleod. Ostensibly the authors wished to bring to a general audience the new and essential scientific concept of energy, but what impelled them to put pen to paper was an article published earlier that year arguing that credit for demonstrating the equivalence of mechanical work and heat, and by extension credit for what had become the general law of energy conservation, should go not to Thomson's close friend Joule but to an obscure German physician by the name of Julius Robert Mayer.

This allegation had come up before. Writing to Thomson in December 1848, Joule had mentioned that "a German of the name of Mayer has set up a claim for the discovery of the equivalent upon the ground that he asserted in 1842 that the heat produced by compressing air was the equivalent to the force employed although he made no experiments to prove it. This is disagreeable to me as it has involved the necessity of writing in reply to the *Comptes Rendus* [the journal of the French Academy of Sciences] but I will not be drawn into a controversy on the subject of priority beyond one rejoinder. I do not want to monopolize. The merit will belong to all those who have worked out the doctrine."

A modest and unassuming man, Joule told Thomson: "I have not the slightest wish to detract from Mayer's real merits, and I hope I have said nothing which may be thought acrimonious or unfair." Mayer responded in turn to Joule, who refused to be drawn. As he explained to Thomson, his view was that while Mayer had undoubtedly proposed an equivalence between mechanical work and heat, he had offered no empirical evidence to back up the assertion. Joule, painstakingly, had done numerous experiments to demonstrate the equivalence, and as he said in another letter to Thomson, "I have not pursued the controversy further because the facts are before the scientific world and I shall be perfectly satisfied with its verdict, whatever it may be."

Joule's reticence had the desired effect. The *Comptes Rendus* published a handful of brief communications, no one else weighed in, and there the matter rested. Until June 6, 1862, that is, when John Tyndall, Faraday's assistant at the Royal Institution, gave a public lecture in which he reiterated Mayer's claim and hinted that credit was not being given because Mayer was both a foreigner and an outsider, not part of the inner circle of reputable scientists. The lecture appeared a little later in the *Philosophical Magazine*, where it came to the attention of Thomson and Tait.

Like Tait, Tyndall had a habit of contentiousness. His grandfather, for obscure reasons, had disinherited his father out of some modest property in County Carlow, Ireland, south of Dublin, leaving the younger Tyndalls to fend for themselves and bestowing on John, it would seem, a permanent sense of grievance against privilege and establishment. Learning mathematics from the books of Professor James Thomson, Tyndall found work as a surveyor first in Ireland and then in England in the railway boom of the early 1840s. He got himself sacked for protesting about pay and conditions on behalf of the junior employees. "I suffer in a righteous cause," he recorded in his diary, and when his father suggested he might be wise to moderate his passions he shot back, "I am not stubborn, but what do you want me to do? Is it to *unsay* what I have said? I have said nothing but the truth. Shall I crawl like a guilty reptile to the knees of Captain Tucker and stain my conscience with a falsehood by telling that I am sorry for what I have done? . . . If I be wrong, you should not have taught me to be *honest*, as honesty and truth have been my guiding light in this transaction." He was 23 at the time, fired with youthful passion for the underdog.

For a few years he bounced from one railway job to another, more than once losing a position amid the numerous legal disputes that raged between companies seeking regional rights and monopolies. By 1847 railway mania was collapsing, and Tyndall found a position as a teacher at Queenwood, a Quaker school in southern England that had started life as Harmony Hall, an educational institution for workers founded by the socialist pioneer Robert Owen. There Tyndall discovered a talent for teaching and an appetite for science, and at the urging of one of his fellow teachers went to the University of Marburg, Germany, to study with the chemist Robert Bunsen. Over the next several years he shuttled between England and Germany, publishing experimental researches mainly in the *Philosophical Magazine*, getting to know Faraday, Huxley, and a few other London luminaries, but maintaining close ties also with the German academic world.

In 1853 he gave a successful popular lecture at the Royal Institution, earning an invitation to give four more at £5 each. Later that year he won permanent appointment to the Royal Institution and remained there the rest of his life. Tyndall scraped together a living from his modest salary, along with fees for lecturing, writing, and translating. In later years he wrote books that brought in extra income. He made a few modest experimental discoveries but was by no means a remarkable or original scientist, tending to throw out qualitative ideas on general grounds rather than work out fine details. Some saw him as an opportunist and self-promoter, though he had an ally in Huxley, who likewise contributed more to science as a disseminator and interpreter than as an innovator. In 1853 Tyndall was told informally he would be awarded one of the Royal Society's annual research medals, but several fellows grumbled that other scientists had greater claims to priority, and Tyndall angrily withdrew his name rather than be the subject of back-room bickering.

Tyndall's enemies may have seen this as a piece of rough justice, since Tyndall had already started to make a name for himself as a man eager to engage in priority battles, usually on behalf of continental scientists whose work was unknown or ignored in Britain. In 1852, at the British Association meeting in Belfast, he made a favorable impression with an account of his recent work on the magnetic properties of certain crystals, but also publicly criticized William Thomson's theoretical opinions on the subject, and for good measure told the assembled scientists that the theory of solar prominences advanced by their countryman Charles Piazzi Smyth had been enunciated earlier by the German astronomer Feilitzsch.

Thomson's ideas on magnetism were evolving during this period, and a few years later he had come around to an interpretation more in line with Faraday's thinking. Tyndall recorded in his diary the dim view he took of this: "Thomson completely backed out of the position which he had assumed in Belfast, and completely disowned the interpretation of his own views as stated in Faraday's lecture. Thomson has in fact backed out of every position he has assumed in regard to the phenomena of diamagnetism and magne-crystallic action. And he has done so, leaving the public to suppose that he had been misconstrued or misapprehended; which tact may generally increase his reputation with the general public, but in the private opinion of me at least does not add a whit to his nobleness."

Thomson could change his mind, of course. Any good scientist must abandon old and incorrect ideas in favor of new and better ones. What rankled with Tyndall, though, was Thomson's habit of speaking as if his earlier views belonged to some forgotten and unimportant era, or hinting that the erroneous ideas he once held were not genuinely wrong but merely inarticulate stumblings toward the more sophisticated opinions he now embraced. Worse, Thomson had a way of picking up all manner of ideas and hints from whatever he read and whomever he spoke to, recasting these bits and pieces into a finished theory, then talking as if he had come up with the whole thing himself. Tyndall saw dishonesty in this, but Tait, who had on occasion seen his own suggestions to Thomson bounced back at him later as Thomson's own, interpreted his friend's character benignly. He is a *"most absolutely honest* man," Tait wrote. "There is no doubt that he cannot distinguish between what he thinks, and what he hears;—for he *never* pays full attention to anything, he is always *also* thinking on something else; so that what he hears gets mixed with what he thinks; and he takes it for his own."

Tyndall, at any rate, was primed for battle when the question of Mayer's priority came to his attention. Tyndall's closest friend described him as sometimes "peremptory, abrupt and dogmatic. . . . He enjoys an intellectual fence for its own sake, and I am not sure that his own dexterity in inflicting sharp lashes is not a source of amusement to him."

In 1825, Faraday had begun a series of Friday evening public lectures at the Royal Institution, which served both to bring scientific innovation to a larger audience and to bring money to the struggling institution. Tyndall, who like Faraday was a lively lecturer and demonstrator of scientific experiments, continued the tradition and in 1862 decided to discourse on heat. Understanding of heat, energy, and mechanical work had revolutionized science thoroughly, but the concepts were little appreciated or understood in the world at large.

Tyndall asked Clausius for advice on the origins of the modern theory, and Clausius sent a bundle of papers, including Mayer's as well as his own contributions. This, it seems, was the first time Clausius had bothered to look closely at Mayer's writings. He was, he wrote to Tyndall, "astonished at the multitude of beautiful and correct thoughts they contain." Tyndall likewise was taken aback to find, in Mayer's little-known paper of 1842, a comprehensive statement of the equivalence of physical action, whether of chemical, electrical, gravitational, or mechanical form, and of the further equivalence of heat to all these phenomena. Mayer had concluded with an estimate of the mechanical equivalent of heat—the number that Joule had worked so hard to find out and which he had first published a year later, in 1843. In another paper came the suggestion from Mayer that meteors falling into the sun provided its heat.

Accordingly, in his 1862 lecture, Tyndall talked of the work needed to raise a body against gravity, the heat released by the same body when it fell and slammed into the earth, the ability of electric currents to make magnets move, the energy generated in a chemical reaction, and so on. He described the universal concept of energy underlying all these seemingly distinct phenomena and asked rhetorically, "To whom, then, are we indebted for the striking generalizations of this evening's discourse? All that I have laid before you is the work of a man of whom you have scarcely ever heard. All that I have brought before you has been taken from the labours of a German physician, named Mayer. Without external stimulus, this man was the first to raise the conception of the interaction of natural forces to clearness in his own mind. And yet he is scarcely ever heard of in scientific lectures, and even to scientific men his merits are but partially known."

Tyndall went on to mention Joule's "beautiful researches . . . quite independent of Mayer" but emphasized again that Mayer had calculated the mechanical equivalent of heat a year earlier. He talked of a meteoric origin for the sun's heat, and how in 1854 "Professor William Thomson applied his admirable mathematical powers to the development of the theory; but six years previously the subject had been handled in a masterly manner by Mayer, and all that I have said on the subject has been derived from him."

Tyndall's fiery sense of justice led him to a vigorous endorsement of Mayer: "Here was a man of genius working in silence, animated solely by a love of his subject, and arriving at the most important results some time in advance of those whose lives were entirely devoted to Natural Philosophy." No doubt Mayer's work had been neglected, and no doubt too

Tyndall's effort to reinstate him was sincere. But in repeatedly stressing the German's priority over both Joule and Thomson, he went out of his way to needle the big names of British science.

This attempt to put Mayer ahead of Joule came to Thomson's attention during the summer of 1862, when Tait was visiting him on the Isle of Arran. In writing a rebuttal they were obliged to face an inconvenient fact: Mayer had indeed published a statement of the equivalence of heat and other forms of energy, along with a calculation of the mechanical equivalent of heat, in 1842, a year before Joule. With Thomson's malleable view of history bolstered by Tait's pugnacity and adaptable sense of logic, their article on energy tried to set matters straight. "We were certainly amazed," they began, "to find in a recent number of a popular magazine, and in an article specially intended for popular information, that one great branch of our present subject, which we had been accustomed to associate with the great name of Davy, was in reality discovered so lately as twenty years ago by a German physician."

Bringing Humphry Davy into the argument was a smart dodge. Davy had allegedly, in the early years of the 19th century, tried an experiment of rubbing two pieces of ice together to see if heat from friction would melt them. It is very hard to imagine how this experiment could have been done convincingly. Davy apparently held the ice by tongs but still, heat could have conducted down them into the ice. Ideally he should have done his experiment under vacuum, so that heat from the air would not melt the ice. In his long account of 1851 on the new theory of heat and work, Thomson had begun by briefly mentioning Davy and also Count Rumford, born in prerevolutionary Massachusetts as Benjamin Thompson, who as a military engineer in Bavaria had noted as early as 1798 the generation of heat by drill bits used to bore out cannons.

Both Davy and Rumford, Thomson said, had provided evidence for heat being a dynamical phenomenon, not an intangible fluid, but it seems that these early ideas only impressed themselves on Thomson's attention after he had belatedly accepted Joule's evidence for the dynamical nature of heat. Nonetheless, in their 1862 article on energy, Thomson and Tait now declared firmly that the relationship of heat and motion "remained a conjecture, unsupported by scientific evidence, until the proof was furnished by Davy. . . . But it is not to be imagined that for all this the pleasant fiction called Caloric was to be abandoned; and consequently, for upwards of forty years after Davy's proof of its non-existence, caloric was believed in, written about, and taught, all over the world."

Breathtakingly absent from this summary is any hint that among those who clung for years to a belief in the "pleasant fiction called Caloric" was Thomson himself. The authors moved swiftly on: "The founder of the modern dynamical theory of the heat, an extension immediately beyond anything previously surmised, is Joule. As early as 1840 we find him investigating the heat generated by electric currents, and in 1841 he published researches which contain the germ of the vast developments of dynamical science as applied to chemical actions. In 1843 he published the results of a well planned and executed series of experiments, by which he ascertained that a pound of water is raised one degree Fahrenheit in temperature by 772 foot-pounds of mechanical work done upon it."

This elaborate recitation is intended to make clear that, although Joule had not published his work until 1843, he had working up to it for a number of years before that. But of course! And so, one may equally well imagine, had Mayer been thinking about the matter for some time before 1842. But that was of no concern to Thomson and Tait, who finally mention Mayer's 1842 paper only to say that in it "the results obtained by preceding naturalists are stated with precision—among them the fundamental one of Davy—new experiments are suggested, and a method for finding the dynamical equivalent of heat is propounded. On the strength of this publication an attempt has been made to claim for Mayer the credit for being the first to establish in its generality the principle of the Conservation of Energy." It was true that Mayer mentioned Davy's alleged experiment, as an example of the general principle he was discussing, but the implication that Davy had prefigured the general idea of the conservation of energy is entirely misleading.

All in all, Thomson and Tait's account is astonishingly dishonest. They are careful not to say anything demonstrably untrue, but pick and choose from scientific history and allow the reader to conclude that Mayer, far from saying anything new, merely summarized what was already well known in 1842. The tendency to interpret the past fluidly is perhaps Thomson's; the belligerence and elision of inconvenient facts are no doubt Tait's. The purpose of the article is not to review the origin of the conservation of energy; it is to place Joule above Mayer by whatever means lend themselves.

Why the great animus against Mayer? Two reasons stand out, one personal and the other philosophical. Thomson had for some years resisted Joule's argument that mechanical work could create heat, but once convinced, he had all the zeal of the convert and praised Joule above everyone else—Joule being the greatest precisely because he had succeeded in changing Thomson's mind. Conversely, to accept that someone else besides Joule had made the connection clear would be to imply that Thomson ought to have seen the truth sooner than he did.

The philosophical reason has more substance. Mayer's 1842 paper contained a great deal more vaporous rhetoric than rational analysis. Although he adduced various examples of physical phenomena, his case rested on a vaguely Kantian assertion that all natural events must have true and proximate causes, that nothing uncaused could happen, and that therefore physical "effects" of any sort could not appear and disappear spontaneously. The language of energy did not yet exist. A thinker as illustrious as Helmholtz, in 1847, talked of the conservation of "force," that term being used interchangeably for the distinct concepts now labeled force and energy. Mayer thus spoke of something being conserved without any precise sense of what that something was.

His calculation of the mechanical equivalent of heat Thomson and Tait attacked vigorously. Take away this numerical estimate and Mayer's paper was nothing but windy prose and metaphysics. Mayer imagined a volume of mercury falling in a glass column, compressing and heating the air beneath it, thus turning the energy of the fall into heat. From simple mechanics and the approximately known properties of air, Mayer came up with a number that was, unhappily for Thomson and Tait, not far from the right answer. Their complaint was that Mayer gave no good reason for assuming that all the energy of a falling weight would go into heating the compressed volume of air. For an ideal gas this would be the result, though whether Mayer knew that is unclear. Air is not an ideal gas—but it's close enough.

It was Joule, Thomson and Tait insisted, who had performed careful experiments to establish that mechanical energy turned into heat with a consistent conversion factor, and from this empirical basis he had argued for the universality of energy. This was how science must proceed: from observation and experiment to theoretical principle. Mayer had it backwards. He had assumed, on dubious metaphysical grounds, a theoretical principle, then had made an unjustifiable calculation of the mechanical equivalent of heat, which by pure luck gave the correct answer. This, above all, Thomson and Tait found intolerable. Mayer had not done the work or propounded good science; he didn't deserve credit for making a lucky guess.

Inevitably, Tyndall had to reply to Thomson and Tait, and Tait could not resist replying again for his side. The *Philosophical Magazine* carried charges and rebuttals and countercharges for several months. Joule wrote briefly, in his levelheaded way, to say that Mayer indeed deserved more recognition than he had received but that "to give to Mayer, or indeed any single individual, the undivided praise of propounding the dynamical theory of heat, is manifestly unjust." For himself he claimed the first "decisive proof" of the dynamical nature of heat, but no more than that. Tyndall responded that he didn't mean to detract from Joule's merit, but defended Mayer against the charge that he didn't know what he was doing. He made some captious remarks about Thomson and Tait putting a scientific question before the readership of *Good Words*, as if the general public was somehow to adjudicate the dispute. Tait responded that the magazine published one scientific article in every issue, and anyway Tyndall was in no position to talk, seeing that he had published essays in *Macmillan's Magazine*, alongside articles on "Water Babies, Sunken Rocks, and Women of Italy." (Of course, Thomson had published one of his serious discussions on the age of the earth in *Macmillan's*; Tait did not mention that.)

Tyndall translated Mayer's papers and arranged for their publication in the *Philosophical Magazine*, of which he had been an editor for some years now. Having earlier stated his admiration and respect for Joule, he now noted that Joule's initial determinations of the mechanical equivalent of heat "were so discordant that nobody attached any value to them" and that Helmholtz (Thomson's great friend—a nice touch!) had in 1847 paid little attention to Joule's work for precisely that reason. Tait responded brusquely that the correct value could, even so, be found in a footnote to Joule's first paper. Tyndall, slightly abashed, admitted that he hadn't noticed this footnote until now, but in any case, what about all the other discrepant numbers in the body of the paper?

These exchanges make immensely tedious reading. Several other names came up: Sèguin and Verdet, both French, and Colding, a Dane, had all made statements sounding like inarticulate versions of a law of conservation of energy before Mayer came into it. Tait, stretching his powers of interpretation to the limit, wanted Tyndall to admit that Isaac Newton himself had understood energy conservation, or rather would have, had he understood heat and light and electricity and magnetism better than he could have done at the time.[4] Tait wrote with belligerence and bluster to cover up the holes in his logic. Tyndall responded with Victorian high dudgeon: "What you have the hardihood to affirm, you certainly must have the goodness to prove or the manliness to retract."

Over about a year the debate fizzled, neither side admitting defeat or acknowledging agreement. Thomson contributed only once. Tyndall on one occasion addressed Thomson rather than Tait, implying none too delicately that he wished to speak to the organ grinder not the monkey. Thomson replied in a brief letter to the editors, objecting to being addressed over Tait's head, saying he did not wish to participate in the discussion but professing full confidence in whatever his colleague said. "Allow me to say I consider a great injury to myself that I should be made even apparently the medium of the statements which Dr Tyndall addresses to me regarding my friend Professor Tait," he concluded, withdrawing from the discussion.

Apart from possible therapeutic benefits to Tyndall and Tait, the controversy achieved little. As Joule had said at the outset, no one person deserves credit for formulating the principle of energy conservation. In vague and speculative form, the idea had been around for some time. As different forms of energy and their interrelationship were understood, the notion grew clearer. Helmholtz, in his 1847 essay, contributed little that was original, but stated a number of points more precisely and sys-

---

[4] By contrast, the historian E. N. Hiebert (1962, p. 105) has remarked: "It is surprising to discover that Newton neither mentioned nor recognized the validity of the principle of the conservation of mechanical energy in any of his works."

tematically than had been done before. Historian of science Clifford Truesdell has argued, on the other hand, that the first incontrovertible and mathematically correct statement of a true conservation law came from William Thomson himself, as late as 1851.

The point is a perhaps an excessively sophisticated one. When Clausius, in 1850, modified Carnot's reasoning by allowing heat and work to interconvert, he had in mind that gas was a collection of atoms or molecules, so that heat was their energy of motion. But he also recognized that there could be forces of attraction or repulsion between these molecules. In that case, when energy was used to compress a gas, say, some of it would go not into a change of temperature but into overcoming intermolecular forces. To accommodate this possibility Clausius invented what is now called "internal energy," which became part of the accounting needed to depict the thermodynamic state of a gas. However, Clausius seems to have decided, on the basis of his molecular picture of a gas, that the internal energy would have the properties it needed to have to make his revised version of Carnot's argument come out right.

Thomson, in his series of papers "On The Dynamical Theory of Heat," took up this question with more care. He saw that he couldn't found a rigorous set of thermodynamic laws if he depended on the wholly unproven hypothesis that a gas consisted of molecules. Instead, he was able to show on quite general grounds, and independent of any assumption about the true constitution of gases, not only that the internal energy must exist but that it had all the correct mathematical properties and relationships to other properties to allow it to serve as a genuine thermodynamic function.

Truesdell argues that this was the real foundation of a rigorously stated law of conservation of energy. Then again, one might say that energy conservation as an overarching concept was established by this time and that Thomson's achievement here was to demonstrate exactly what form the first law of thermodynamics, as a particular instance of energy conservation, must take.

These are academic arguments, in both the good and bad senses. That such niceties remain debatable today illustrates the hopelessness of trying to apportion credit among the many scientists who contributed to the formulation of the laws of energy conservation and thermodynamics.

Tyndall's estimation of Mayer was excessive. Tait's position, both variable and extreme in that he would give pride of place sometimes to Joule, or to Davy, or to Newton, but never to a German, was absurd. Thomson's opinion on the matter remains enigmatic. He praised Joule unceasingly, rarely pushed hard for his own case (unless writing with Tait), and even after the spat with Tyndall could be generous to Mayer. In his 1868 lecture "On Geological Time," he made passing reference to "Mayer, the great German advocate of the modern theory of heat, who did so much to urge the reception of the idea of an equivalence between heat and mechanical power." This reinforces a suspicion that the harsh words against Mayer in "Energy" came mostly from Tait, admittedly with Thomson's acquiescence. Thomson didn't care much for history, his own or anyone else's. He was inclined to say, and probably even believe, whatever seemed appropriate or congenial at the time.

***

Through these years Margaret Thomson remained an invalid, often housebound though accompanying her husband on summer trips to Arran or to Kreuznach, where the waters failed to cure her. Late in 1862 she suffered a fall that did her no physical harm but left her shocked and nervous. Poems from later years record episodes when death, her angel, fluttered closer. From August 20, 1866:

> *Is it, then, thine, this clasp importunate,*
> *O death? I will submit;*
> *I cannot struggle with a power so great,*
> *But yield me, as is fit.*

Struggle, submit, yield: she survived anyway. In 1868 the Thomsons spent some weeks at Bellagio, on Lake Como in northern Italy, before retreating over the Alps to Bad Kissingen, whose waters had been recommended for Lady Thomson, as a change, presumably, from the waters at Kreuznach. The same doctor told Thomson his lame leg would benefit from immersion, but as much as he hoped for a miraculous cure for his wife, he had little time for dubious treatments on his own account. As he told his sister Elizabeth, "I have commenced trying [the waters], but my faith is not so great in their efficacy." Even so he wrote to Helmholtz that "my wife has been feeling much better and able to walk since she came here, and it seems as if she had derived real benefit from the waters."

Thomson maintained constant hope for improvement in Margaret's condition, despite all evidence to the contrary. Friends and colleagues testified occasionally to Margaret's lively intelligence and charm. Her poems spoke otherwise. From Bellagio in 1868, from the sun and the warmth and the soothing, gorgeous landscape of Italy, she concluded an ode to death thus:

> *All have failed but thou, O death!*
> *To thy promise be thou true,*
> *E'en though Despair suggest to Faith*
> *That there is cause to doubt thee too.*

In the second half of 1869 her health became decidedly worse, to the point where her doctors would not allow travel at all. That winter Thomson took his wife to the Scottish coast at Largs, away from the smoke and grime of their Glasgow residence, where they still lived in one of the old college houses squeezed among the city's slums. He refused to teach the spring classes, so that he could stay with her in the fresh sea air. The university engaged a substitute teacher for that session. Margaret's health declined steadily, but Thomson took her occasional rallies as slight cause for optimism. Though Margaret's sister and others assisted him, he spent many hours at his wife's bedside. He could sleep apparently at will and got up at three every morning to take over nursing duties. Margaret Thomson died on June 17, 1870, bringing to a close the 17 years of nameless ill health she had endured since overtaxing herself on her honeymoon tour.

Thomson's commitment to his wife had been one of uncomplaining devotion rather than passion. Her loss created an absence in his life—an absence not so much of emotion as of occupation. His immediate solution was work and travel. Although the Atlantic expedition of 1858 and the tussle with Whitehouse had demonstrated the virtues of his mirror galvanometer, Thomson had long sought to remedy its one flaw: It needed an operator to stand by at all times and record the flickering motion of the light beam as it registered incoming signals. Electromechanical devices of the old Morse type had evolved into machines that punched marks onto a paper tape, but these instruments demanded bigger electric pulses than long oceanic telegraphs could provide. Thomson tried for a while to build a device that would record electric sparks as scorch marks on paper strips, but failed to produce a reliable or sensitive system.

In 1867 he took out a patent on a new instrument, which he called his siphon recorder. A fine stream of ink passed down a narrow glass needle, which Thomson adapted from needles used to administer vaccines. As in the mirror galvanometer, this "siphon" was connected to a small magnet that moved from side to side in response to electrical signals. A paper tape ran beneath the tip of the needle. The important innovation in the siphon recorder was that Thomson devised a system to electrically charge the ink in its reservoir, so that electrical repulsion constituted an internal pressure driving the ink through the needle. This was an electric rather than a mechanical pump, pushing a finely controllable ink jet from the end of the siphon.

This new technology was ingenious but delicate. It took Thomson two years of fiddling to make a prototype good enough to install at a telegraph station, where it could be tested alongside the mirror galvanometer. In fact, one might argue it took around 120 years of development for the siphon recorder to graduate to a trustworthy piece of technology. The modern inkjet printer, which became affordable and dependable only in the late 1980s, uses essentially the trick that Thomson dreamed up in the 1860s.

By the summer of 1870, following Margaret's death, Thomson was eager to take to the seas again to try out the latest version of his siphon recorder. The operators of the French Atlantic cable, which ran from Brest to St. Pierre, had been using siphon recorders with mixed success for about a year. A long cable from Falmouth, on the southwestern coast of England, to Bombay began transmitting in April 1870, and the Prince of Wales, among other distinguished guests, saw messages arrive in London on one of Thomson's siphon recorders.

Thomson spent several weeks in Cornwall tinkering with his siphon recorder at the English end of the India cable. From there he wrote optimistically to his wife's sister that "the days of signalling by the 'spot of light' are numbered, and a luminous electrified pen will succeed the mirror." This was premature. Though the siphon recorder enjoyed some success, it was never the unqualified triumph that the mirror galvanometer had been a decade earlier. Telegraph companies hoped it would allow a reduction in manpower, since it recorded messages without an operator, but its mechanism proved in practice so delicate that a technician generally had to stand by to keep it running correctly. More than two years later, Thomson was sent copies of reports from telegraph company men testifying to the siphon recorder's balkiness. From Bombay, in September 1872, complete with misspellings: "The signals are, comparitively, not so distinct on Recorder as on Mirror. . . . Will you kindly point this out to Sir William Thompson to see if he can remedy it." From Aden on the Arabian Peninsula: "I am sorry to say however the Recorder has only worked here with varying success, and in no way to justify any one in reccommending for you a staff reduction. . . . Close connection with the Instrument does not dispel the opinion that I formed when I first saw it i.e. that as a Telegraph Instrument it had too many things to get out of order." The correspondent explained that in damp conditions the ink stopped flowing, and they had to dry the machine at the kitchen fire. He continued: "We loose in speed by it and the signals not being too distinct the speed is still further reduced by repititions being necessary. . . . I hope when Sir William Thomson's assistant Mr. Leitch returns here and gives us the benefit of his advice and experience that I shall be able to report that we are making progress."

Mr. Leitch, who had assisted Thomson in Cornwall in 1870 and then headed out to the Far East, unfortunately died of dysentery in Alexandria at the end of 1872, though not before making the siphon recorder at Suez work tolerably well. After a year of trials in Bombay, operators concluded that the recorder worked better than the mirror galvanometer from October to May, but that during the monsoon season the constant humidity drained away electric charge from the ink, and they had to return to the mirror instrument instead. From balmier Mediterranean climes, on the other hand, came a ringing endorsement: "At Malta, the Mirror is a thing of the past. . . . The Recorder has undoubtedly tended to lessen greatly, the number of errors." Happily for the telegraph company, this report also noted a reduction in staff numbers at Malta because of the recorder's satisfactory functioning.

The siphon recorder, despite its delicacy, added to Thomson's reputation and to his wealth. The devices were made in Glasgow by his long-time instrument maker, James White, who formed a company with Thomson as senior partner. Thomson asked for, and received, licensing fees of £1,000 a year from each company using the siphon recorder—several times his professorial salary.

Thomson never went in for ostentatious displays of affluence, but he showed a taste for some of the things that money could buy. Late in the summer of 1870, a few months after his wife's death, and having spent some time again at sea on cabling business, he decided he needed a boat of his own. It was no mere weekend pleasure craft he chose. During the course of a trip to London on telegraph business, he consulted his old colleague, the Glasgow mathematician turned lawyer Archibald Smith, a sailing man himself, about a vessel he had heard of. Smith wrote a few days later: "You quite take away my breath by your plans for a schooner of 120 tons." The craft in question was at Cowes, on the Isle of Wight, and after seeing it Thomson described it to his brother James, in tones reminiscent of the letter he had written years ago to his father justifying the purchase of the "funny" at Cambridge: "It is the *Lalla Rookh*[5] 126 tons vessel of 17 years old but of oak & very strong and in perfect condition, also a very good model & said to be a fast sailer, that I have bought."

Thomson engaged a captain and crew and sailed his new boat around the coast to Glasgow. After a few local jaunts he put it up for the winter and started planning nautical excursions for the following year, fondly imagining his scientific friends would eagerly join him. From his seafaring connections sprang new work and interests. In September 1870, as he was purchasing the *Lalla Rookh*, the Royal Navy was conducting trials of the 7,000-ton H.M.S. *Captain*, the prototype of a modern design of gunship intended to carry British seapower and political might around the world. But in a sharp storm at Vigo Bay, on the northwestern tip of

---

[5]The name is from Thomas Moore's popular but now largely forgotten narrative poem of 1817, *Lalla Rookh* ("Tulip Cheek"), recounting the travails of a princess of that name who journeys from Delhi to Kashmir in search of the man she is to marry.

Spain, the *Captain* capsized with the loss of all aboard, almost 500 men, including its designer. The Admiralty, counter to long-standing habit, thought it wise to obtain technical expertise in its investigation of the catastrophe and accordingly asked Thomson as well as his colleague William Rankine (who was by this time professor of engineering at Glasgow) to serve on a committee of inquiry. The two contributed an analysis of the dynamical stability of ships against rolling due to wind and waves. They concluded that the *Captain* carried such a quantity of turreted weaponry that it became top heavy and therefore unstable even in moderate seas. Once it tipped more than nine degrees away from vertical, they estimated, it could not have righted itself. Thomson's work on this committee (he traveled from Glasgow to London every other week throughout the spring of 1871) contributed to the establishment of elementary principles of physics in ship design and to criteria that any new design must meet. As always this was, for him, no distraction from real science but a vital aspect of what science must do. Well into the second half of the 19th century the point needed still to be made that scientific laws were not restricted to the laboratory and the ivory tower but had an essential role in the industrial and military prowess of Great Britain.

Thomson, still only in his mid-40s, was racing incessantly about Britain by sea and rail, consulting on cable ventures, meeting with Admiralty officials, dealing with lawyers and businessmen on patent rights and licensing fees, and darting off to Oxford or Cambridge or some other academic institution to discuss pure science. C. G. Knott, a student of Tait's and later his biographer, recalled an occasion when Thomson stopped in at Tait's lab in Edinburgh on his way to London in order to get some experimental data he needed. "Full of impatience and excitement Thomson kept moving to and fro between the slabs on which the instruments stood, suggesting new combinations and jotting down in chalk on the blackboard the readings we declared. Tait stood by, assisting and at the same time criticizing some of the methods. At length Sir William went to the further side of the lecture table and copied into his note book the columns of figures on the blackboard. After a few hasty calculations he said: 'That will do, it is just what I expected.' Then off he hurried for a hasty lunch at Tait's before the start for London where during the next week he was to give expert evidence in a law case. As they withdrew, Tait looked back at us with a laugh and said '*There's* experimenting for you.'"

Amid all this activity and traveling Thomson was still, from November to May each year, professor of natural philosophy at Glasgow University, with a complement of courses to teach. Telegraph and other business he mostly managed to conduct during the long summers (when cabling voyages invariably took place), but when the Admiralty summoned him to assist Her Majesty's government on matters of national importance university lectures took second place. By this time his nephew James Thomson Bottomley, born in 1845 to his since deceased sister Anna, had trained as a physicist himself and had become his assistant. Bottomley took classes with increasing frequency as official duties called Thomson elsewhere.

Throughout 1870 Thomson was largely absent during a momentous period in the university's history. After years of planning, the university relocated to a new building, designed in high Victorian style by George Gilbert Scott, who had been responsible a few years earlier for that most magnificent of Victorian buildings, St. Pancras railway station in London. The new Glasgow University rose up in a more salubrious part of the city, west of the old site. As long ago as 1845, James Thomson had written to his son in Cambridge to say that the railway company had agreed to buy the old college site for £30,000, in order to build a new railway station for Glasgow, and would kick in another £70,000 to help the university move. But the railway boom came to a bust a few years later, and the deal fell through. In 1863 the City of Glasgow Union Railway Company made a similar proposal, now offering £100,000 for the old site. Much more was required to build a new college, but with a combination of government money and funds raised by public subscription (in those days commercial men took civic pride in the great city universities), ground was broken for the new university in 1867, with the move completed in time for the start of the 1870-1871 session.

That year, however, Thomson attended his dying wife in Largs and rarely came to Glasgow. During the summer he worked on his siphon recorder, in Cornwall and London, and bought the *Lalla Rookh*. He did not attend the inauguration of the new buildings, saying he was still officially in mourning. He taught the first part of the 1870-1871 session, but during the second part was frequently away for the *Captain* inquiry.

Out of a mix of loyalty, affection, and pragmatism, Thomson remained a Glasgow man. By 1869 Cambridge University had finally worked its way around to seeing the value of experimental physics and approached Thomson, who had been teaching the subject at Glasgow for almost a quarter of a century, about coming down to Cambridge to instruct students there in the novelties of experiments and data. His wife's illness prevented him from making any commitment, though clearly the prospect tempted him. The Duke of Devonshire (who had been second wrangler in his day and was then chancellor of the university) offered money for the construction and endowment of a physical laboratory at Cambridge in memory of his great-uncle, the accomplished amateur scientist Henry Cavendish, who among other things had calibrated Newton's inverse square law of gravity by measuring the force between two large masses.

Cookson, Thomson's tutor of three decades ago and now master of Peterhouse, tried to draw him down from Scotland. But Thomson had begun to settle into his spacious laboratory in the new Glasgow buildings, and the prospect not only of moving but of starting an entire new Cambridge course, in a laboratory yet to be built or equipped, did not attract him. He confessed to Cookson an anxiety about taking on new responsibilities, especially when set against "the great advantages I have here in the new College, the apparatus and assistance provided, the convenience for getting mechanical work done" (the latter referring to his long-standing relationship with the instrument maker James White). Thomson, in short, was comfortable at Glasgow and had built a life combining scientific work, teaching, inventing, business, advising, and so on, which he could not imagine disrupting. Cambridge would make new demands, and the purer academic environment of that place was, at this point in Thomson's life, far less attractive, narrower, and less thrilling perhaps than it had seemed to him at the age of 20. He stayed in Glasgow, and in the course of his life refused the Cavendish professorship three times. Cambridge appointed James Clerk Maxwell instead, who had for the previous five years abandoned professorial life to be a country squire in the Scottish lowlands, at his modest Galloway estate, although he never ceased to work at mathematics and physics.

Maxwell, though he stands as the greatest theoretical physicist of the

19th century, had been instrumental in working out the practicalities of the BA system of electrical units. Probably it was good for Cambridge that they got him and not Thomson, who as he admitted to Cookson had too full and complex a life to devote himself entirely to the creation from scratch of a wholly new laboratory. Possibly it was not such a good thing for Maxwell, though. Diligent as always, he set the Cavendish Laboratory on the road to becoming for a considerable stretch of time the premier experimental venue in the world. He spent an inordinate amount of his time collecting and editing the scientific papers of Henry Cavendish, when more papers from James Clerk Maxwell would have bestowed greater benefit on the world. In 1879, still producing works of enormous profundity and promise, Maxwell died of intestinal cancer, at the age of 48.

***

"I am very glad Maxwell is standing for the professorship," Thomson had written to Stokes in March 1871, perhaps partly out of relief that he could set the matter behind him. Thomson and Maxwell knew each other but were never particularly close. Maxwell, a loner with a cryptic and "pawky" Scots sense of humor, lived a largely detached life. Jemima Blackburn, the wife of Thomson's Cambridge friend and Glasgow colleague Hugh Blackburn, professor of mathematics in succession to his father, was a cousin of Maxwell's. She was well known in her lifetime for her drawings and watercolors of nature, especially of birds, and made the acquaintance of a number of Victorian luminaries outside the scientific world, notably Anthony Trollope and William Thackeray. She had early encouragement in her artistic endeavors from the great critic Ruskin. The Blackburns entertained numerous visitors and parties at their homes (they had a house in Glasgow and a rambling mansion on the Moidart peninsula, on the west coast of Scotland, looking out toward the islands of Rum, Muck, and Eigg). Thomson, especially before he was married, spent a great deal of time with his friends, and so too for a while did James Clerk Maxwell. When the teenaged Maxwell attended the Edinburgh Academy, before going to Cambridge, he lived in Edinburgh with his father's sister Mrs. Wedderburn, Jemima's mother. He and Jemima played and made toys together. Jemima recorded a few charming sketches of

James, both in Edinburgh and at his father's small estate, where the Wedderburns visited.

As an adult, though, James Clerk Maxwell was far less at ease in the boisterous Blackburn household than was William Thomson, who had a splendid time amusing the numerous Blackburn children. Thomson was a bluff, engaging, straightforward fellow, but Maxwell was by nature an ironist and an observer. A watercolor by Jemima shows Maxwell sitting silently to one side, looking away, while Hugh Blackburn and Thomson lark about with the youngsters. Maxwell's withdrawal from social life was completed when he married a woman whom, according to Jemima, no one liked, possibly not even Maxwell himself. Katherine Dewar was several years older than her husband, and the daughter of the principal of Marischal College in Aberdeen, where Maxwell was professor from 1857 to 1860. In Jemima's words, "The lady was neither pretty, nor healthy, nor agreeable, but much enamored of him. It was said that her sister had brought about the match by telling him how much she was in love with him, and being of a very affectionate tender disposition [Maxwell] married her out of gratitude. Her mind afterwards became unsettled but he was always most kind to her, and put up with it all. She alienated him from his friends, and was a suspicious and jealous nature." As a cousin and childhood friend displaced by a wife, Jemima had reason to be irked, but similar accounts come from other acquaintances. A student at Aberdeen described Maxwell as "the most delightful and sympathetic of beings" but said he had a "terrible wife." There is a Cambridge anecdote that she fetched her husband away from a social gathering once by announcing loudly, "James, it's time you went home, you are beginning to enjoy yourself." The son of one physicist said that "Mrs. Maxwell, although she no doubt had her points was, to put it bluntly, a difficult woman." Another Cambridge physicist advised new arrivals that Maxwell, when he was a professor there, was always eager to talk to his colleagues at the laboratory, but warned them they should not call on him at home.

Maxwell, like Thomson, belongs to the rank of distinguished scientists who managed only second place in the wrangler competition. Maxwell lost in 1854 to E. J. Routh, who went on to become a moderately well known applied mathematician and a notable coach of more wran-

glers. Already acquainted with Thomson through the Blackburns, Maxwell had studied closely Thomson's early work on the analogies between
electric lines of force and heat flow and wrote to him in February 1854
asking for advice on how to proceed. "Suppose a man to have a popular
knowledge of electrical show experiments and a little antipathy to
Murphy's *Electricity* [a textbook of the day that did not impress Max-
well], how ought he to proceed in reading & working so as to get a little
insight into the subject wh may be of use. . . . If he wished to read
Ampere and Faraday &c how should they be arranged, and at what stage
and in what order might he read your articles in the Cambridge Journal?"

In the mid-1850s Thomson had published his innovative papers on
electricity and magnetism, the main achievement of which had been to
cast in mathematical form Faraday's qualitative ideas about lines of force
and the state of "electrotonic" tension that Faraday believed to pervade
electrified space. Digesting these works, Maxwell conceived that
Thomson had more or less evolved a full theory of electromagnetism, but
being a busy man of diverse interests had not yet got around to assem-
bling it all in publishable form. By the end of 1854 Maxwell was telling
Thomson that he had "been rewarded of late by finding the whole mass
of confusion beginning to clear up under the influence of a few simple
ideas." In this endeavor he was, he said, "greatly aided by the analogy of
the conduction of heat, wh: I believe is your invention at least I never
found it elsewhere. . . . This is a long screed of electricity, but I find no
other man to apply to on the subject so I hope you may not find it
difficult to see my drift."

A little less than a year later, in September 1855, Maxwell was eager
to start publishing his own ideas and discoveries but remained cautious
about encroaching on Thomson's territory. "I would be much assisted by
your telling me whether you have not the whole draught of the thing
lying in loose papers and neglected only till you have worked out Heat or
got a little spare time," he wrote. "As there can be no doubt that you have
the mathematical part of the theory in your desk all that you have to do is
to explain your results with reference to electricity. I think that if you
were to do so publicly it would introduce a new set of electrical notions
into circulation & save much useless speculation. I do not know the
Game-laws and Patent-laws of science. . . . I certainly intend to poach

among your electrical images, and as for the hints you have dropped about the 'higher' electricity, I intend to take them."

In fact, Maxwell was mistaken. Thomson had no theory, in loose papers or otherwise, waiting to be published. In 1855 Maxwell wrote the first of his great papers on electromagnetic theory, entitled "On Faraday's Lines of Force." This was not yet a theory of electromagnetism. Maxwell had worked out a full mathematical treatment of lines of force on Faraday's model, but did not link these notions to the movement of electric charges or variations in magnetic fields.

As different as were their personalities, Maxwell and Thomson diverged too in their views of mathematical science. Maxwell aimed to capture electromagnetism in mathematical form, finding relationships between the spatial distribution of charges and magnets and the fields they produced, their variation with time, and (as Faraday had so triumphantly demonstrated) the generation of currents by moving lines of magnetic force. This grand synthesis Maxwell eventually created, but not for another 10 years, and by that time Thomson resisted the theory Maxwell gave to the world. Thomson had the same goal but wanted to get there by a different route. For him, mathematics was to be trusted only if it emerged from a tangible physical model. As early as 1847 he had published a paper in which he tried to portray electric and magnetic influences by analogy to the behavior of solid materials with specific properties. He imagined an elastic material with some rigidity, capable of being deformed by external forces but also resisting such forces and returning to its normal disposition when unstressed.

As with his earlier analogies between electric lines of forces and lines of heat flow according to Fourier, Thomson believed that mathematical similarity betokened an underlying physical connection, in the sense that different phenomena appeared to follow the same kinds of law. Such thinking bolstered his entire approach to electromagnetism. He wished to discover what properties a physical solid would need—density, rigidity, elasticity—in order to provide a complete and consistent model for the behavior and interrelationship of electric and magnetic effects. His early work didn't do this. It suggested three different analogies for electric forces, electric currents, and magnetic forces. But it was a step in the right direction, Thomson believed. Find a plausible physical solid that contained an analogy to all electromagnetic phenomena at the same time, and you will have found a full and satisfactory theory of electromagnetism. Such a theory would not, in the modern sense, give much clue as to the elementary nature of electromagnetism, but that was entirely in keeping with Thomson's distaste for "metaphysics." Find the right analogy, obtain the correct equations, and you have found all you can need or want, so he fiercely believed. This belief never left him. Maxwell may have started out thinking along the same lines, but his genius led him to see, after many years of cogitation, that the mathematical laws governing electromagnetism must be *sui generis*, laws that correctly accounted for electromagnetic phenomena and interrelationships, without any necessary connection or analogy to other parts of physics.

Indeed, Maxwell's finished theory of 1865 contained elements that Thomson disliked precisely because they had no analogy to the behavior of any known kind of physical solid. Maxwell built a theory of mathematical relationships between charges, currents, and what we now recognize as the electromagnetic field. But Thomson saw this as abstract mathematics, divorced from direct physical meaning. There was no model, no tangible, palpable elastic substance underlying the quantities Maxwell defined. He accepted that Maxwell had found an important link between the propagation of electromagnetic influences and the speed of light, but he saw that as an isolated triumph within a theory that was, to him, no theory at all.

Such reservations were not unique to Thomson. This was, after all, something new in physics. Maxwell's was the first modern field theory, rendering in precise mathematical form Faraday's extraordinary vision of electric and magnetic influences pervading space. In 1855, according to his niece, Faraday felt isolated in his views on electromagnetism. "How few understand the physical lines of force!" he said to her. "Thompson [sic] of Glasgow seems almost the only one who acknowledges them. He is perhaps the nearest to understanding what I meant." Then Maxwell took up the cause, acknowledging his initial debt to Thomson, whose early analogy between heat flow and electric tension was, he said, "one of the most valuable of these truly scientific, or *science-forming* ideas." Subsequent work too he "considered as a development of Thomson's idea."

But over the years Maxwell drifted further from Thomson's conception of a theory of electromagnetism, while remaining always true to Faraday's vision. He wrote to Faraday in 1857 that "as far as I know you are the first person in whom the idea of bodies acting at a distance by throwing the surrounding medium into a state of constraint has arisen, as a principle actually to be believed in. . . . Nothing is clearer than your descriptions of all sources of force keeping up a state of energy in all that surrounds them, which state by its increase or diminution measures the work done by any change in the system." This "state of constraint" or "state of energy" was Maxwell's way of reaching toward what we now understand as the electromagnetic field.

Faraday, who had long complained of difficulties with memory, was by this time in his late 60s and beginning to lose his mental powers. He stopped writing to friends because he could not finish a sentence without losing his train of thought. Reluctantly, he resigned his lifelong position from the Royal Institution in October 1861, telling the board members: "I have been most happy in your kindness, and in the fostering care which the Royal Institution has bestowed on me. . . . My life has been a happy one and all I desired. . . . I desire therefore to lay down this duty; and I may truly say, that such has been the pleasure of the occupation to me, that my regret must be greater than yours can or need be."

In his final years, Faraday became silent and motionless. He lived with his wife at Hampton Court, in a "grace and favour" apartment at Queen Victoria's disposal. In one of his last letters to an old friend, he had written of his sustaining faith: "I am, I hope, very thankful that in the withdrawal of the power and things of this life,—the good hope is left with me, which makes the contemplation of death a comfort—not a fear. Such peace is alone in the gift of god, and as it is he who gives it, why shall we be afraid? His unspeakable gift in his beloved son is the ground of no doubtful hope; and *there* is the rest for those who like you and me are drawing near the latter end of our terms here below." He died in 1867, never knowing of the way in which Maxwell had succeeded in casting his vision of the electromagnetic field into a mathematical form that other scientists would slowly accept. After Maxwell's death in 1879 it was not until 1888, when Heinrich Hertz succeeded in creating and detecting radio waves, that the scientific world began to embrace wholeheartedly Maxwell's views.

Even then, Thomson refused to see the physics in Maxwell's elegant mathematics and worked endlessly on alternative theories that would produce the right results on a quite different physical basis. He did not succeed, but neither, even at the end of his life, did he accept that Maxwell's theory would do.

***

From the time of Newton and Galileo until the startling innovations of the early 20th century, modern theoretical physics emerged by evolution more than revolution. Around 1860, soon after they first met, Thomson and Tait conceived the idea of writing a textbook, in several volumes, in order to lay out the mathematical principles of "natural philosophy" that they both saw as the model for a finished science of the inanimate universe. At this time there was nothing we would now recognize as a general textbook of physics. There were volumes of mathematics, in Cantabrigian or French style (although the two were gradually becoming closer), but these were essentially axiomatic exercises in applied philosophy. Ludwig Fischer, Thomson's undergraduate rival who had become a professor at St. Andrews, wrote to him in 1855 asking, in a postscript, "Do you know of any elementary work on Mechanics starting with the idea of 'mechanical energy' or 'work'?" No such book existed. Thomson and Tait wanted to start from the new conception of energy, in all its generality, and show how the rules of physical interaction followed ineluctably from the crucial law of conservation. Not only did this simplify the concepts of mechanics, but it allowed mechanical laws to connect to all other branches of physics, as long as the concept of energy was held foremost.

But their textbook was to be furthermore a practical compendium of physics, allowing the student who mastered its techniques to apply them not simply to idealized laboratory examples and Cambridge tripos questions but also to electric telegraph cables and steam engines and ship design. At end of 1861, Tait was writing to Thomson with the broad outline of a plan and concluded: "I fancy that we might easily give in three moderate volumes a far more complete course of Physics Experimental & Mathematical than exists (to my knowledge) either in French or German. As to English, there are *none*." A couple of weeks later he listed his idea of chapter titles, moving from basic kinematics and dynamics, to hydrostatics and hydromechanics (that is, the behavior and motion of fluids), to the properties of matter, on to sound, light, and heat, then to electricity and magnetism, culminating in a final discussion of that essential principle, the conservation of energy. The modern student will recognize this list as forming the backbone of that time-honored subject, classical physics. One may easily get the impression that classical physics had been around for dusty generations before the beginning of the 20th century and the emergence of quantum mechanics and relativity. In fact, it is a product of the late 19th century, and Thomson and Tait's textbook marks as well as anything its formal appearance. Thomson and Tait were not simply the first to think of putting all these subjects between one set of covers. Rather, they were the first to see these subjects as subparts of a single discipline, elements of a conceptual whole.

Concept was one thing, execution another. As Thomson gallivanted around Britain and further afield on scientific, commercial, and government business, and organized sailing expeditions on the *Lalla Rookh*, it was Tait's task to get the book written. In what turned out to be a woefully premature pronouncement, he advised Thomson on Christmas Day in 1861 that completion of the first volume by the following May seemed altogether feasible. "Let us apportion our work, and fall to. An average of three or 4 (or less) hours a day would give us the volume in 6 weeks," he announced. The most attractive aspect of Tait's character is that he could be as abrupt and unappeasable toward Thomson, for whom his admiration bordered on besotment, as he was toward enemies such as Tyndall. Thomson began by proposing more subjects and suggesting three or four volumes of experimental matters alone. Tait expressed his alarm at this, not least because of "the *expense to the students*, especially the Scotch ones. We may mulct & bleed Oxford and Cambridge & Rugby &c &c to any extent, but how about our own classes? What we want *at once* is not the fame of authorship, but the supply of a want of elementary teaching."

By the middle of January they were still haggling about the list of contents. Tait's difficulties with his coauthor became evident: "I will shortly send you the *revised* headings, that you may see whether they correspond with your ideas, which I confess I have but vaguely gleaned from your notes." Thomson sent Tait bits and scraps and sketches, which

Tait assembled into some form of continuous text, and returned these products to Thomson, who rather than refining and polishing began adding whole new sections and offshoots in the margins. Thomson in any case did not respond promptly, and Tait had to plead constantly to get anything out of his coauthor: "I wish you would send back my sketch of the Chap. on Prop[erties] of Matter with your amendments &c, & I will have it written as soon as is consistent with care and completeness." The next day, on another aspect of the book, he urged *"at all events act speedily."*

To no avail. More than two years later, in May 1864, Tait was writing thus: "D^r T, *Do* look alive with my MSS. It should have been all in type this week." They had taken to addressing each other as T and T′, apparently a shorthand begun by Maxwell, who also referred to them sometimes as the archiepiscopal pair, having discovered to his amusement that the archbishops of York and Canterbury at that time were also Thomson and Tait. (Archibald Campbell Tait, the archbishop of Canterbury, had attended Edinburgh Academy some 15 years before Tait and Maxwell.)

A month later Tait became more peremptory still: "I wish you would go ahead. I am getting quite sick of the great Book. . . . If you send only scraps and these at rare intervals, what can I do? *You have not given me even a hint as to what you want done in our present Chapter about Statics of Liquids and Gases!* . . . . Now all this is very pitiable: I declare you did twice as much during the winter as you are doing now. I sent you a great bundle of proof sheets only ten days ago, but you have taken no notice of them whatever. You proposed certain preposterous problems which I could not be bothered working out." Each year from 1861 Tait urged Thomson to make haste so they could have at least a small installment of their text printed by September, ready for the incoming class of university students. Each year the book failed to materialize.

Thomson published scientific papers and reviews at a great rate throughout his life, but his productivity came from working on so many projects at once that he could easily pick up another when he tired momentarily of one. Stokes, as editor of the *Philosophical Transactions* of the Royal Society in London, had encountered the difficulty: "You are a terrible fellow and I must write you a scolding. The vol of the Phil. Trans. ought to have been out by the 30th of Novr and here's your paper won't be ready for a month yet." Thomson responded to this and other scoldings with invariable good humor, which may not have pleased his correspondents, since he carried on as dilatory as before, only more amused.

Tait found, as others had found before him, that it was impossible to chide Thomson to any effect. In his obituary notice for Tait many years later, Thomson recalled that "the making of the first part of 'T and T'" was treated as a perpetual joke, in respect to the irksome details for the interchange of drafts for 'copy', amendments in type, and final corrections of proofs. It was lightened by the interchange of visits between Greenhill Gardens, or Drummond Place, or George Square, and Largs, or Arran, or the old and new College of Glasgow; but of necessity it was largely carried on by post. Even the postman laughed when he delivered one of our missives, about the size of a postage stamp, out of a pocket handkerchief, in which he had tied it to make sure of not dropping it on the way." All this to-ing and fro-ing may not have appeared to Tait quite as much of a joke as it did to Thomson. At any rate, having existed in a sort of samizdat form for years, circulating among the undergraduates of Edinburgh and Glasgow as proof sections in various states of completion, the *Treatise on Natural Philosophy* by Sir William Thomson and P. G. Tait appeared, at over 700 pages long, in October 1867.

After its difficult gestation, the book did well. Reviewers liked it, Helmholtz quickly arranged for a German translation, and reprintings and revisions were soon in demand. Undergraduates bought it at a smart pace, although J. M. Barrie remarked that it was "better known in my year as the 'Student's First Glimpse of Hades.'" Commenting in *Nature* on an updated version published in 1879, Maxwell identified the greatest virtue of the *Treatise* as the liberation of arcane mathematical propositions in abstract dynamics into the world of practical scientists and even engineers. "The credit of breaking up the monopoly of the great masters of the spell, and making all their charms familiar to our ears as household words, belongs in great measure to Thomson and Tait," Maxwell wrote in his idiosyncratic way. "The two northern wizards[6] were the first who,

---

[6]The "Wizard of the North" was Sir Walter Scott.

without compunction or dread, uttered in their mother tongue the true and proper names of those dynamical concepts which the magicians of old were wont to invoke only by the aid of muttered symbols and inarticulate equations."

As late as the 1870s, in other words, Maxwell found it necessary to remark on the gap between practical physics and mathematical reasoning. Thomson and Tait provided the means to bridge that gap. That the book appeared at all was entirely due to Tait, but the philosophy behind it was Thomson's. Tait at heart was a formalist. He invested many years of effort in promoting a novel kind of mathematics called "quaternions," a cousin to modern vector analysis. Quaternions, Tait claimed, offered a marvelous degree of compactness in writing down complex mathematical equations. He proudly explained to the Cambridge mathematician Arthur Cayley how he had transformed one of Thomson's mathematical arguments: "Three *pages* of formulae can easily, and with immense increase of comprehensibility, be put in as many *lines* of quaternions."

This quixotic enthusiasm drew few converts, however, least of all Thomson. Tait compared quaternions to a pocket map, containing a prodigious amount of information in compact form. To this Cayley dryly replied that, as with a pocket map, the thing had to be unfolded again to be of any use. Maxwell, in a characteristically backhanded remark about one of Tait's quaternionic papers, noted the "remarkable condensation not to say coagulation of style, which has rendered it impenetrable to all but the piercing intellect of the author in his best moments." Tait strove to introduce quaternionic notation into T and T', but Thomson breezily paid no attention. "We have had a thirty-eight year war over quaternions," he explained long afterward to a colleague. "Times without number I offered to let quaternions into Thomson and Tait if he could only show that in any case our work would be helped by their use. You will see that from beginning to end they were never introduced."

Thomson never approved of mathematical formalism for its own sake. Although Cayley disdained quaternions, he was still too rarefied a mathematician for Thomson's taste. Writing to Helmholtz in order to convey his compliments to another German scientist who had succeeded in working out a long, difficult, but practical problem, Thomson remarked: "Oh! That the Cayleys would devote what skill they have to such things instead of to pieces of algebra which possibly interest four people in the world, certainly not more." More bluntly, he told Glasgow students that "the art of reading mathematical books is judicious skipping." For scientists, in other words, the point of mathematics was to solve physical problems. Remarkably, Thomson and Tait's 1867 *Treatise on Natural Philosophy* was the first textbook to bring to undergraduates a comprehensive summary, mathematical but at the same time practical, of physics in all its scope and variety.

***

While goading Thomson toward completion of their joint textbook, Tait had also managed to write a book of his own, prompted by the controversy with Tyndall over the origins of the law of conservation of energy. His *Sketch of Thermodynamics* appeared in 1868. Ostensibly, Tait aimed to provide a thorough and rational account not simply of the two laws of thermodynamics but also of their genesis. He praised Joule as the originator of the first law and made light of Mayer's claims. But he also had a second agenda, relating to the second law. Clausius, in 1865, had come up with the name "entropy" and phrased the second law of thermodynamics as the rule that entropy never decreases. In what were called reversible processes, entropy stayed the same; in irreversible processes (the working of real engines, friction, conduction of heat without production of work, and so on), entropy grew. The notion of reversibility goes back to Carnot. Thomson, in 1852, had made a start on understanding the thermodynamics of irreversibility. Two years later Clausius had written down the modern definition of entropy but didn't yet extend its utility to all processes, reversible and irreversible. When he finally came up with the name entropy, Clausius recognized a physical concept that had emerged, in a tangled and difficult way, during the intervening years. Tait set himself to untangle this history, and he came to the new and remarkable conclusion that his friend Thomson had actually done all the hard work in 1852 and that Clausius's subsequent contributions were at best minor clarifications.

At least that is what his published *Sketch* declared. His correspondence with Thomson reveals a different story. In his 1852 paper, Thomson had imagined a series of steps constituting a reversible cycle, and for each step he calculated the heat going in or out of the system, divided by the temperature at which the exchange took place. The sum of these increments, he showed, was zero. This is the germ of a formal definition of entropy. However, Thomson did not draw any special attention to this quantity but only used it in the course of obtaining other results.

In letters to Thomson in January 1868, Tait castigated him for providing an inadequate proof of the fact that the sum of the heat transfers divided by temperature was zero around a reversible cycle. It "is no proof at all—not even a *chain* of reasoning, merely a set of detached links! How you let it be printed in such a state I can't imagine. Everybody sees you had the proof in your eye, but whether you or the printers omitted a leading step I can't of course tell."

When he came to write his *Sketch*, though, Tait suppressed his private reservations and declared blithely that the general definition of what became entropy was, after all, clearly stated in Thomson's 1852 paper. He sent drafts of the relevant sections to both Clausius and Helmholtz. Clausius evidently responded, and critically, because Tait later wrote to Helmholtz professing no wicked intent: "Is it fair to ask you whether you think with Clausius that my little pamphlet will only do me harm? . . . I wish to avoid strife and to produce a useful little text-book; but, if Clausius is right, I had better burn it at once." The letter from Clausius to Tait apparently does not survive, but there is a fragment from Tait to Thomson, of the right date, referring to an angry missive from an unnamed German: "I enclose a letter just rec'd from him, which rather startles me—you having told me how meek and mild he appeared to you. Dummheit [stupidity] and Hinterlist [cunning, trickery] are pretty strong—and I don't at all like his application of 'auffällig' [egregious] to myself."

This may well have been from Clausius. Helmholtz tried in a more gentlemanly way to dissuade Tait: "For my part I must say that I have a great aversion to all priority quarrels. . . . If then you divest your writing of its polemical garb it will in my opinion be thankfully received and will have more influence than with this polemic."

Calm consideration not being Tait's cup of tea, he went ahead and published his *Sketch of Thermodynamics* "in all its individuality," as his biographer put it. In writing the book, Tait had corresponded a good deal with Maxwell, who absorbed Tait's views on Clausius and Thomson and the second law and promoted them in his own *Theory of Heat*, published in 1871. Clausius now protested in the pages of the *Philosophical Magazine* that Tait and Maxwell had managed to bestow all the credit for working out the concept of entropy on Thomson, leaving him only with the honor of providing the name. He argued, cogently, that Thomson did not make the important step of extending his analysis to irreversible as well as reversible processes. Another tedious dispute followed. Tait responded by ignoring the main point and making some nitpicking remarks about Clausius's original definition of entropy.

Under this provocation Clausius then embarked on his own book about thermodynamics, which predictably met with Tait's disapproval. Thomson's response to this was indirect and curious. He arranged for the publication in the *Philosophical Magazine* of a supposedly personal letter, beginning "My dear Thomson," in which Tait evaded Clausius's arguments and reinstalled Thomson as the inventor of entropy. Thomson then added a short note of his own to say, without elaboration, that he agreed with Tait's assessment. Thus he allowed the idea to get into print that he had come up with the concept of entropy, without ever quite saying so explicitly. Tait's method was to take every statement by Clausius strictly at face value, whereas in scanning carefully Thomson's published works on thermodynamics he allowed himself to infer what Thomson really meant to say, although he hadn't quite managed to say it clearly at the time. By going along with Tait, Thomson was colluding in a significant reinterpretation of history. Perhaps, Tait being so persuasive and insistent, Thomson really came to believe his friend's rereading of his own papers. Perhaps, with hindsight, when the laws of thermodynamics had become more or less obvious, he imagined that he had known the truth all along.

Many years earlier, in his dispute with Whitehouse over the Atlantic cable, Thomson had remarked that his theory of the telegraph was "like any theory, merely a combination of established truths." This, if he really believed it, marked his great flaw as a scientist. To ponder the wider significance of the quantity that eventually became entropy would to him, perhaps, have been going beyond the established truths. Thomson was content to fit empirical knowledge into a system of elementary theoretical rules. Clausius (and also Rankine, in his obscure way) went further and perceived a universal second law of thermodynamics and a general conception of entropy. Tait, with Thomson's implicit agreement, scorned Clausius and Mayer and others because they made large statements on insufficient evidence. So they did, and that is precisely why they deserve credit for seeing where the truth lay even though their evidence was incomplete and their reasoning loose. It must have been galling to Thomson to understand, looking back, that he could have gone further. Tait, who came on the scene after the early confusion had been swept away, could see this even more clearly. Unable to understand why his brilliant friend Thomson had not taken his ideas to their seemingly obvious conclusion, Tait decided that in fact he must have but simply hadn't bothered to write it all down explicitly. But brilliance and imagination are not the same thing.

***

In 1871 the British Association convened in Edinburgh for its annual meeting. Thomson accepted the invitation to serve as president and delivered the meeting's keynote address. In keeping with tradition, he paused to mark the passing of a number of notable scientists that year and summarized recent progress in his own field, physics. But presidency of the BA was a forum from which he could address all of science, and Thomson did not shy away from pronouncing on a number of other issues. He restated his restrictive views on the age of the earth and the sun, scolded the geologists again for not paying enough attention, and then moved on to biology.

The inanimate sciences, Thomson maintained, offered a model for scientific inquiry in general. "The essence of science, as is well illustrated by astronomy and cosmical physics, consists in inferring antecedent conditions, and anticipating future evolutions, from phenomena which have actually come under observation. In biology the difficulties of successfully acting up to this ideal are prodigious. The earnest naturalists of the present day are, however, not appalled or paralyzed by them, and are struggling boldly and laboriously to pass out of the mere 'Natural History stage' of their study, and bring zoology within the range of Natural Philosophy."

Having praised biologists for taking at least baby steps on the way to becoming real scientists, Thomson moved on to criticisms of Darwinian theory. In a nutshell, he would admit the possibility of natural selection but would not go so far as to endorse evolution in its full scope. By natural selection he conceived of the idea that creatures with slightly different qualities might be more or less well suited to their conditions. Some would prosper, others would suffer. This had a nice mechanistic ring to it, which appealed to Thomson. But if evolution meant the appearance of wholly new kinds of creatures from those already existing, or even more radically the appearance of life when none had existed before, then he would not go down that road.

Criticism of this sort he had perhaps learned from his friend Fleeming Jenkin, whose 1867 review of the *Origin of Species* had made a number of intelligent observations hampered by a crabbed vision of what evolution involved. In essence, Jenkin had conceived of a fixed population within which a limited number of genetic elements (as we would now call them) shuffled about from one generation to the next. He offered an example illustrative of his time. Jenkin imagined a white man shipwrecked on an island inhabited by savages. The unquestionably superior qualities of this Crusoe would lead to him being acclaimed king, acquiring many wives, and so on. But by virtue of his very desirability, his innate advantages would then be diluted and dissipated among his numerous offspring. Such would be the fate of any superior individual within a larger population, Jenkin argued; better-adapted individuals would always be overwhelmed by the common herd. Thus he believed advantage, in Darwin's theory, would always be squelched by what statisticians call regression to the mean.

Thomson took a similar line. He and Jenkin apparently believed that genetic elements would be randomly reassigned at each generation, and so failed to grasp the idea that a small genetic advantage, if it is passed down from one generation to the next, can gradually come to dominate a population. They therefore balked at the idea of new permanent traits, still less new species, arising. He quoted with approval Darwin's famous sentence about the "grandeur in this view of life" as the result of selection acting upon and enlarging some original stock. But then he deliberately omitted Darwin's mention of the "origin of species by natural selection" because, as he said, "I have always felt that this hypothesis does contain the true theory of evolution, if evolution there has been, in biology." On a related point Thomson was adamant: "Dead matter cannot become living without coming under the influence of matter previously alive. This seems to me as sure a teaching of science as the law of gravitation."

No clear argument emerges from Thomson's summary. He would not accept that life can come into being from inanimate matter without the agency of some higher power, but neither would he say that the Creator assembled life on earth directly in its present form. He accepted that natural selection can work in creating some of the variety of life, but he leaned toward the view that human origins are a question apart. He would not relinquish the role of a Creator and suggested that Bishop Paley's old argument from design had been too lightly abandoned. In short, he wanted to have a scientific account of life, but only up to a point. In Thomson's view the evidence of creation and design was all around us. And yet he did not want to say that there is *only* creation and design, because Darwin's theory of variation and selection appealed to him as a rational mechanism acting on living organisms.

In the end he finessed the difficult question of where creation ends and natural selection takes over. He noted that material from elsewhere rains constantly on to the earth in the form of meteors. He observed that when a barren lava flow, after not too many years, becomes covered with vegetation, we take it for granted that life has blown in from elsewhere, rather than spontaneously originating on the cooling rocks. Thus he introduced his new suggestion: "Hence and because we all confidently believe that there are at present, and have been from time immemorial, many worlds of life besides our own, we must regard it as probable in the highest degree that there are countless seed-bearing meteoritic stones moving about through space. If at the present instant no life existed upon this earth, one such stone falling upon it might, by what we blindly call natural causes, lead to its becoming covered with vegetation. . . . The hypothesis that life originated on this earth through moss-grown fragments from the ruins of another world may seem wild and visionary; all I maintain is that it is not unscientific."

The notion that life originated elsewhere in the universe didn't solve the problem of its origin, only pushed the question offstage and allowed

Thomson to imagine that whatever happened subsequently on earth adhered to his mechanistic view of science. His audience reacted with a mixture of puzzlement and amusement. Huxley commented that unless Thomson believed that life came to earth in the form of elephants and acorns and crocodiles and coconuts, a largely evolutionary explanation for the present array of species remained necessary. Churchmen were disappointed that Thomson didn't denounce Darwin altogether. Biologists and geologists were skeptical of the idea of viable germs of life flying about through the empty reaches of space.

Nevertheless this proposal by Thomson is characteristic of his thinking. He believed in the universal and encompassing power of scientific reasoning and felt no hesitation in applying the certain rules of mathematical physics in areas beyond the realms in which he had made his reputation. Mathematical physics, indeed, was his model for science in general. Yet when powers of scientific analysis led him inexorably toward a fundamental question—inanimate origin of life or creator?—he abruptly became timid, and drew a veil over the hard dilemmas that his firm belief in science threw up before him.

5

COMPASS

The *Lalla Rookh*, during Thomson's first winter of ownership, underwent repairs and modifications, and with the guidance of Mrs. Tait was fitted out with draperies, sheets, tablecloths, and dinnerware, sufficient to make it a floating residence for half a dozen people. Thomson playfully applied scientific and mathematical principles to the matter of equipping his vessel. The question of fabric for the bedsheets, he explained to Mrs. Tait, "has, after anxious consideration and consultation with naval experts, been decided in favour of linen. The cotton fabric seems to be too hygroscopic to be suitable for sea-going places." He wanted the towels and the larger bath sheets to be made of the same material, objecting to the practice that "sometimes bath sheets are made thicker (apparently with the idea of maintaining a constant proportion of thickness to length or breadth) which is a mistake." Thomson lavished money on his new toy: 12 pairs of sheets, 5 dozen towels, $3^1/_2$ dozen table napkins of double damask, 10 tablecloths, and whatever quantity of kitchen towels, cloths, and dusters Mrs. Tait deemed necessary. She was to place orders for all these furnishings and advise Thomson of whatever else should be acquired.

Of Mrs. Tait herself, little is known except that she put up with P. G. Tait for many years and bore four sons and a daughter. Tait's biographer,

C. G. Knott, spends a paragraph describing how Tait had become friendly at Cambridge with two Belfast brothers, William and James Porter, who in their years were third and seventh wranglers respectively. Both tutored for a while at Peterhouse, William eventually becoming master of the college, while James after a time entered the church. A third brother, John, had a distinguished career in the Indian Civil Service. After leaving Cambridge and before returning to Edinburgh, his hometown, Tait was for six years a professor in Belfast, where he maintained close connections with the Porter family. During this time, Knott relates, he "married one of the sisters of his Peterhouse friends." Which one, what her name was, and on what basis Tait made his selection, Knott fails to say.

By the spring of 1871, at any rate, Thomson was looking forward to sailing adventures as soon as the Glasgow teaching session ended. He urged Helmholtz to come to the British Association meeting at Edinburgh in early August then join him in a scientific party to cruise the Hebrides and western isles on the *Lalla Rookh*. He proposed to invite not only Maxwell and Tait (if he could be lured away from the golf course) but also Huxley, his opponent on evolution and the age of the earth and sun, and even Tyndall, with whom he and Tait had clashed over the allotment of credit for the laws of thermodynamics. Tyndall had encountered Thomson at Dundee, when the BA met there in 1867. "Thomson met me in the Kinnaird Hall; blocked my passage, smiled and stretched out his hand. I grasped it, expressed in a word my gratification at meeting him, and walked in. Shook hands with Tait afterwards at St. Andrews. They were very cordial to me," Tyndall recorded in his diary. Thomson easily put any past unpleasantness behind him; probably he could not conceive that anyone might bear a grudge. Even Tait softened a little. A few years later a German astronomer by the name of Zöllner published a *Treatise on Comets* in which he attacked all and sundry, not only Thomson and Tait but also Helmholtz and Tyndall. Tait approached Tyndall with the most comradely offer he could manage, which was that they should march into battle together. "There will be a splendid row, which is some consolation," he wrote.

But Tyndall would not be drawn. "Whether it is that the fire of my life has fallen to a cinder, the book has produced singularly little disturbance in my feelings," he replied. "Ten years ago I should have been at the throat of Zöllner, but not *now*. I would rather see you and Clausius friends than Zöllner and myself. Trust me C. is through and through an honest high-minded man." Rebuffed, Tait found an opportunity to resume hostilities a few years later. Writing the life of his predecessor at Edinburgh, J. D. Forbes, he disinterred a dispute, dating back to the 1840s, over a theory of glacier motion in which Tyndall and Forbes had been on opposite sides. The subsequent sniping spilled over into the pages of *Nature*. In September 1873, Tait mocked "the flow of word-painting and righteous indignation which Dr Tyndall so abundantly possesses." Tyndall shot back, describing Tait as "this man whose blunders and whose injustice have been so often reduced to nakedness, without ever once showing that he possessed the manhood to acknowledge a committed wrong." The following week the editor referred to further communications he had received from both parties, which he chose not to print as the exchange had "assumed somewhat of a personal tone."

In Belfast the following year Tyndall served as president of the BA and gave an address that excited controversy and repudiation. Going back to Democritus, Epicurus, and Lucretius, he praised those thinkers throughout history who had striven for rational, material explanations of natural phenomena, rather than resorting to a "mob of gods and demons." He mentioned Giordano Bruno, burned at the stake by the Church for his heresy in promoting Copernicus and his sun-centered universe. Coming up to date he praised Darwin and the modern biologists who sought to explain the origins of life through science; he even suggested that human beings and their intelligence might have an origin in material processes. He had no specific explanation of how this could come about, but he urged his audience to think of the question rationally, rather than resorting to vague and unquestioning invocations of divine intervention.

This was blasphemy in some quarters. Thomson wrote to his brother-in-law, the Rev. King, that Tyndall's suggestion was "especially inappropriate." Stokes said Tyndall was surely wrong: Atoms have no emotion or thought, so how could life made only of atoms acquire such things? Maxwell, typically, produced a poetical satire, beginning:

In the very beginnings of science, the parsons, who managed things then,
Being handy with hammer and chisel, made gods in the likeness of men . . .

He then mocked Tyndall's supposition that collections of atoms could do
things that single atoms could not:

For by laying their heads all together, the atoms, as councillors do,
May combine to express an opinion to every one of them new.

(This is a clever play also on Lucretius, who argued in *De rerum natura*
that atoms move at random "for surely [they] did not hold council . . .
flexing their keen minds. . . .")

To the mechanistic thinkers of the late 19th century it seemed of
course impossible, contrary to common sense as well as reason, that inanimate matter could of its own accord turn into sentient creatures. The
problem is still not solved: Most scientists today would agree with
Tyndall's proposition that intelligent life can, somehow, arise from inanimate origins, though a number of unimaginative philosophers cling stead-
fastly to the old view. Tyndall was not, in his specific achievements, a
great scientist, yet in his general views he was forward looking and
imaginative.[1] Thomson was the opposite. A brilliant scientist and solver
of problems, he could not or would not look very far forward, because he
did not like to speculate where he had no solid ground beneath him.
Therefore he believed that life, especially human beings, came from a
divine impulse, and that was that. The combination of technical acuity
and imagination in one mind is a rare thing indeed.

***

During these years Thomson adopted an increasingly capacious
mode of living. His work on the committee investigating the loss of the

---

[1]Another noteworthy speaker at the 1874 BA meeting was the all-around Victorian
sage and pontificator Herbert Spencer, who talked of what is now called social Darwinism—the idea that civilization is itself a manifestation of evolutionary progress. This
drew another smart verse from Maxwell:

The ancients made enemies saved from the slaughter
Into hewers of wood and drawers of water.
We moderns, reversing arrangements so rude,
Prefer ewers of water and drawers of wood.

Captain took him to London in the summer of 1871, but now he shunned the train and instead sailed on his new yacht through the Irish Sea and around Cornwall to Dartmouth, Weymouth, and Southampton on the south coast of England, taking the train to London and staying there at the Athenaeum club when necessary but living on his yacht as much as he could. His voyaging had a certain evangelical quality. He persuaded friends and family members to accompany him, cheerily confident they would share his enthusiasm for life afloat. He took his nephew James Bottomley and brother-in-law David King with him as far as Penzance, where King left to be replaced by the other brother-in-law, Bottomley's father. King said that he would have enjoyed *Lalla Rookh* more had she remained on the slip at Greenock, and William Bottomley opined (after a tough stint against a stiff east wind) that the best part of yachting was going ashore. Thomson took these remarks as jests and assured Mrs. Tait that everyone had had a splendid time.

He threw himself into sailing with the same energy he used to attack any scientific, technical, commercial problem that came his way. He appeared on deck in the middle of the night to make sure the watches were at their posts. He urged his captain and crew on in conditions they thought dangerous. "You will not rest till you have your boat at the bottom," Captain Flarty muttered to him once as they fought gale force winds. This he never did, although he managed to run the *Lalla Rookh* aground some years later, an incident he amusingly recounted to a friend as an experiment showing that wood yielded more easily than rock. Despite, or perhaps because of, his devil-may-care attitude, Thomson was greatly loved by his crew. He talked to the men without pretension, worked as hard as they did, and was eager to learn the arts of navigation. Inevitably, having learned, he wanted to improve and rationalize the science of sailing. When a break from business in London presented itself, he took off for Lisbon and, sailing across the Bay of Biscay, began experiments on sounding devices. Traditionally, sailors threw a weighted line over the side to judge the depth of the water, but for an accurate sounding the ship had to come to a full halt. Thomson set about devising easier, faster, and more accurate methods.

Back in London, between work for the Admiralty and additional duties as examiner for the India Telegraph Service, he arranged a reunion with several Cambridge friends he had not seen for a quarter of a century. They got together and played a little music, but Thomson found the occasion oddly unsettling. "It was a strange reunion, like a return from another world. . . . It can never again be what it was, and it is too full of sadness for the present." The past—Thomson had no time for it. In the few quiet moments he could find, usually when he was alone on the *Lalla Rookh*, he struggled to compose his presidential address for the British Association's 1871 Edinburgh meeting, in which he resumed his attack on the geologists and biologists and proposed his cometary idea for the origin of life.

Huxley and Tyndall declined Thomson's invitation to a postmeeting jaunt on the *Lalla Rookh*, as did Maxwell, who was preparing to take up his new appointment as Cavendish professor in Cambridge and who was in any case the exact antithesis of the jolly boating fellow. Other duties had prevented Helmholtz from attending the BA meeting, but he came over from Germany later in August to go sailing with Thomson. First, though, he went to St. Andrews where, as Tait had avidly proposed, he might "learn (at its headquarters) the mysteries of GOLF!" Helmholtz failed to succumb. He wrote to his wife: "Mr. Tait knows nothing here besides golf. I had to go along with him. My first swings succeeded, but after that I hit only ground or air. Tait is a curious kind of savage—exists here, so he says, only for his muscles, and only today, Sunday, when he dared not play, though he didn't go to church either, would he be brought around to rational matters." In explaining the strange game to his wife, Helmholtz's scientific precision deserted him: He reckoned that each hole was about one English mile long and that the players walked 10 miles during a round.

Helmholtz went on to stay briefly at Thomson's new house in Glasgow, while Thomson, out at sea somewhere, arranged where the sailing party should meet. This was a faculty house, part of the new university buildings. Following Margaret's death, Thomson had not got around to making the place presentable. Helmholtz described to his wife the unfinished rooms, uncarpeted and unpainted, and the furniture stacked here and there, waiting to be set out. It produced in him "an indescribably sad impression, as if no one cared about the place." In the dining room he came across "an exceedingly fine and expressive portrait of her, and below it the couch where she used to lie, and her coverlet. I was very sad and could hardly restrain my tears. It is very sad when men lose their wives, and their life is left desolate." Helmholtz had lost his first wife 12 years earlier, and a couple of years later had married a considerably younger woman. The evidence of Thomson's loss brought back memories. To him the empty house evoked Thomson's now empty life—except that Thomson had been so busy equipping his new yacht, sailing to the south coast and to Portugal and back, and attending meetings in London that the state of his Glasgow house had drifted from his mind. Before long he would have to return to Glasgow to start the new session. The house could wait until then.

Helmholtz joined Thomson on the *Lalla Rookh* at Inverary, on Loch Fyne, where they saw highland games before sailing back to Glasgow and thence to Belfast where they picked up James Thomson. Assorted nephews and sisters-in-law joined and left at various places. They recrossed the Irish Sea and sailed about the west coast of Scotland until they reached the Blackburns' house at Rushven, on the Moidart peninsula. Helmholtz found this "a lonely property, a very lovely spot on a bay between the loneliest mountains." Jemima Blackburn's animal and bird drawings impressed him. She painted a little watercolor of him and the Thomson brothers observing birds out at sea. The Thomsons, a couple of nephews, and the boisterous Blackburns and their children constituted a "friendly and unconstrained" party. Helmholtz was taken aback by Thomson's habit of abruptly withdrawing from the games and conversations, to sit with his green notebook and make calculations. "How would it be if I accustomed the Berliners to the same proceedings?" he asked his wife, with puzzlement or perhaps envy. Oddest of all, Helmholtz thought, was that after Thomson had assembled his guests at dinner aboard his yacht the night before they were to set off toward Skye, he immediately disappeared to his cabin to work at some problem in his notebook. The dinner conversation faltered, and Helmholtz went off to stroll up and down the gently rolling deck with as much "unsteady elegance" as he could muster. He observed, of himself or Thomson or both, that "a husband who is no longer in his first youth feels uncomfortable when he wanders about in the world, all by himself, without higher guidance, and I think that if the world were peopled with men only, it would not be particularly beautiful, but would be very practical, and not at all refreshing."

For Thomson nothing could be more refreshing than immersing himself in practical questions. He loved the sea as much as he loved natural philosophy, perhaps more. The six winter months in Glasgow he taught and lectured and worked on scientific matters. The summer months he arranged to spend as much time as possible on the *Lalla Rookh*, sailing as often as he could but relying on it too as a refuge from the endless demands on his time. Problems of navigation gave his intellect a full range of scientific and practical matters to attack. Merely being at sea presented interesting questions. He had long corresponded with Stokes about the numerous physical and mathematical issues arising in the motion of fluids: streamlines, waves, turbulence, eddies, rotation, and so on. Underneath all these phenomena lay Newton's simple laws of motion, time-tested and elementary, but in fluids these laws manifested themselves in an enormous variety of ways. Fluids, treated by mathematical physics, could be idealized as incompressible or more realistically given some degree of compressibility. They had density, viscosity, elasticity, all of which could change with temperature. Thomson loved this kind of thing. He had no fear of mathematics and no particular aesthetic sense of it either. If he found that a model of some problem wasn't yielding the full range of observed behavior, he would happily pile on more complications. There is an old joke about a theoretical physicist asked to come up with improvements in milk production on a dairy farm, who after months of secretive analysis announces that he has solved the problem in its entirety, at least for the ideal case of a perfectly spherical cow. Thomson wouldn't have understood the joke. What do you mean, a spherical cow?

While sailing, he and Helmholtz had indulged in a sort of competition to see who could correctly explain the behavior of various waves and ripples they observed from the deck of the *Lalla Rookh*. When Thomson had to leave the yacht for a while to attend to some problem ashore, he said jocularly, or perhaps not, "Now, mind, Helmholtz, you're not to work at waves while I'm away." A later commentator disparaged Thomson as someone who "had immense intellectual strength, but was deficient in intellectual taste." How could a man capable of founding thermodynamics and laying the groundwork for modern electromagnetic theory fritter away his time and mental energy in explaining the ripples on the Sound of Mull, or tinkering with devices to measure the depth of the ocean from a moving boat? But to Thomson a problem was a problem was a problem. Whatever puzzle came before him engaged his interest. In a sense his contributions to thermodynamics and electromagnetism were the aberration, his interest in telegraphy and navigation the more characteristic examples of his talents. By resolving certain contradictions between Faraday and the French electrical theorists, and between Carnot and Joule's views of heat and work, he had found a way forward, illuminated the path that theoretical physics must take. But having gotten over the immediate difficulty, Thomson's attention turned elsewhere. It was others, in the end, who completed the journey. Thomson was a practical thinker, a resolver of difficulties, not a metaphysician.

***

On his first voyage with the *Lalla Rookh* through the Bay of Biscay, Thomson had experimented with a sounding device consisting of a 30-pound lead weight attached to a reel of thin piano wire that could be spun with ease from the stern of the ship. The design owed something to his experience with the machines that played out telegraph cables. An accurate sounding demanded that the ship come to a halt, as before, but Thomson's idea was that the depth could be taken much more quickly than with a weight on a hemp rope, so the interruption to progress would be minimal. His first attempt almost came to grief because of an elementary difficulty that he was "much ashamed" not to have thought of beforehand. The wire unspooled nicely, and he got a rapid and accurate depth—2,600 fathoms, agreeing with the chart. But when about a third of the wire had been retrieved, the reel showed alarming signs of strain and began to buckle. Even though the tension on the wire was at most 50 pounds, Thomson realized, the effect was additive: Each turn of the reel added that much tension, so that if the whole length were wound in, over 100 tons of pressure would squeeze the reel.[2] The crew had to stop and haul in something like a mile of thin wire by hand. Thomson devised a

---

[2]Try winding a length of dental floss around a finger. Even with light tension, you can cut off the blood supply and cause pain very easily.

secondary pulley that would take up the tension and allow the wire to be retrieved, another trick that came from cabling expeditions.

Two years later, sailing to Brazil on the cable ship *Hooper* with his old telegraph colleague Fleeming Jenkin, Thomson found that he could take "flying soundings" with reasonable accuracy. With a light wire unreeling freely, he assumed that the weight dropped vertically from the point of release while the ship steamed on. The ship's speed being known, application of Pythagoras's theorem provided the depth from the distance traveled and the length of wire played out. A table converting length of wire and ship's speed into depth relieved the sailor from needing to know his square roots.

Thomson took out a patent on his sounding machine in 1876, but soon afterward started work on a different system. Instead of a passive weight, he attached a simple pressure gauge to the end of the wire. Essentially this was nothing more than an open-ended glass tube with some air in it, weighted so as to keep the open end facing down. As it descended and pressure increased, water advanced up the tube, squeezing the air into a smaller volume. To mark the water level, Thomson tried dyes that got washed away as the water rose, but eventually he settled on a reactive chemical that changed color on contact with water. When the pressure gauge was retrieved, visual inspection revealed how far the water had made it up the tube, which indicated the greatest pressure it had experienced. This yielded directly the maximum depth attained.

This pressure-recording device was nothing new, and the patent for the chemical marker belonged to a T. F. Walker. But Thomson, with his piano wires and pulleys (adapted from cabling machinery), supplied the means to raise and lower the device easily and reliably. Again, Thomson showed his knack for putting together disparate elements, solving some practical difficulties, and coming up with a working system—an empirical counterpart to his theoretical achievements in thermodynamics and electromagnetism. A series of patents in 1880, 1883, and 1885 stamped Thomson's name on the system, and although bureaucracy proved sluggish and reluctant, he succeeded in getting his device adopted by the Royal Navy and other navies. Only with the advent of echo-sounding sonar devices in the early 20th century did new technology supplant his basic design. Of course, there were licensing fees, royalties, and consulting opportunities. Thomson spent many long days aboard the *Lalla Rookh* making sure his system was both reliable and practical for the average sailor. He took pride in being meticulous as well as ingenious, and when the thing was ready he was equally assiduous in maintaining his legal rights and establishing his income.

The 1873 trip to Brazil had another satisfactory outcome for Thomson. To repair electrical flaws in cable coiled aboard the *Hooper*, the expedition paused for a couple of weeks on Madeira, one of the Canary Islands off the west coast of Africa. There Jenkin and Thomson made the acquaintance of Charles Blandy, a local businessman and prominent islander. (To this day the Blandy company produces madeira and other fortified wines.) Blandy had two daughters, who learned Morse code from the visiting technical men. The women signaled with a lamp from their house to the *Hooper*, anchored a mile and a half distant.

Jenkin always remembered Madeira with a pang of alarm. While they were out one day riding on the steep island hills, Jenkin's horse darted unexpectedly and almost pushed Thomson's over the cliff edge. "No harm was done," R. L. Stevenson relates, "but for the moment Fleeming saw his friend hurled into the sea, and almost by his own act; it was a memory that haunted him." Jenkin idolized Thomson almost to the point of worship and could hardly bear to think he might have killed him. Thomson, however, never mentioned the incident and apparently thought no more of it. The moment passed, and he was safe. It was past; no reason to dwell on it.[3]

Thomson, by contrast, left Madeira with a budding romance. Anecdote has it that when the *Hooper* steamed from the island to lay a cable to Brazil, Thomson's attention was drawn to a white cloth fluttering from a window of a house overlooking the port. Peering through his eyeglass, Thomson interpreted the Morse code flapping: "Goodbye, goodbye, Sir William Thomson."

---

[3]Thomson was equally blithe about the occasion when he almost killed his friend Helmholtz, who was visiting his Glasgow laboratory. Showing off the sturdiness of a heavy rotating iron disc in some magnetic experiment, Thomson whacked it with a hammer, which sent it flying across the room. It took off Helmholtz's hat but not, luckily, his head. They both seemed to think it was quite a joke. Helmholtz merely reported to his wife that Thomson had done his hat in. (Königsberger, 1906, ch. 9)

The message drew Thomson back the following May, as soon as the Glasgow session had ended, in the *Lalla Rookh*. He and Frances Anna Blandy, known as Fanny, were married on June 24, 1874, in the British Consular Chapel on Madeira. Thomson was two days shy of his fiftieth birthday. His new wife was in her mid-30s. Though little is known about either his first wife or his second, they were clearly very different. Margaret Thomson had been, even before her years of ailing, a refined, sensitive, artistic soul. Fanny Thomson was, like her husband, cheery and gregarious, and unlike him socially accomplished and elegant. She was a capable, outgoing, practical woman. She loved to organize dinner parties, attending to flowers, seating arrangements, menus, and all manner of diversions. She would gently kick her husband's shin under the table if he seemed about to reach into his pocket for a green notebook.

Thomson's chance encounter the previous year evidently awakened feelings that his frenetic activity had concealed even from him. The day after Fanny accepted his proposal he wrote to his sister: "When I came to Madeira in the *Hooper* it had never seemed possible that such an idea could enter my mind, or that life *could* bring any happiness. When I came away in July I did not think happiness possible for me, and indeed I had not begun even to wish for it. But I carried away an image and an impression from which the idea came. . . . Hope grew stronger till yesterday, when I found that I had not hoped in vain. . . . When you know Fanny you will be able to really congratulate me. Even now I think you will be glad for my sake." No one ever remarked of Thomson, at any time in his life, that he seemed to be an unhappy or melancholic or brooding kind of man, but his long years with Margaret had been mostly toil and worry. Fanny was a bright soul and charmed Thomson's brother and sister and all the nieces and nephews. She traveled with him frequently and made her own social arrangements while he attended scientific or business meetings. Thomson's increasing wealth and reputation made him the center of a widening circle of notable acquaintances in business and politics as well as science, but Thomson himself had little time for purely social matters. Fanny gave him a life appropriate to his circumstances, and he gratefully participated.

Later that year the newlyweds bought a piece of land at Netherhall, on the Ayrshire coast near Largs, where Thomson had so often spent summer months. He undertook the design and construction of a splendid house.[4] He supervised carpenters and masons, with the result that everything took longer than it need have done, as he could never resist the urge to improvise. After the house was finished he used it as a venue for experiments in domestic science. It was probably the second house in Great Britain to be equipped with electricity (the first was Cragside, near Newcastle, built by the industrialist Sir William Armstrong, who had a dynamo installed at a waterfall on his property). Thomson at first used large storage batteries of a French design, similar to a modern lead-acid car battery, which he wrote enthusiastically about to the *Times*. By the early 1880s he was experimenting with generators running from the domestic gas supply and running an impressive variety of incandescent lamps as well as electrical experiments.

With his new wife, new yacht, and new house along with all his teaching, research, and commercial activity, Thomson displayed a seemingly infinite capacity for doing things. A green notebook accompanied him at all times. Even before his marriage he had begun to rely on a string of assistants, former Glasgow pupils, who helped him in the lab and with his writing. He explained to one colleague, "as I have so many engagements, and so much laboratory work, that I am kept constantly standing and walking about, I can seldom sit down to write anything, and am obliged to do nearly everything I wish in black and white by dictation." Early in 1874 he gave his presidential address to the Society of Telegraph Engineers, then in its third year. He spent a couple of weeks working with his assistant in odd moments and came up with four minutes' worth of material. When it came time to deliver his lecture, he started with the few sentences he had painfully composed, then winged it. A stenographer took down his talk and the lecture reads loosely but is cogent and lively. C. G. Knott on one occasion was with Thomson and Tait in Edinburgh and had agreed to write up Thomson's remarks to the Royal Society of Edinburgh for publication in *Nature*. He had difficulty sum-

---

[4]S. P. Thompson refers to the house itself as Netherhall, a practice many later writers have picked up on. No doubt in those days a letter addressed to Thomson at Netherhall, Largs would reach the recipient. But Netherhall is the name of a small community, not Thomson's house.

marizing Thomson's largely ad lib presentation and approached him the next day for further enlightenment. Thomson stared perplexedly into space for a while, struggling to recollect, then had a sudden thought: "Oh, I'll tell you what you should do. Just wait till the *Nature* report is published—that fellow always reports me well."

For all the evidence of the dissipation of his intellect into innumerable half-completed researches, Thomson showed an unswervable ability to pursue technical and practical projects to a fine state of perfection. In a letter to the *Times* and in contributions to the British Association, he had suggested that each lighthouse should signal with a distinctive Morse code pattern, as he had found that sailors coming across a light were frequently so unsure of their location that they didn't know which hazard they were near. He badgered naval men and civil servants whenever he had an opportunity, and his system was eventually adopted.

Since the late 1860s, Thomson had busied himself with an analysis of tide heights at various ports. This was important information for the Admiralty as the Royal Navy stationed itself across the globe, and the British Association had taken on tide prediction as an official project, with Thomson taking a lead role. It had been established that tides at any location could be analyzed into a series of harmonic components, each component having a certain period and a magnitude, from which tides could be predicted with good accuracy. This entailed complex mathematical analysis of measured tides and further mathematics to make a prediction. Around 1876, Thomson devised his Tidal Harmonic Analyser, essentially a mechanical calculator, based on an invention by his brother, and using a set of cogs of appropriate sizes to mimic the components. With the machine correctly set, for a specific location, anyone could predict future tides by cranking the handle.

The Analyser was another characteristic invention. The mathematics it embodied came from others, notably Laplace in France and Airy in England. The germ of the mechanism came from James Thomson. But it was William Thomson who combined the theoretical and practical elements, recast the mathematics into amenable form, developed his brother's innovation into a more general calculating device, and produced a working machine that did exactly what it was supposed to do, in a way that demanded no expertise on the part of the operator. It led him also into a petty dispute in which Thomson showed his increasingly inflexible assertion of his own priorities and interests.

Since about 1872, Edward Roberts of the Nautical Almanac Office had assisted Thomson on the tidal prediction project by performing the tedious but routine calculations needed to obtain the magnitudes of harmonic components from observations at various ports around the world. When the mechanical calculator came to be built, Roberts had the responsibility of working out such details as the correct numbers of teeth for the gears. In 1879 he composed a short paper for the *Proceedings* of the Royal Society, with the title "Preliminary Note on a New Tide-Predictor." This came to Stokes, in his editorial capacity, who learned indirectly that Thomson wished the title to be changed to "Preliminary Description of Sir William Thomson's Tide Predictor Constructed for the Indian Government." Stokes then related to Thomson how "utterly surprised" he was that this "very mild and unobjectionable" change caused Roberts to fly into a huff and refuse to have the paper published in its new form. In the end it was published with its original title, but then Roberts began to speak of the Roberts tide predictor and claim that the important part of the invention was his, acknowledging Thomson only for one or two useful hints. As Thomson explained the matter to Stokes, on the other hand, the design was due to him, except that while riding on a train from Brighton to London, a Mr. Tower with whom he was traveling had suggested driving the machine with a chain-and-pulley mechanism originally devised by Charles Wheatstone for his old letter-printing telegraph receiver. "That is the very thing for me," Thomson had instantly said.

The only innovations Roberts introduced into the predictor were bad ones, Thomson said, which he had to take out again. He wished, in the end, he had had the machine made by James White in Glasgow, a superior instrument maker and a man he could trust. He engaged White anyway to then start work on a predictor with additional improvements, so there would be no doubt who was the true inventor. This little flap amounted to nothing much, but any of the hesitation and deference Thomson had shown 20 years earlier in dealing with Whitehouse's claims over telegraph theory and instrumentation had long since vanished. Thomson picked up clues and hints wherever he could find them and relied on assistants and engineers and technical men to help him refine his ideas and turn them to practical use. But when a finished product was ready for display to the world, there was no question whose name would be attached to it.

***

The sounding machine, the tide predictor, the signaling code for lighthouses—estimable innovations all, but inferior both in importance and in the magnitude of bureaucratic struggle they entailed to the central element of Thomson's career as a marine philosopher. The essential navigational device, passed down from antiquity, perhaps originally from China, was the compass. At the beginning of the 19th century boats were almost wholly wooden. First, metal rivets made an appearance, then a few strengthening iron beams were incorporated into hulls. Brunel's steamer, the *Great Britain*, launched by Prince Albert in 1843, was the world's first fully iron-clad ocean-going ship. Commercial shipping rapidly changed from wood to steel. The Royal Navy followed suit, as slowly as it could decently manage. Iron ships still relied on compasses, but iron had magnetic properties of its own. Ships ran aground because their compasses no longer pointed north but were deflected by the iron hulls and superstructure that carried them. The problem was not new in Thomson's day, but the solutions devised thus far he found unsatisfactory. Naturally, he could do better.

Maritime legend has it that a Portuguese captain, João de Castro, noticed in 1538 that his compass needles twitched when heavy iron cannons were moved about the deck. Two and half centuries later William Wales, an astronomer sailing with Captain Cook's *Resolution* in the south Pacific in the 1770s, noticed the irresolute behavior of the ship's compasses but failed to see the cause. Wales reported that the compass direction drifted depending on the ship's course, its latitude, and the placement of the compass on the ship, but concluded somewhat obtusely that there must be something wrong with the compasses. Finally, on Royal Navy expeditions to Australia in 1798 and 1801, Matthew Flinders systematically studied compass deviations and began to understand their origin. He noticed particularly that the departure of a compass needle from magnetic north changed sign when the ship crossed the equator—that is, it erred to one side in the northern hemisphere and to the other in the south.

Flinders had more time to ponder this problem than he might have liked. Unaware, sailing the south seas, that France and England were at war, he was captured by the French in 1803 and remained a prisoner of war for several years. With ample opportunity for reflection, he came to the conclusion that compass deviation was related to the "dip" of the terrestrial magnetic field. If the earth is pictured as a giant bar magnet, its magnetic poles coinciding approximately with the geographical poles, then lines of magnetic force will emerge vertically at the poles and curve around the planet, becoming parallel to the earth's surface at the equator. The angle between the magnetic field and horizontal is the dip.

War being somewhat more gentlemanly in those days, at least for officers, Flinders was able to publish his findings in the *Proceedings* of the Royal Society while still imprisoned. His analysis attracted the interest of scientists but not naval men. Freed in 1810, he proposed correcting a compass by placing an iron bar adjacent to it, finding the correct position by trial and error. His thinking was that the iron in any ship, though it was scattered about in some complex pattern, would act to a first approximation like an iron rod at some fixed location relative to the compass. A properly placed "Flinders bar," as it eventually became known, would cancel the magnetic distortion produced by the ship, leaving the compass to measure the true magnetic field of the earth.

Flinders died in 1814, at 40, having been unable to submit his correction bar to practical scrutiny. Six years later Peter Barlow of the Royal Military Academy tried a similar scheme involving an iron plate positioned near the compass. Barlow's system went into practice on a number of ships but didn't succeed widely, in particular because a plate that corrected compass error north of the equator was found, perplexingly, to magnify it in the southern hemisphere. In the meantime other scientifically inclined navigators had confirmed and extended Flinders's original analysis of the variation of errors with a ship's course and latitude, without so far coming up with any systematic account of the cause or a practical method to deal with the deviations.

There arose the practice of "swinging" a ship to quantify compass deviation. In a suitable harbor, a captain would use geographical landmarks to align his ship north, south, east, and west, and at a dozen or more compass points in between, so as to measure the discrepancy between the known direction and the compass indication. This procedure yielded a table of corrections which the navigator then applied, out in the open sea, to get true direction from compass reading. This assumed that the necessary correction, measured in one place, would work anywhere on the globe, which Flinders and others had already shown not to be the case. The Royal Navy made it official policy to use correction tables rather than the Barlow plate and other unreliable devices, with the recommendation to captains that they should swing their ships regularly, especially in the course of long voyages. This itself was no easy matter: Swinging a ship was time consuming and required some independent way of establishing compass directions. Nor was the use of correction tables as straightforward as it might appear. Mistakes happened when a navigator subtracted a tabulated correction from a compass reading, instead of adding it—not as absurd as it sounds, as the problem is much like puzzling out whether to put clocks forward or back when going from summertime to wintertime, without the benefit of a handy mnemonic, and with the rule being different from one ship to another.

By the mid-1830s, compass deviations were poorly understood but were undeniably getting larger as ships used increasing amounts of iron in their construction. No satisfactory mechanical correction existed, and correction tables had a way of confusing all but the most sophisticated seamen. Ships ran aground with staggering frequency—hundreds of British vessels, naval and commercial, were lost every year—and the common sailor learned a great distrust of compasses of any kind (which, making matters worse, were often poorly built and unreliable from the outset, apart from the question of deviations).

In 1835, on his own initiative, Captain Edward Johnson of the Royal Navy investigated compass deviation on the *Garryowen*, a 130-foot-long iron paddle steamer with an enormous funnel 28 feet high. He placed compasses at many points around the boat and swung it to measure deviation at different locations. Imaginatively, he also set a number of compasses around the harbor where he was swinging the *Garryowen* and found that these too suffered deviations changing with the orientation of the ship. This was a crucial though dismaying discovery. Flinders, Barlow, and others had taken it for granted that an iron ship passively distorted the magnetic field passing through it. Johnson now concluded that in addition an iron ship had some permanent magnetism of its own, which is why it affected compasses nearby. He speculated that as iron was heated and cooled and shaped and hammered, immersed all the while in the earth's magnetic field, it acquired permanent magnetism that was then built into the ship under construction.

The British Admiralty finally lumbered into action, forming in 1837 a Compass Committee to address this "evil so pregnant with mischief"— namely, the dismal performance of compasses on Her Majesty's warships. (Queen Victoria came to the throne that same year, which perhaps provided a little fillip of opportunity for change and reform in the realm of officialdom.) The Committee investigated all aspects of compass performance, including basic design and quality control as well as deviation. Captain Johnson served on the Committee, which the following year asked Astronomer Royal George Airy to investigate deviation on the paddle-steamer *Rainbow*. Airy, a man of great mathematical skill but not altogether conversant with magnetic phenomena, concluded that compass deviation on the *Rainbow* came almost wholly from the ship's permanent magnetism, and worked out how to compensate for this interference by positioning two small bar magnets, one on the fore-aft line through the compass, the other laterally—athwartships, in nautical language. Airy's method was simple. The ship was swung to point north, and one magnet was placed to make the compass point north as well. The other magnet was positioned similarly by swinging the ship east-west. (In fact, Airy proved that a single magnet would correct the compass, but working out its location required a difficult mathematical analysis.)

The technique seemed to work reasonably well, and commercial shippers (who showed more enthusiasm than the Admiralty for solving the problem, being acutely aware that days lost in passage from navigation errors translated into lost business) hired Airy to install his correction system on a number of ships. But the vessels he corrected were mainly of wooden construction, although with a growing number of iron components. As the amount of iron in ships grew larger, compass deviations grew too, and Airy's method proved insufficient. In particular, when a ship corrected in England went south of the equator, its compass frequently became less trustworthy than if it had not been corrected at all. The reason is that permanent magnetism is not the whole story. A mass of iron distorts a magnetic field that passes through it, in a way that depends on geometry and orientation. The simplest way to deal with this phenomenon is by thinking of the iron developing an induced magnetism in response to the external field. An iron ship, therefore, acts on a compass in two ways: There is a fixed, permanent, or hard magnetism and a variable, induced, or soft magnetism that depends (as Flinders had long ago found) on the ship's position relative to the earth's magnetic field—in other words, on its latitude and heading.

As if this were not complication enough, there is also heeling error. When the ship tilts to one side or the other, out of vertical, its geometry relative to the earth's magnetic field changes, and so the induced or soft magnetism changes too. As Airy had calculated, the hard magnetic error in a compass can be fixed with a couple of permanent magnets, suitably placed.[5] Likewise, the soft error can in principle be compensated by placing soft iron correctors, whose induced magnetism counteracts that of the ship as a whole. But the heeling error, which became significant only for fully iron-built ships, introduces additional complication in placing the correctors. Heeling error, moreover, is linked to the variation of error with latitude, since both depend on the angle with which the earth's magnetic field passes through the ship.

Although Airy was able to earn handsome fees for correcting compasses (£100 or more per ship, compared to his annual salary as astronomer royal of £500), he showed no enthusiasm for going into this line of business as the magnitude of the task became more apparent. He acknowledged the importance of soft as well as hard magnetism and experimented with adding suitable iron correctors as well as magnets, but then largely lost interest. Such complications, on the other hand, were posi-

---

[5]Even this is not quite true. In a newly built ship, each iron component has some permanent magnetism, and these interact slowly with each other, just as bar magnets thrown randomly into a bag will alter each other's magnetism over a period of time. It was found that an iron ship's hard magnetism changed over the first few months, gradually settling into a more or less permanent pattern.

tively an attraction to William Thomson. It is not clear when he first took a serious interest in compass correction. A letter to Stokes as early as 1850 tells of him going to Borley Rectory in Suffolk to visit the Rev. W. W. Herringham, an old Cambridge friend, where he planned "to see 'the Retribution' *swing*, for det^g the dev^n of his compass." On the 1858 Atlantic voyage, the mass of cable aboard the *Niagara* had upset the compasses enough that it steamed away from the midocean rendezvous on the wrong bearing and had to change course and follow one of the smaller ships instead. Thomson, however, recorded no recollection of this incident.

In 1871 the editor of *Good Words* once again called on his friend for an article on some technical or scientific subject, and Thomson, proud new owner of the *Lalla Rookh*, decided to write about the nautical compass. His account, a general history of the compass, appeared in 1874 and did not offer any great novelty. But that same year brought the death of his old friend Archibald Smith, Glaswegian, London lawyer, and sometime mathematician. Thomson wrote an obituary notice of him for the Royal Society. Smith was just over 60 years old when he died, and Thomson blamed his demise in part on exhaustion. While working at the law during the day, he devoted his evenings to mathematics, and the particular problem that consumed so much of his energy was a complex and detailed analysis of compass deviation. This was official labor, undertaken for the Admiralty's Compass Committee. When Airy's compensation system proved inadequate, the Navy settled on a policy of mathematical correction but in a more sophisticated way than before. The soft iron deviation varied, as was now clear, with the ship's bearing, because of the dependence on angle between ship axis and magnetic field. The old system of swinging a ship to get a single set of corrections was utterly inadequate, but Smith (elaborating an earlier treatment by Poisson) devised a method for determining a ship's soft iron properties from a prescribed set of swings, which could then be transformed into a mathematical formula to derive tables of compass corrections, individualized to each ship, and varying according to the ship's bearing.

This was a fantastically intricate business. Every year ships' captains had to perform a complex set of swings to obtain the necessary data. These numbers went to London, where Smith's mathematics transformed them into an array of correction tables that were sent to the ship concerned for a navigator to apply to the ship's compass. The mathematics, to Thomson's eyes, was refined, elegant, and powerful, but as a system to allow an ordinary sailor to chart a course, the method was beyond impossible. The Admiralty rather preferred it this way. It wouldn't do for plain sailors to know how to steer their ships; that was a task for officers. Only a select few could master the art of compass correction, which obviated the danger of untutored seamen going astray or worse through blind trust in a poorly corrected compass. On the other hand, the mathematical correction method offered so many chances for error that it was far from foolproof, even in expert hands.

Between sailing his own ship, writing for *Good Words*, and studying Smith's handiwork, Thomson found a new challenge to latch on to. His demotic instinct rebelled against the mathematical system in favor of the kind of mechanical compensation that Airy had tried, which in principle put navigation in the hands of everyday sailors. In 1879 he wrote another article to summarize his progress but admitted the task was enormous. "When I tried to write on the mariner's compass," he told his readers, "I found I did not know nearly enough about it. So I had to learn my subject. I *have* been learning it these five years, and still feel insufficiently prepared to enlighten the readers of *Good Words* upon it when I now resume the attempt to complete my old article."

Apart from the difficulties presented by the hard and soft magnetism of iron ships, Thomson discovered, there was a woeful history of compass design that ignored elementary matters in dynamics. A compass needle is a magnetized rod, which tries to align itself with the earth's magnetic field. Centuries ago, compass designers had learned to mount the needle on a support or card that floated in a bowl of water—not a practical solution for a ship rocking around in a violent sea. Instead, a compass card balanced on a pivot—a so-called dry card—was set in a bowl and mounted in gimbals (a 16th-century innovation) that allowed it to rotate freely on perpendicular horizontal axes.

A single needle sitting on the diameter of a card does not work well. In a vessel rolling side to side, this needle experiences a purely dynamical influence that tries to line it up with the axis of the ship, so that the needle twists about its own long axis. Mariners often made their compass needles heavier, thinking they would be more stable. But the heavier the needle, the more it responds to dynamical rather than magnetic forces. A long, heavy needle on a fiercely rolling ship will reliably and stably point toward the bow, regardless of where the ship is heading.

The Compass Committee settled on a design in 1840 that dealt fairly well with these problems. On the compass card four needles were mounted parallel, some distance apart, and placed in such a way as to the make the card symmetric in its dynamic properties (that is, it would rotate with equal stability about any diameter). "By a happy coincidence," as Thomson put it, this arrangement of needles also had beneficial magnetic properties. A compass in an iron ship experiences the earth's magnetic field, distorted by the body of the ship. To a first approximation, the distortion pulls a needle out of true, to one side or the other. But there are also higher-order disturbances, with more complex geometry—harmonic components, essentially, of the distortion. With multiple needles correctly placed, the most important of these higher order effects disappear, because they exert equal but opposite pulls on the different needles, leaving the compass card as a whole unaffected.

This card, of mica-covered paper 7.5 inches in diameter, with two needles 7.3 inches long and two 5.4 inches long, formed the heart of the Admiralty Standard Compass. Considerable thought had been expended on the design of the pivots, the compass bowl in which the card was suspended, the gimbals, and so on. It was, by deliberate choice, an uncompensated compass. It came with instructions to place it on a ship's midline, as far away as possible from any large iron structures. Ships' masters received detailed education on how to swing their vessels and how to use Smith's correction tables. In 1842 the Admiralty set up a Compass Department, under Johnson, to oversee everything from manufacture to testing to installation to maintenance.

The Admiralty Standard Compass was therefore not merely an instrument but an entire system, with detailed rules and regulations and, inevitably, an attendant bureaucracy. In its day it represented an enormous leap forward in quality and reliability, and many other navies around the world adopted it. During the mid-1840s hundreds of the new compasses were ordered and installed.

The lurch into modernity signified by the Admiralty Standard Compass did not pull the rest of the Royal Navy along in its wake. In particular there was resistance to the commissioning of fully iron-built steam ships. At first, legitimate questions arose about the reliability of steam technology and the soundness of riveted metal hulls, but these subsided as commercial shipping interests forced the rapid development of better engines and more robust ships. During the 1840s steam and iron drove out wood and sail in the merchant marine, but now the old guard of the Admiralty resisted for sentimental reasons. Some sections of the press chimed in too, declaring that the splendidly rigged wooden ships that had built and now safeguarded the British Empire should not be thrust aside to make way for ugly, smelly metal boats.

Consequently, the Admiralty Standard Compass became part of entrenched naval practice when iron ships had not yet put in an appearance, so that compass errors were generally small and mathematical correction worked reasonably well. By the early 1850s, when the Admiralty could no longer resist steam and iron, the Standard Compass system was inviolate, although difficulties in dealing with much greater compass errors were by then becoming apparent. Deviations of one or two degrees, as in the older ships, posed no insuperable difficulty. Deviations of 10 or 20 or more degrees, which were beginning to be found, were another story altogether.

Captain Johnson had long ago seen these difficulties coming and urged a reconsideration of Airy's correction methods. But he died in 1853. His successor, Captain Frederick Evans, had no scientific training but by dint of great effort mastered the mathematical correction of Archibald Smit—and having mastered it, resisted any hint of change. In any case Airy's method had failed to prove its worth. In 1854 Airy had compensated the compasses of the *Tayleur*, an iron-clad passenger vessel. On its maiden voyage, one day out of Liverpool, it ran aground in heavy seas with the loss of 290 out of 538 people. At the British Association meeting in Liverpool later that year, with the tragedy still fresh in the public mind, the Rev. Dr. William Scoresby, formerly captain of an arctic whaler, later churchman and amateur scientist, charged that Airy's compensation methods were not to be trusted because a ship's magnetism could change. Just as banging and hammering during construction imparted permanent magnetism to a ship's iron components, Scoresby said, so violent motion in a storm could do the same. This was a highly dubious contention, even allowing for the rudimentary understanding of magnetism at the time, and Airy argued against it. But to nonscientific observers, including Captain Evans and many Admiralty figures, Scoresby's alarms were disturbing. Evans oversaw a test of Airy's compensation methods on a smaller vessel, but despite a good result remained unconvinced and stuck to Archibald Smith and his mathematical tables.

The loss of the *Tayleur* and the ensuing controversy at the 1854 BA meeting caused a ruckus among insurers, who became reluctant to offer coverage for new ships. The BA in response formed the Liverpool Compass Committee, which reported in 1860 with a broad recommendation in favor of compensation rather than mathematical correction, though it was clear both systems had drawbacks. Evans conceded that some basic correction by magnets or soft iron had merit, though he hesitated at "the placing of so dangerous a tool as a moveable magnet in the hands of the untrained navigator." He and Smith produced a series of immense Admiralty manuals on the mathematical correction of compasses, which found favor with many navies around the world. Even a simplified version, though, was too forbidding for the average mariner.

In the meantime iron ships were getting bigger, and in what Thomson, with a jab at Darwinism, called "a process of 'Artificial Selection' unguided by intelligence" compasses got bigger too. Highly decorated compasses with long needles looked mighty impressive on the gleaming bridges of new ocean liners, but the bigger the needle, the greater the magnetic force needed to keep it aligned. Dynamical problems resurfaced. The big compasses tended to be unstable or else unresponsive. Worse, a bigger needle needs a stronger magnet to correct it, to the point where secondary magnetic interactions between needles and compensators added to the local magnetic distortions. As one naval history put it, "between 1850 and 1880 ships were, therefore, sailing about with unsteady compasses and heavy deviation tables, and the officers were blaming the compasses instead of mastering the real enemy—inadequate compensation."

William Thomson, coming across all this in the 1870s, found a problem to relish. A handful of elementary and unarguable physical principles dictated the dynamics and magnetic behavior of compasses. The interaction among the earth's magnetic field, the hard and soft magnetism of a ship, and magnetized needles entailed huge mathematical complexity but no scientific novelty. At the end of it, there was a practical issue to be dealt with, and the solution should be one that ordinary seamen, not just mathematically trained navigators, could grasp and use. In 1876, after several summers of experimenting on the *Lalla Rookh*, Thomson took out the first of several patents for a compass design of his own. Like his sounding machine and tide predictor, Thomson's compass contained little that was truly original. It was, like the Admiralty Standard, intended as a complete system, but unlike the official instrument it was intended for the use of ordinary sailors. Thomson, one of very few knowledgeable sailors who could actually appreciate the exquisite mathematics of Smith's correction methods, shunned the mathematical system in his own compass design.

As early as 1874 Thomson had written to Evans suggesting that a lighter compass card, properly suspended, would be more stable in heavy weather, and with guns firing, than the Admiralty Standard Compass card. Evans's cool reply, on top of the difficulty he had experienced in getting his sounding machine tested, caused Thomson to remark later that "innovation is very distasteful to sailors. I have a semi-official letter to that effect." This wariness set the tone for subsequent battles.

In his 1876 compass, Thomson turned the compass card into a light aluminum ring, 10 inches across. Putting all the weight at the edge gave the ring an extraordinarily long period of natural oscillation, almost 60 seconds in the prototype, whereas the Admiralty Standard card had a period of perhaps 20 seconds. As ships got larger, they rolled more slowly, and Thomson perceived the importance of giving the compass card a slower oscillation than the ship itself—otherwise in heavy seas the compass would tend to rock in synchrony with the ship and become magnetically unresponsive.

Within this ring Thomson then suspended eight short, slender needles on silk threads. His card was lighter and magnetically more sensitive than the Admiralty Standard, but at the same time had greater dynamical stability. Its magnetic delicacy made it easier to compensate. As well as using magnetic and soft iron correctors more or less as Airy suggested, Thomson included a Flinders bar (something the Liverpool Committee had strongly endorsed) because it largely took care of heeling and related errors. Characteristically Thomson designed a complete compass mounting or binnacle[6] in which all the correctors could be positioned in a prescribed and restricted fashion, so that any sailor with a few hours of training could compensate the compass in a reliable way. Some naval historians have charged that Thomson gets too much credit for merely putting Airy's method into practice, but as well as adding the Flinders bar Thomson took pains to devise a system in which compensation would be both straightforward and trustworthy. Captain Evans of the Compass Department had complained with some justice that Airy's prescription was too loose to yield consistent results; there was insufficient guidance on the size and strength of magnets and correctors, how close they should be to the compass, and so on. In Thomson's compass the correctors were designed along with the card and so had optimal properties and placement to harmonize with both the magnetic and the dynamical attributes of the compass card. Thomson "enunciated no new principles but was the first to combine successfully all the requirements in one compass and binnacle," according to one history of navigation, and this was no small achievement.

Airy was unimpressed. One of Thomson's students took an early, admittedly rather crudely mounted compass to him at the Royal Observatory but Airy, after looking at it for a while, just said, "It won't do." The mandarins of the Admiralty were likewise indifferent. The superintendent of compasses by this time was William Mayes, who had sailed with the *Agamemnon* on Atlantic cabling expeditions. Evans had by now ascended higher but maintained overall control of compass matters. Complaints were coming in, with increasing frequency, from commanders who found that the Admiralty Standard Compass performed poorly in ships moving at speed or when firing guns. But Mayes, with Evans's backing, resisted all thoughts of change or innovation. Their job, they said, was to implement policy set out in Admiralty documents.

By the late 1870s Thomson had succeeded only in getting grudging permission from the Admiralty to install one of his compasses, at his expense, on some suitable occasion yet to be determined and on a ship

---

[6]Archaically "bittacle," from Latin *habitaculum*, a small dwelling, probably referring originally to the cabin in which a compass was kept.

yet to be selected, for the purposes of testing. He was becoming wily, though. Through his long association with cabling, his work with the Admiralty on other matters, and by his frequent sailing around Britain on the *Lalla Rookh*, Thomson was a familiar and cheery figure among naval officers around the country. He made his compass available informally to a number of captains, a strategy that Admiralty men regarded as a low trick and which may have hardened opposition. However, he began to win allies. Favorable reports on his compass trickled in, and a base of support grew among mid-level naval officers. The Thomson compass encountered problems, particularly a tendency to become unstable under rough conditions. Thomson refashioned the gimbals and the card suspension. Some Admiralty officials regarded this as a variety of cheating. Into the 1880s the Thomson compass made inroads in commercial shipping and was gaining a few crucial promoters in the Navy. Evans and Mayes and their Compass Department, however, remained steadfast in their determination to hold on to the now 40-year-old Admiralty Standard.

***

Energy, a word originally born of science, has emerged into everyday language. Entropy has come part way, used occasionally in nontechnical contexts with varying degrees of accuracy and appropriateness. Some words never make it out of the scientific lexicon: enthalpy, diamagnetism, quaternion. And some miscarry within science itself. In 1884 Thomson delivered a series of lectures at Johns Hopkins University in Baltimore, in the course of which he amused his audience with a novel and curious terminology: thlipsinomic, platythliptic, plagiotatic, cybotatic, euthythliptic, and more besides. No modern physics student will recognize these words or could guess what they refer to. A physicist today reading Thomson's Baltimore lectures, as they became known, might well be able to follow the author sentence by sentence, equation by equation, but the purpose of the whole intricate exercise would seem opaque. Celebrated as they were in Thomson's lifetime, the Baltimore lectures stand as an elaborate monument to a forgotten cause, like one of those architectural follies wealthy Victorians liked to put up in some bosky corner of the estate to surprise visitors.

Thomson first visited the United States in 1876, when he acted as a judge in the technical instrumentation section of the Centennial Exhibition in Philadelphia. Sailing across the Atlantic with Lady Thomson on the S.S. *Russia*, he had kept the crew and passengers entertained with constant experimenting on his compass and a new version of his sounding machine. Returning on the *Scythia*, he claimed to have found a previously unknown shoal somewhere in mid-Atlantic, where he plumbed a depth of only 68 fathoms at a point where the charts said 1,900 fathoms. Oceanographers have inexplicably failed to rediscover this shelf.

Among the dazzling array of inventions he saw in Philadelphia, Thomson particularly noticed an automatic telegraph receiver and an electric pen presented by the 29-year-old Thomas Edison, whose remarkable career was just beginning. Edison had started as a junior telegraph operator, teaching himself electricity and some engineering. Just as telegraphy was the first commercial technology to make use of the science of electricity, so the telegraph industry offered a route into technical careers for those, like Edison, who did not tread an academic path.

Thomson also saw a liquid compass by E. S. Ritchie of Boston, in which a card and needles floated on water in an enclosed vessel instead of being suspended in air on pivots and gimbals. The U.S. Navy had already adopted the liquid compass, a decision that was to prove wise in years to come. Ritchie sent one of his devices to the Compass Department in London, but Evans brushed off this innovation, all the more easily, no doubt, as it was not only radical but foreign. Thomson likewise showed no great enthusiasm for Ritchie's compass. He had already designed his first dry-card compass and wasn't about to be deflected from his purpose by an entirely different design.

Most astonishing, though, was Alexander Graham Bell's telephone. Thomson heard "marvellously distinct" the words "to be or not to be" spoken through the device, along with a selection of items read from a local newspaper that were not so easy to make out. He brought back a pair of telephones from Philadelphia to show off at the British Association meeting in Glasgow later that year but had some difficulty with Bell's primitive microphone and couldn't get the apparatus to perform. (Edison's "button" microphone of the following year, using compacted carbon powder whose resistance varied with applied pressure, made the telephone into a far more practical instrument.)

After Philadelphia the Thomsons went on a whirlwind train tour taking in Niagara Falls, Toronto, Montreal, Boston, and Newport. At the BA meeting in Glasgow, Thomson spoke enthusiastically of "the originality, the inventiveness, the patient persevering thoroughness of the work, the appreciativeness, and the generous open-mindedness" that he had seen on display in Philadelphia. He noted sharply that in America "every good thing deserving a patent was patented" and told his audience that the "onerous" British patent system was "far behind America's wisdom in this respect" and that if the British and European patent laws were not amended "America will speedily become the nursery of useful inventions for the world." The *New York Times* saw fit to reproduce these glowing remarks.

Eight years later the British Association organized its first meeting abroad, in Montreal. Thomson was part of a large contingent from Britain, and he and several others went on to Philadelphia to attend a meeting of the American Association for the Advancement of Science, a counterpart to the BA founded in 1847. In both Montreal and Philadelphia the local newspapers splashed accounts of the visiting luminaries on their front pages. Science was the driving force of the age, and its achievements, most recently the telephone, seemed little short of miraculous. "To see such men is a privilege," declared the Montreal *Gazette* of the city's eminent visitors. "The meeting of the British Association . . . has been one of the happiest events in our history and one from which much and manifold good may be reasonably expected." The paper devoted dutiful pages reporting to Montrealers the arcane discussions of the visiting savants.

Philadelphia, more cosmopolitan and confident, was not quite so overawed, but still the *Inquirer* ran a long account of the many famous men coming to town for the AAAS meeting, notable among them "Sir William Thomson, England's great mathematician and electrician." Thence Thomson went on to Baltimore, where in a story on events at Johns Hopkins the *Sun* announced that "the great event in the year's work will probably be the lectures by Sir William Thomson . . . considered by many scientists second only to Newton." Daniel Coit Gilman, president of Hopkins, had written to Thomson in 1882 inviting him to deliver a series of lectures on whatever subject he cared to choose, telling him that this "would give a strong impulse to the study of Physics in this country." Although Thomson, after his earlier visit, had praised the technical inventiveness of the Americans, academic science was in a rudimentary state. Johns Hopkins was at that time probably the nearest to a European research institution, with departments of science, graduate students, and noteworthy professors.

Gilman asked Thomson to start with a general talk to a large audience but emphasized that the point of his lecture series would be to introduce a select group to the most advanced topics and pressing questions. He passed on advice from Wolcott Gibbs, chemistry professor at Harvard, who believed that "the very best and most effective—most stimulating—course would be one on the *obscure* and *difficult* points in our modern physics. For instance on the difficulties we meet in the wave theory of light, in the atomic and molecular theory of matter, in electricity, as regards the want of any physical theory whatever." Gibbs wanted "a really vigorous showing up of our shortcomings, especially if supported by new views such as Thomson could and would bring forward. . . . Every professor of physics in this country would want to hear such a course."

Thomson settled on the wave theory of light as his theme. Maxwell's theory, according to which light was a form of electromagnetic radiation, was by then two decades old but not yet widely understood or accepted. Heinrich Hertz's laboratory demonstration of radio waves was four years in the future. Maxwell's one undeniable success, in Thomson's estimation, was the connection he found between the speed of light and the propagation of purely electric and magnetic phenomena in a vacuum. To modern thinking, this alone almost demonstrates the fundamental correctness of Maxwell's theory. As far as Thomson was concerned, it was a pregnant quantitative prediction that the current evidence supported but by no means proved beyond doubt.

In particular, Thomson found Maxwell's theory deficient because it had nothing to say on precisely what constituted electric and magnetic phenomena, on what light was, or on how these effects passed through a vacuum and interacted with matter. Maxwell proposed certain general concepts—electric and magnetic fields—and showed mathematically how they related to each other. He showed that these interlinked fields could sustain oscillations that traveled at a fixed and finite speed. Defining the strength of electric and magnetic forces according to their respective inverse square laws are two constants known respectively as the permittivity and permeability of the vacuum. These two constants are linked in a simple way, Maxwell showed, to the predicted speed of electromagnetic radiation. The speed thus calculated was suggestively close to the speed of light, according to the best available data.

Thomson accepted the importance of the speed prediction but in other respects disliked what he saw as the abstract nature of Maxwell's theory. How do electric and magnetic effects propagate through space, and what physical mechanism determines the permittivity and permeability? Surely what we call empty space must be a medium of some kind, if it is to support wave motion. What constitutes this medium and how does a wave motion manifest itself? Sound waves in air, every physicist knew, were pressure waves, the air bunching up and spreading out alternately. What, Thomson wanted to know, was the corresponding picture for electromagnetic waves? On these issues Maxwell was silent. He simply labeled the vacuum by certain parameters, characterized electromagnetic fields by certain mathematical functions, and proposed relationships among these things. It was a start, Thomson agreed, but it was not yet physics as he understood the term.

When one thought of light and electromagnetism and their behavior in matter, moreover, further difficulties and complications arose. To some extent, Maxwell could model various materials—insulators, conductors— by using appropriately adjusted values for the permittivity and permeability. But this merely glossed over the fundamental questions, Thomson believed. Why was one material a conductor and another an insulator? Some materials responded to a magnetic field by becoming magnetic themselves, in the same sense as the applied field. Others developed magnetism that *opposed* the applied field. Why? Again, Maxwell's theory allowed these phenomena to be given mathematical labels, but that didn't explain anything. Thomson wanted to know what went on in inside a material when an electromagnetic influence pervaded it. He wanted models that would explain and predict the behaviors that Maxwell's theory merely accommodated and labeled.

Finally, there were phenomena that Maxwell's theory failed to address at all. Newton long ago had shown that pure white light could be split by a prism into an orderly rainbow of colors. In 1814 the German astronomer Joseph von Fraunhofer found that hundreds of dark lines crossed the spectrum of light from the sun, and in the mid-1800s Robert Bunsen and Gustav Kirchhoff showed how these characteristic lines, appearing at certain fixed wavelengths of light, indicated the presence of individual chemical elements. Thus was born the science of spectroscopic analysis.[7] But what was the physical mechanism by which some substance snipped out a handful of little sections of incoming light, rather than absorbing evenly across the whole spectrum? Maxwell had no answer. Thomson couldn't accept a theory of light with nothing to say on so elementary an issue.

He began his series of lectures on October 1. A reporter for the Baltimore *Sun* dropped by to see the celebrated scientist in action and wrote that "the lecturer is a man tall, though somewhat stooping, with kindly eyes, gray hair, and broad high forehead. He speaks easily, but has a habit of constantly twitching his hands while addressing an audience." Thomson had turned 60 a few months earlier but was still a slender, lively man, the limp from his shortened leg exaggerating the impression of constant activity.

He intended his lecture series as an extended collegial seminar, engaging his audience of about 20 in discussions that led to consultation of books and papers, augmented by hasty overnight calculations. The course of one day's discussion fed into the next day's agenda. He had his topics in his head but prepared little for each session. The English physicist Lord Rayleigh (who was born into the upper crust as John William Strutt and acquired his title when his father died) attended about half of the Baltimore lectures and remarked to his son years later: "What an extraordinary performance that was! I often recognized that the morning's lecture was founded on questions that had cropped up when we were talking at breakfast." This spontaneous disorderliness pleased Thomson's audience, Rayleigh thought, more than a set of carefully prepared talks would

---

[7]This afforded another opportunity for P. G. Tait to embark on a crusade. He protested vehemently that Bunsen and Kirchhoff had unfairly taken credit for spectroscopy from his friend Balfour Stewart.

have done. "They were very much impressed and he got some of them to do grinding long sums for him in the intervals," he recalled. Writing at the time to his mother, though, Rayleigh gave a somewhat less sanguine view. Baltimore had been a success, he said, although the "lectures were quite in the usual Thomsonian style, a sort of thinking out loud in an enthusiastic incoherent manner." J. J. Thomson (no relation), discoverer of the electron, remarked of William Thomson that "he has been known to lecture for an hour before reaching the subject of the lecture. It was only very rarely that he prepared either a speech or a lecture. There was, to the few who were already interested in the subject he was talking about, generally both charm and interest in these diversions."

Ostensibly, Thomson talked in Baltimore of the wave theory of light. In the printed version of his lectures (they were stenographed and reproduced), he appears to talk at great and often mystifying length about waves in fluids and solids with various presumed characteristics, and even more enigmatically of the oscillations of imaginary mechanical constructions that he asked his listeners to ponder. There were apparently featureless spheres, inside which lay concealed, like Russian dolls, smaller spheres linked to the adjacent ones with springs (and zigzag springs, mind, not the usual spiral sort). There were geometric arrays of rigid rods, joined in such a way that they could rotate and pivot in a restricted fashion. There was a flywheel on an axle inside a sphere; but that wasn't complicated enough, so Thomson proposed two flywheels inside the sphere, on a split axle that could pivot about its midpoint.

For one example Thomson devised an actual model, which became known as the "wiggler." On a steel wire suspended from the ceiling half a dozen short wooden slats were attached, like the rungs of a ladder. At the ends of the slats weights were placed, with bigger weights on the higher slats. The whole array could be oscillated by a pendulum attached at the bottom. By varying the frequency of oscillation Thomson showed a great variety of motions of the wiggler, with some slats going one way and some the other, at different frequencies, and perhaps with one in the middle remaining stationary.

These bizarre toys, in one way or another, were supposed to represent molecules of matter, or more specifically the way molecules interacted with light. Spectroscopy made it clear that matter responded not just to light in a general sense but to particular frequencies of light, both absorbing and emitting at these preferred points of the spectrum. Thomson asked his audience to imagine that a molecule must be some sort of machine with a complex array of internal vibrations and oscillations. Light could set the thing going (in which case a particular frequency would be absorbed, as light energy went into the molecule) and once it was vibrating the toy molecule could emit light again—perhaps at the same frequency but in general (as in the wiggler) at some other frequency that had been also set going.

Thomson also discussed at length the propagation of light through the unknown medium—the ether—that sustained electromagnetic oscillations. The problem was to get the right physical characteristics for the ether, so that light behaved as it was empirically known to behave. He told his audience of a favorite demonstration. In his Glasgow laboratory he had almost filled a glass jar with water, thrown in some corks (which naturally floated), poured in a two-inch layer of wax (Scottish shoemaker's wax, to be precise), and then scattered a few bullets on top. This arrangement sat there, doing nothing at all as far as the eye could tell. But after six months, as Thomson fondly explained, the corks and the bullets had vanished from sight. After a year, the corks had emerged on top of the wax, having floated slowly upward, while the bullets had sunk through and dropped to the bottom of the jar. A familiar medium, he concluded, could seem solid by the hour, but fluid by the month or year. This was relevant in understanding light, because the ether, to sustain the extremely rapid vibrations of light waves, must in some sense be rigid—just as a chunk of hard metal, when struck, will ring at high pitch, whereas a more pliable block of wood would give a dull thud. On the other hand, the ether must also be forgiving and tenuous, for the simple reason that slow-moving solid objects (such as the earth) apparently passed freely through it, with no hindrance.

Therefore the ether, as an initial basis for contemplation, was something like a wax—hard on short timescales, soft with respect to slow changes. Of course no real wax truly mimicked the required physics of the ether. The point was that such things were possible, broadly speaking, and therefore conceivable. He mentioned Burgundy pitch and Trinidad pitch and Canada balsam as having properties interestingly different from his shoemaker's wax. He talked of glycerin. The ether was *like* a wax or a jelly, he insisted; it was a matter of coming up with the appropriate characteristics which, Thomson admitted, he had not yet been able to do.

In the printed version of his 20 Baltimore lectures, Thomson occupies more than 200 pages with seemingly endless calculations of the vibrations and oscillations of increasingly rococo arrangements of waxes and jellies, rods and springs, passive weights and gyrostats, in all combinations. It is Thomson's scientific style at its acme. Take a handful of ingredients, all uncontroversial and with well-understood properties. Combine them, to discover what kinds of behavior they present and how the phenomena might relate to the passage of light, or the interaction of light with matter. If the modeled phenomena do not have the required range and complexity to simulate their empirical counterparts, then add something: More springs! Another flywheel! Another set of hinged, frictionless rods! Thomson displayed no doubt that the strategy was correct. His response to any difficulty was to add more bells and whistles. There was no hesitation, no going back.

The curious vocabulary—thlipsinomic, platythliptic, plagiotatic, and the like—came up as Thomson cataloged the entire range of physically allowable behavior of three-dimensional solids with density, compressibility, and viscosity. There is simple compression, as when a pastry cook squeezes a lump of dough into a ball. There is shear, as when the cook, with the heel of the hand, pushes dough *across* the board to create an elliptical disk. And there is torsion, or rotation, as when a rope of dough is twisted to make a spiral. Pastry dough, however, is what physicists call an isotropic material. Its properties do not have any directionality. By contrast there are layered materials, such as mica or graphite, which may slide easily when pressure is applied in one direction, but buckle or crack under a force applied perpendicularly.

In general, a substance that isn't isotropic responds to an applied stress with a distortion at some angle to the applied force. To account for the full range of strains produced in a maximally nonisotropic solid, 21 independent coefficients are needed. Thomson's thermodynamics and engineering colleague William Rankine (who had died in 1872 at the age of 52) had been a man of classical learning and had devised a set of 21 names for these coefficients, according to the geometrical relationship of stress and strain they each denoted. Rankine, Thomson explained in one of his numerous Baltimore digressions, had a particular obsession with the way English had acquired an erroneous pronunciation of Latin and Greek words by absorbing them through French and campaigned quixotically for reform. He "was the last writer to speak of cinematics instead of kinematics," Thomson said. "Cyboid is a very good word, but I do not know that there is any need of introducing it instead of Cubic. . . . Rankine was splendid in his vigour, and the grandeur of his Greek derivatives. Perhaps he over did it, but I do not like to call it an error." In Rankine's system one had to say Kikero instead of Cicero, which Thomson admitted was too much, and in the end he preferred a more conventional language. "Platytatic" and "platythliptic," for example, became "sidelong normal" and "sidelong tangential."

Whatever antique charm these words may once have possessed, the concepts they stand for have left no mark on modern physics. Thomson's aim, described but not achieved in his Baltimore lectures, was to find an ether, characterized by the correct set of values of the 21 coefficients, that would support oscillatory light waves with precisely their observed properties and relation to electric and magnetic stresses. Maxwell's theory, of course, did just that. But there was no physics, or not enough, in Maxwell, as Thomson saw it. He summarized his central objection: "I never satisfy myself until I can make a mechanical model of a thing. If I can make a mechanical model, I can understand it. As long as I cannot make a mechanical model all the way through, I cannot understand, and that is why I cannot get the electromagnetic theory. . . . That is why I take plain dynamics. I can get a model in plain dynamics, I cannot in electromagnetics."

Thomson, in other words, did not literally think that space was filled with some version of Scottish shoemaker's wax or Canada balsam or that matter was really composed of tiny spheres concealing springs and gyrostats. Unless, however, he could reproduce the mathematics of any physical phenomenon in terms of some directly appreciable mechanical model, he did not believe he had explained anything. Maxwell, early on, had used mechanical models similarly in order to arrive at the eventual form of his theory. He imagined, in one famous discussion, magnetic effects propagating through space analogously to the way rotation would pass among spheres rolling against idler wheels interposed between them. But as he explored further Maxwell found that adhering to strict mechanical pictures limited his ability to understand the links between electric and magnetic phenomena, and he learned to rely on mathematical laws alone, even if they represented physical entities that had no immediately perceptible mechanical counterpart. In the end he abandoned mechanical pictures and, as one historian put it, presented his "Dynamical Theory of the Electromagnetic Field" in 1865 "stripped of the scaffolding by aid of which it had first been erected."

Maxwell's is the modern strategy. Pictures and analogies of all and any kind are frequently useful in drawing up ideas for new theoretical constructs, but in the end those constructs stand or fall by their internal mathematical consistency and their empirical usefulness. Thomson's insistence that every theory must be reducible to a suitable arrangement of simple Newtonian ingredients limits the imagination far too much, and for no good reason. It is the ultimate expression of a "mechanical" view of the universe. The obviousness of this position was self-evident to Thomson, who neither would nor could provide deeper justification. On this subject he was the last holdout.

***

Thomson adjusted and modified his compass and other nautical instruments with just as much ingenuity and resourcefulness as he fiddled with ether models. For such technical inventions his strategy was sound. During the 1880s word of Thomson's compass spread around the Royal Navy, and ships' masters began to carry them surreptitiously, against official policy. One man in particular become a crucial and outspoken booster. John Arbuthnot Fisher, born in 1841 in Ceylon (now Sri Lanka), became a midshipman in the Royal Navy at the age of 13, having passed an entrance examination that consisted, his biographer reported, of "writing out the Lord's Prayer, and jumping over a chair, naked, in the presence of the doctor; after which he was given a glass of sherry as evidence of his having become a naval officer." By the age of 18 he had command of his own ship, on which he oversaw a technical advance, the firing of guns by electric impulses coming from crude batteries of zinc and copper plates immersed in vinegar.

Fisher was a blunt, outspoken man whose career teetered constantly on the edge of insubordination but was carried off with sufficient brilliance that he ended up, years later, as First Lord of the Admiralty. He was the kind of officer loved, if feared, by his men, respected warily by his fellow officers, and barely tolerated by the Admiralty. He was a reformer and an enthusiast for scientific innovation. In his memoirs he raged against mindless official resistance to any kind of change. "We still have ancient Admirals who believe in bows and arrows," he fumed. "Didn't the Board of Admiralty issue a solemn Board Minute that wood floated and iron sank? So what a damnable thing to build iron ships!" The merest hints of change, he complained, were reliably opposed by "some Commander Knowall . . . Admiral Retrograde . . . and then some old 'cup of tea' writes to the *Times* . . . these carbonised cranks who wield the pen, actuated by the wrong kind of grey matter of their brain."

For a few years he taught gunnery school in Portsmouth, drilling young seamen in the mastery of another new technology, the torpedo. He succeeded by force of personality as much as pedagogic skill. One student, asked to explain why π was equal to 3.14159, supposedly wrote that it was "the most suitable number Captain Fisher could think of."

Given command of the *Northampton*, Fisher first encountered Thomson and his compass in 1879. In cold weather and wearing a thin overcoat, Thomson spent hours on deck adjusting his compass, while young officers sent to assist him came and went shivering. Fisher told him at some point to come in from the cold, but Thomson assured him: "No, thank you, I am quite warm. I've got several vests on." He then explained to Fisher his theory, allegedly acquired from the Chinese, that many thin layers were better than a few thick ones. Thomson, like Fisher, was robust, indefatigable, always interested in new ideas, and impatient with the past. They became great friends and allies.

Two years later Fisher, still only 40, became captain of the *Inflexible*, the largest ship in the Royal Navy. Invited by Thomson to dine one evening at the Royal Society, he saw Joseph Swan's new incandescent light[8] and immediately decided he needed them for his ship. His con-

---

[8]The English inventor Swan came up with his lightbulb at about the same time Edison devised his, but Edison's superior commercial sense and his mastery of electric systems won the day.

stant requests for modifications and improvements (he stirred up trouble by insisting on more toilets) caused him, so he said, to be "regarded by the Admiral Superintendent of the Dockyard as the Incarnation of Revolution."

The electric lighting on the *Inflexible* ran at first on a 600-volt system, and on one occasion a sailor got a bad shock from touching a poorly insulated wire. Fisher asked Thomson, who happened to be on board, to take a look. "He diagnosed the matter as 'a nasty little leak, but not likely to be dangerous to life'," Fisher recalled. "Just then the cable slipped through his hand and the bare wire touched his finger. He leapt into the air, and his immediate second diagnosis was 'Dangerous, very dangerous to life. I will mention this to the British Association.'" In fact, a man was electrocuted not long after in a similar accident. The potential was cut to 60 volts.

Thomson's utter lack of embarrassment at changing his opinion so immediately impressed rather than irked Fisher as the sign of man, like himself, capable of adapting unhesitatingly to circumstances. They were both pragmatists. Fisher understood very well the intricate system of swinging a ship and obtaining correction tables, but inveighed against its complexity, its lack of practicality, and especially against the dim-witted bureaucracy that had caused it to survive long past its useful lifetime. As a young man Fisher had found his way to a new ship in the eastern Mediterranean by hitching a ride, or whatever the nautical expression may be, on a tramp steamer heading out from Italy. The captain of this old vessel had been plying about the Mediterranean his whole life and told Fisher, who was curious about the man's informal navigational practices, that he generally got about successfully by knowing his "lamp-posts"—the lighthouses—and by having his engineer tell him how many turns the engine had made, from which he could figure how far they'd sailed.

"Well," Fisher asked, "what do you do about your compass? Are you sure it's correct? In the Navy, you know, we're constantly looking at the sun when it sets, and that's an easy way of seeing that the compass is right."

"Well," the old captain explained, "what I does is this. I throws a cask overboard, and when it's as far off as ever I can see it, I turns the ship round on her axis. I takes the bearing of the cask at every point of the compass, divides by the total number of bearings, which gives me the average, and then I subtracts each point of the compass from it, and that's what the compass is wrong on each point. But," he concluded, "I seldom does it, because provided I make the lamp-post all right I think the compass is all right."

In his way, as King Edward VII told him many years later, Fisher was a socialist. He loathed the British class system and the privileges accorded to the genteel members of society. "We fight God when our Social System dooms the brilliant clever child of the poor man to the same level as his father," he wrote (he had been brought up by his maternal grandfather, who he said was swindled out of his painstakingly acquired means). The compass correction system, so far as Fisher saw it, was a microcosm of aristocratic elitism and conservatism.

Having become a champion for Thomson's compass, Fisher took no small pleasure in battling the Compass Department at every opportunity. "It was an immense difficulty getting the Admiralty to adopt [Thomson's] compass. I was reprimanded for having them on board. I always asked at a Court-Martial, no matter what the prisoner was being tried for, whether they had [Thomson's] compass on board. It was only ridicule that got rid of the old Admiralty compass. . . . But what most scandalised the dear old Fossil who then presided over the Admiralty compass department was that I wanted to do away with the points of the compass and mark it into the three hundred and sixty degrees of the circle (you might as well have asked them to do away with salt beef and rum!) . . . the 'Old Salts' said at that time, 'There he is again—the d—d Revolutionary!'"

The "dear old Fossil" in question was either William Mayes or his superior and mentor Captain Evans, who, like Fisher, had gone to sea as a midshipman at the age of 13 and worked his way to the top. Both men were Navy through and through and knew no other life. Evans's devotion manifested itself in an unremitting determination to adhere to tradition and obey official regulations to the smallest of the small print. Even Captain W. E. May, a historian generally sympathetic to the Compass Department and somewhat hostile to Thomson, described Evans as "pig-headed and self-opinionated. Once in the Compass Branch he had his orders and he meant to stick to them." Fisher was no doubt opinionated too, but he was flexible and adventurous; his commitment was to a mythical naval history of heroic deeds and courageous individuals saving the day over the pedantic objections of desk-captains in thrall to their rulebooks.

Not only Fisher but other captains began to rely on the Thomson compass and press for its official adoption. In 1883 the Admiralty relented a little, approving the use of the compass though only in a subsidiary relation to the Admiralty Standard. Still, according to Fisher, Mayes was plotting against the good cause. One day in 1885 he had been talking at the Admiralty with a captain who complained that his Thomson compass had been so poorly located, with respect to the iron structure of his ship, that he could hardly use it. By chance Mayes appeared just at that moment, and Fisher roundly declared to his colleague: "I can state from long experience that Capt Mayes may be relied upon to use every exertion to place Sir Wm Thomson's Compass in the worst possible position." "The result of this speech," Fisher later told Thomson, "was most gratifying—I am convinced that the proper way to treat Capt Mayes is to deliberately and calmly insult him."

In the meantime Thomson used the patent courts to fight off various competitors, with a determination that was often more thorough than admirable. His own compass, with a couple of exceptions, was the result of his putting together a variety of ingredients from numerous sources. He particularly defended his compass card—aluminum ring, small needles suspended on threads, light but with long period—as his chief innovation, but other compass builders with slightly different layouts could claim, with some justice, that these ideas had been floating about for some time before Thomson put them all in one card. He initiated a number of legal challenges against competing designs and won all of them, in one case by appeal to the House of Lords after a lower court had gone against him. Thomson had financial resources and friends in high places; the justices of Great Britain had no expertise in deciding technical questions; and Thomson's reputation and sometimes hectoring manner in court overcame the opposition. Some years later Thomson appeared as an expert witness in a patent case involving electric wiring systems. A clever barrister with some technical knowledge seized on a small error Thomson made in his testimony to push the case for his side. After he had come down from the stand, Thomson was told of his slip-up and somehow blustered his way back on to the stand where he began to deliver an impromptu technical lecture. As J. J. Thomson recounted the anecdote, the barristers objected to the judge: "'My Lord, what has this to do with the case?' 'I don't know! I don't know!' said the judge, and Thomson went on."

At the end of 1887 the Admiralty appointed a new superintendent of compasses, Captain Ettrick W. Creak F.R.S., a man both scientifically knowledgeable and forward looking. He did not doubt that the old Admiralty Standard Compass, now coming up to its 50th birthday, had been kept on long past its natural lifetime. But that did not make him eager to take on the Thomson compass. The U.S. Navy, back in the 1860s, had decided to go with liquid compasses of the kind that Ritchie exhibited in 1876 in Philadelphia. The trick had been to design a chamber in which a compass card could float stably on water, under conditions of constant pressure, and with this and other improvements accomplished the liquid compass had far fewer of the dynamical problems associated with a dry card balanced on a pivot.

The Royal Navy had recently commissioned a number of fast torpedo boats, and in these, when they moved at high speed, or in rough conditions, or with weaponry firing, both the old Admiralty Standard and the Thomson compass proved unstable and useless, while a simple liquid compass remained level. Creak argued for the adoption of a range of compasses, including an improved Admiralty Standard, the Thomson compass, and liquid compasses for torpedo boats and for gunnery positions on other ships.

The ponderous mass of the Admiralty Board, however, prodded relentlessly by Fisher and others, had by this time finally turned from its old course enough to embrace the Thomson compass, and having belatedly and inelegantly come around, their Lordships were not inclined even to consider another technology. Although Creak mustered evidence in favor of the superiority of the liquid compass in some circumstances, the Board decided, on November 19, 1889, to make the Thomson compass the sole official compass of the Royal Navy.

For Thomson this was the final victory and vindication. At the time of the decision, Thomson was staying in London with Fisher, and when he returned to Glasgow at the end of the month he told his sister Elizabeth that "much mean and underhand work has been brought to light." Some 60 letters from ships' captains concerning the Thomson compass had allegedly been squirreled away at the Compass Department, suppressed by Evans and Mayes, and it was Fisher's irresistible force that had brought them into the open. A few of these letters contained critical remarks but fully 51 (so Elizabeth King wrote to her daughters) "spoke in terms of unbounded admiration and appreciation. . . . I believe this has been going on for years, and that Admiral Fisher has been instrumental in exposing the abuse. . . . Uncle William does not want it talked of."

In his history of the Compass Department, however, A. E. Fanning tells a different story. The 51 letters existed but were old. When the board met to make its final decision, Creak (being an honest man, Fanning says) dug out these old testimonials and presented them along with 24 more recent reports, which described difficulties with the Thomson compass as well as the virtues of the liquid compass. "His compass . . . was excellent for many applications, but for the requirements of the Navy of the 1890s its introduction was a retrograde step," Fanning concluded. It was, after all, more than 13 years since Thomson had first taken out a compass patent, and although he had made many modifications since then, the fundamental design remained the same.

In 1883 Creak, then an assistant at the Compass Department, had written to Thomson congratulating him on the award of the Copley Medal by the Royal Society. "I may single out amongst the many practical results of your researches the benefits you have conferred on Navigation. Foremost amongst Navigational instruments comes your compass, and your steady advocacy of that instrument against adverse forces has made me—perhaps one among many—long since review the position I had taken up and thank you for having made me think the matter over again increasingly to the advantage of your conclusions. You can review the position of your compass as regards the Navy with pride and satisfaction." Years later, though, after he had retired, Creak confided to a friend: "When the Thomson compass was first introduced as Standard Compass on board I felt it my duty to try and make it a success. It was, however, in many respects the bete noire of my existence." It was not until after Creak had retired that his successor was able finally to introduce liquid compasses into the Royal Navy, Thomson having more or less retired from the scene by this time. Thomson's compass, though, could still be found on merchant vessels and other ships well into the middle of the 20th century.

Scientific biographers of Thomson, if they have paid any attention at all to his compass innovations, have generally taken the matter to be a sorry saga of dim-witted naval administrators resisting marvelous innovations from a superlative scientific mind. Writers sympathetic to the Navy, on the other hand, portray Thomson as a man of undoubted talent and enthusiasm, with some genuine knowledge of the sea, who managed to parlay a handful of modest ideas in compass design into a commercial monopoly for his own manufacturing concern, using his reputation as a bludgeon in the law courts to beat down even small claims of originality from others, and persuading the Admiralty and the law to overlook both the deficiencies of his own design and the virtues of his competitors'.

The truth, inevitably, seems to lie somewhere between these extremes.

6

KELVIN

Monday, June 15, 1896, a week shy of the longest day of the year, and Glasgow remained sunny and pleasantly warm well into the evening hours. Flowers and electric lighting (still something of a novelty) brightened the lecture rooms and hallways of the university buildings. Upwards of 2,000 distinguished visitors roamed the campus, spilling out onto the spacious lawns where they were serenaded by the pipers of the Gordon Highlanders. Scholars from around the country mingled with prominent Glaswegian businessmen and politicians. Stokes, now Sir George Gabriel Stokes, came from Cambridge. The astronomer Simon Newcomb was there, representing the National Academy of Sciences in Washington, D.C. From Princeton University came Professor Woodrow Wilson. The Prince of Wales, detained elsewhere by a prior engagement, sent his apologies. Almost every college and university in Great Britain sent one or more representatives, as did institutions from across Europe and North America.

In the library the visitors could marvel at an array of electric and mechanical devices, commercial and scientific instruments, all the product of one man's inventive powers. Upstairs, courtesy of the Eastern, Anglo-American, and Commercial Cable companies, telegraph equipment and siphon recorders stood ready. Congratulatory messages ticked in from around the world. One, sent from within the university, took seven minutes to travel via Newfoundland, New York, Chicago, San Francisco, Los Angeles, New Orleans, Florida, Washington, New York, and Newfoundland, arriving at the library, where it was presented to Lord Kelvin. He composed a short reply of thanks and sent it back around the same route. It looped around the western hemisphere in only four minutes.

Sir William Thomson had become Lord Kelvin (to be precise, Baron Kelvin of Largs) four and a half years earlier, in Queen Victoria's New Year's Honours list of 1892. He was the first British scientist to be raised to the peerage, but his ascent into the upper reaches of nobility did not spring from his purely scientific achievements. His commercial success and personal wealth exemplified Victorian entrepreneurial virtue and contributed to Britain's economic and technological prowess. His telegraphic and marine navigation systems served in support of the empire. Lately he had made political forays on behalf of the Liberal Unionist party. A Scot of Irish origins, he vehemently opposed home rule for Ireland on the grounds (which his father would heartily have endorsed) that it would inevitably lead to religious quarrels and sectarian politics. He was not politically sophisticated, but he was plain spoken and direct, and his reputation guaranteed that his voice was heard.

Accepting the peerage, Thomson had to choose a title for himself. Lord Cable! Lord Compass! his nieces suggested. Lord Tom-Noddie would suit him, Thomson joked, in his simple way. His sister Elizabeth, more soberly, came up with "Kelvin" a couple of days later, only to find that Fanny and William Thomson had already had the same thought. Kelvin is the name of the small river that runs beside the university into the Clyde; it connects the academic world with the open sea.

The 1896 Glasgow celebration marked another milestone. William Thomson, later Sir William, now Lord Kelvin, had been a Glasgow professor for 50 years since taking up his post at the age of 22. The only position that might have drawn him away was the Cavendish chair at Cambridge, which he had in the end refused three times: at the outset, when Maxwell was appointed; again when Maxwell died in 1879, and the chair eventually went to Lord Rayleigh; and once more when Rayleigh resigned after five years to return to his estate at Terling, in Essex, in order to set up his own laboratory and work in peace. Approached then about leaving Glasgow for Cambridge, Thomson, then 60 years old and just returned from giving the Baltimore lectures, had written: "I am afraid it cannot be—alas, alas—The wrench would be too great. I began taking root here in 1831 [when his father came to Glasgow with his young family], and have been becoming more and more fixedly moored ever since. . . . To make a new departure . . . would be a life's work again."

The Cavendish professorship went to a much younger man. Joseph John Thomson, invariably known as J. J., was a 28-year-old physicist from Manchester, second wrangler in 1880, to whom William Thomson's old friend and colleague James Joule was a distinguished but by then frail and impoverished man, supported by a government pension after he had lost money through failed investments. Introducing his young son to Joule one day, J. J.'s father had said: "Some day you will be proud to be able to say you have met that gentleman." Late in 1884, "to my great surprise and I think to that of everyone else," J. J. became Cavendish professor. He remained there for 35 years and built the Cavendish into the world's preeminent institution for experiments in the new physics of the late 19th and early 20th centuries. He discovered the electron in 1897. His colleagues and students pioneered the investigation of radioactivity and atomic and nuclear physics. This lay in the future. But William Thomson's final refusal of the Cavendish chair and J. J. Thomson's appointment marked the end of one kind of physics and the beginning of another.

At his jubilee Kelvin was lauded over three days with banquets and speeches testifying to his "pre-eminent service in promoting arts, manufactures, and science," his contributions to "the improvement of natural knowledge," his "triumphs . . . in the advance of scientific theory and experiment," his "splendid discoveries . . . and valuable scientific inventions, which have . . . conferred signal benefits on the whole civilized world," and more, much more, in the same vein. In reply, Kelvin began conventionally enough, thanking the city and university for their long loyalty to him, thanking his numerous colleagues—"friends and comrades, day-labourers in science"—for their congratulations and for the work they had all done over the years. But then, echoing Newton's famous phrase about the small boy playing on the seashore, he went on: "When I think how infinitely little is all that I have done I cannot feel pride; I only see the great kindness of my scientific comrades, and of all my friends in crediting me for so much. One word characterizes the most strenuous of the efforts for the advancement of science that I have made perseveringly during fifty-five years; that word is *failure*. I know no more of electric and magnetic force, or of the relation between ether, electricity, and ponderable matter, or of chemical affinity, than I knew and tried to teach to my students . . . fifty years ago."

It was a startling moment in an emotional evening. A great-niece of Kelvin's, granddaughter of his sister Elizabeth, wrote that the word failure "seemed to ring through the hall with half-sad, half-yearning emphasis. Some of the people tried to laugh incredulously, but he was too much in earnest for that." Kelvin moved swiftly on, to talk of the joy of experimental discovery, of the innumerable inventions and marvelous devices that scientific study had brought into being in the second half of the 19th century. This was more than adequate compensation, he told his audience, for the "philosophical failures" he spoke of. As people rose in turn to offer their own words of praise, Kelvin "seemed nearly to break down for a moment, but got through, and everybody said he never spoke better," his great-niece reported. "There was something pathetic about it all—a sort of wonder that people should be so kind to him, and a wish that he had done more to deserve it all."

Then they all sang "Auld Lang Syne."

***

What Kelvin called failure is, in the standard histories of science, a progression of remarkable triumphs. By 1896 thermodynamics was largely settled, and Maxwell's theory of electromagnetism had gained experimental support and widespread acceptance. These, with Newtonian mechanics, formed the core of classical physics, a body of knowledge that held center stage for just a decade or two before the unexpected discoveries of the 20th century began to push it to the background. From 1895 to 1897, the years bracketing Kelvin's jubilee, the first of those new discoveries had put in an appearance: X rays, radioactivity, and J. J. Thomson's identification of the electron. Physics, far from being wrapped up, still had the capacity to surprise and perplex. Even so, physicists would have called this a time to take satisfaction in what had so recently been achieved. In 1846, when William Thomson took up his Glasgow position, neither heat nor energy, nor light nor electricity or magnetism, were understood except in a rudimentary way. Fifty years on, profound mathematical theories encompassed all these phenomena. Yet Kelvin talked of failure.

Natural philosophy had not gone as Kelvin had hoped. It turned into physics, for one thing, a title he disliked. During an 1862 lecture he had quoted Johnson's definition—"Naturalist. A person well versed in Natural Philosophy"—and had said that "armed with this authority, chemists, electricians, astronomers, and mathematicians may surely claim to be admitted along with merely descriptive investigators of nature to the honourable and convenient title of Naturalist, and refuse to accept so un-English, unpleasing, and meaningless a variation from old usage as 'physicist.'" Certainly he would rather be a student of natural philosophy than of physics, a subject he believed was becoming too abstract, too mathematical, and too isolated from the rest of science. Of mathematics itself, Kelvin had no fear; he had been a mathematical prodigy. But in the closing years of the century mathematical formalism was driving out, as Kelvin saw it, physical realism. He had still not reconciled himself to the elegant but spartan electromagnetic field theory of Maxwell. In his Baltimore lectures a dozen years earlier he had promoted his endlessly intricate attempts to construct mechanical models of the ether, a tangible physical medium that would carry electromagnetic influences. In 1896, Kelvin still pursued this increasingly lonely quest.

Kelvin also, and in similar isolation from the mainstream, cultivated his own view of atoms and molecules. As long ago as January 1867, only a few months after the successful conclusion of the Atlantic cable venture, the newly minted Sir William Thomson had presented to the Royal Society of Edinburgh a long account of what he called "vortex atoms." His ideas, as always, combined novelties gleaned from other sources. In particular he referred to the "magnificent display of smoke-rings, which he recently had the pleasure of witnessing in Professor Tait's lecture-room" and to a theoretical analysis of fluid motion from his old friend Helmholtz.

Tait had taken a wooden packing box, cut a circular hole in one end, and replaced the opposite end with a taut cloth. He filled the box with smoke from a piece of smoldering phosphorus, and by striking the cloth sharply with the flat of his hand, he could produce smoke rings up to a foot across and an inch in thickness. These rings, sailing gracefully across the room, were "pungent and disagreeable," Thomson said, but wonderfully suggestive. He watched as two rings grazed up against each other: They met, quivered, then bounced away intact, like rubber rings. This put him in mind of "the clash of atoms" implied by the new kinetic theory of gases, in which the motion and collision of atomic entities were presumed to account for the overall properties of a gas.

Kinetic theory was then beginning its long and eventually triumphal ascent. The idea that matter consisted of small, hard atoms had ancient roots, but the modern theory, arising in the middle of the 19th century, owed most to the efforts of Clausius, Maxwell, and Ludwig Boltzmann in Vienna. If a gas consisted of tiny atoms speeding about and colliding constantly with each other, as kinetic theory held, then the overall properties of the gas ought to follow directly from consideration of the behavior of the atoms, as dictated by simple Newtonian mechanics. This was simple in principle but enormously complicated in practice, since there were trillions upon trillions of atoms in an ordinary volume of gas. Perhaps the greatest triumph of kinetic theory was Boltzmann's derivation of a statistical formulation of entropy from the collective motions of atoms.

Thomson did not altogether object to kinetic theory, but he found it inadequate and restrictive, as indeed in a number of ways it was. An atom must clearly be more than an inert lump, with no qualities except mass and velocity. Atoms absorbed and emitted light at characteristic frequencies, not uniformly across the spectrum. This was the foundation of spectroscopy. Why was one material transparent and another opaque? Kinetic theory did not address such matters. Its defenders would argue that they were using an idealized model to tackle a specific issue—the derivation of the large-scale thermodynamics of gases from the microscopic dynamics of atoms. But Thomson never liked to deal with idealizations and limitations. If a model didn't explain everything he wanted to explain, he would add to it somehow. In this case, what he needed was a model in which atoms had some sort of structure, some array of intrinsic properties, by which he could try to understand the interaction of atoms with light and other electromagnetic phenomena. The vortex atom looked like a good bet.

Tait's experiments with smoke rings arose from his translation of an 1858 paper by Helmholtz that discussed rotatory motion in fluids. Helmholtz had defined, for a fluid with some arbitrary set of internal motions, a quantity he called the *Wirbelbewegung*, or vortex motion, which he showed was conserved. That is, the collective rotational motions of an idealized frictionless fluid might behave in a hugely complicated way, the fluid stirring about this way or that, but their total magnitude measured by Helmholtz's prescription would remain constant. (A cup of tea, stirred with a spoon and then left alone, will of course come to a standstill after a time. This is mainly because of friction between tea and cup.)

In particular Helmholtz had shown that a "vortex ring"—a toroidal or doughnut-shaped volume of spinning fluid—was stable. Vortex rings could not appear out of nowhere, nor could they vanish. Thomson seized on this mathematical theorem and built on it a tentative atomic theory. Permanent existence was a basic criterion for any structure that might qualify as an atom, but vortex rings had much more going for them than that. Like the atoms of kinetic theory, they would interact with each other in ways determined purely by dynamics, although in a far more complicated fashion. The collision of two rings was a difficult though "perfectly solvable mathematical problem," Thomson wrote. "Its solution will be the foundation of the proposed new kinetic theory of gases."

Beyond that, the vibrations and oscillations of vortex rings, which Thomson had amused himself with in Tait's laboratory by poking at a smoke ring with his finger, had the capacity to explain spectroscopy. Each atom, pictured as a ring with some set of possible oscillations determined by its structure, would interact with light at a characteristic set of frequencies. Finally, if a vortex ring was a stable rotatory motion of the light-transmitting ether itself, then the physical attributes of the ether should completely determine the interaction of a vortex atom with light.

Here in principle, Thomson asserted, was the foundation of what we might now call a grand unified theory of light and matter. "Helmholtz's rings are the only true atoms," Thomson confidently declared. Working out a full theory would not be easy: "Even for a simple Helmholtz ring, the analytical difficulties which it presents are of a very formidable character, but certainly far from insuperable in the present state of mathematical science." This sort of exercise, like his Baltimore models of the ether, suited Thomson perfectly. Underlying it all was simple Newtonian mechanics, applied to a certain medium. Vortex atoms were purely dynamical constructions, and consequently all their properties followed from dynamical laws alone. Matter was a dynamical phenomenon. Light was a dynamical phenomenon. The laws of electricity and magnetism, formulated by Maxwell in what Thomson regarded as a suspiciously abstract style, would turn out to be the dynamics of the ether. Maxwell himself, speaking at the British Association meeting in 1870, endorsed Thomson's proposal as a project worthy of serious investigation and said that if it succeeded the constitution of the physical world would be "nothing but matter and motion."

Thomson latched onto this marvelous, intoxicating vision without reservation. The striking spectroscopic properties of sodium—the bright double line that gives sodium lights their lurid yellow hue—had caught Thomson's interest years ago, and he was quick to suggest that "the sodium atom . . . may very probably consist of two approximately equal vortex rings passing through one another like two links of a chain. It is . . . quite certain that a vapour consisting of such atoms, with proper volumes and angular velocities in the two rings of each atom, would act precisely as incandescent sodium-vapour acts—that is to say, would fulfil the 'spectrum test' for sodium." In the space of two sentences Thomson's enthusiasm for vortex atoms took him from "very probably" to "quite certain" without a second thought. Thermodynamics was already a branch of mechanics. Now light and electromagnetism and the properties of atoms would all likewise reduce to dynamical theorems and proofs. This would be a theory of everything, for its day. There was nothing else to explain.

***

Even by the standards of the 19th century, before the mills and factories of the burgeoning academic industry had processed scholarly and scientific prose into the passive-voiced porridge it has mainly become today, it took an odd author indeed to begin a work of mathematical physics thus:

> The following story is true. There was a little boy, and his father said, "Do try to be like other people. Don't frown." And he tried and tried, but could not. So his father beat him with a strap; and then he was eaten up by lions.
>
> Reader, if young, take warning by his sad life and death. For though it may be an honour to be different from other people, if Carlyle's dictum about the 30 millions[1] still be true, yet other people do not like it. So, if you are different, you had better hide it, and pretend to be solemn and wooden-headed. For most wooden-headed people worship money; and, really, I do not know what else they can do. In particular, if you are going to write a book, remember the wooden-headed. So be rigorous; that will cover a multitude of sins. And do not frown.

The paper then discussed some propositions concerning wave motion in electromagnetic theory. Its author, Oliver Heaviside, was certainly different from other people, but he made no pretence of being wooden headed. Born in 1850 in the London slums (around the corner, he would say, from the blacking factory where the young Charles Dickens spent a harshly formative period), Heaviside escaped his origins by becoming a telegraph engineer—the same route Edison took with such great success. But Heaviside was at heart a mathematician and a theorist, and he singularly lacked the personal skills by which men get on in business.

He applied for membership in the new Society of Telegraph Engineers, but was informed that mere telegraph clerks did not qualify. "What would Edison say if he were here now?" Heaviside later remarked. "I was riled. I had already had one of my inventions tried in a rough experimental way by the [Post Office] with success. . . . So I went to Prof. W. Thomson and asked him to propose me. He was a real gentleman and agreed at once. . . . So I got in, in spite of the P.O. snobs." Having proved his point by obtaining membership, he attended no meetings and never paid his dues, with the result that he was kicked out some years later.

Beginning in the mid-1870s, he began a lengthy project to formulate a comprehensive theory of signal transmission by the electric telegraph

---

[1] Thomas Carlyle, a Scot, replied "thirty millions, mostly fools," when asked about the population of England.

according to the full Maxwellian theory of electromagnetism. This was the subject Thomson had begun, 20 years earlier, with only a limited understanding of electric phenomena at his disposal. Heaviside's treatment was mathematically sophisticated, but practical too, and led to new principles for the design of long telegraph cables, whether overland or undersea. The appearance of the telephone at this same time made new demands of electrical theory. It was no longer enough to get indeterminate but recognizable blips down a cable. Telephony over any distance demanded an output that faithfully reproduced the input. Heaviside's theoretical analysis supplied a sound basis to the new technology, but his ideas were at first firmly resisted by the British Post Office. Heaviside's invariable response to opposition was sarcasm of a creative and eccentric flavor, which won him no allies.

As the scope of electrical technology blossomed, the Society of Telegraph Engineers transformed itself in 1888 into the Institution of Electrical Engineers. As well the telephone, industrial and domestic electricity were on the rise. Systems for wiring, insulation, and connection were tried out and patented. Thomson himself devised and then marketed through his instrument company one of the first electricity meters.

In his address as inaugural president of the IEE, Thomson made special mention of Heaviside's new treatment of telegraphy. But his praise came on the back of a hesitant and grudging nod toward Maxwell: "Maxwell's 'electro-magnetic theory of light' marks a stage of enormous importance in electro-magnetic doctrine, and I cannot doubt but that in electro-magnetic practice we shall derive great benefit from a pursuing of the theoretical ideas suggested by such considerations. In fact, Heaviside's way of looking at the submarine cable problem is just one instance of how the highest mathematical power of working and of judging as to physical applications, helps on the doctrine, and directs it into a practical channel."

Heaviside had used Maxwell's theory to help him understand the telegraph better. Thomson saw this achievement exactly backward. He believed Heaviside's investigation of the telegraph would illuminate Maxwell's theory and remedy what he regarded as its flaws.

By this time Heaviside had abandoned with disgust his connection to the unappreciative telegraph industry and had gone to live with his parents in Devon, barely surviving on their meager resources. He had a brother living nearby but hardly ever visited "because he thought the cart-men shouted abuse at him." Around this time he was put up for membership in the Royal Society, a process that resembled entry into the baseball hall of fame. Names were proposed and seconded, a secret ballot was taken, some succeeded, others did not, and those who failed could campaign again the next year, until they got in or got the message.

Heaviside was proposed by Oliver Lodge, another young physicist making a name for his work in electromagnetism, and seconded by Thomson. As he explained to Thomson, his attitude was awkward: "I have to give you my best thanks for your consideration in offering to second Oliver Lodge's F.R.S. proposal. As he has probably told you, I am somewhat cranky on the subject; rather than be passed over, I would prefer never to be nominated; so he has suggested postponement." To Lodge he wrote enigmatically: "You may judge of the intensity of my feelings as to possible rejection by the fact that I have so good a man as you for my proposer and no less than Sir W. T. for seconder, and still I am not happy. (I had a wicked mammy, a more than brutal pappy; they kicked me, strapped me, flogged me, whacked me. *Still* I was not happy!)" Such was Heaviside's strange, pugnacious humor; he was living with his parents (nursing them, so he claimed) when he wrote this.

An arrangement was made. Lodge, Thomson, and others worked behind the scenes, apparently, to rig the balloting in 1890 and guarantee Heaviside a place, and so he became F.R.S., with or without his full compliance. He never traveled up from Devon to go to meetings in London.

Crankiness apart, Heaviside was now working hard in his isolation, going from a specific treatment of telegraphy to a more general and theoretical reworking of Maxwell's electromagnetism. The Nobel laureate Leon Lederman has joked that the essential criterion for an acceptable "theory of everything" in modern physics is that the necessary equations should fit on a T-shirt. Science and engineering students may occasionally be seen wearing T-shirts with Maxwell's equations on them, these being, until about the 1970s, the nearest thing to a theory of everything that physicists had thus far devised. In Maxwell's own time, however, no one would have worn a T-shirt bearing his equations, and not just because Victorian gentlemen didn't wear T-shirts. What we now regard as

Maxwell's equations in their standard form—four concise laws, cryptic to the uninitiated, encapsulating the links between electricity and magnetism—are due to Oliver Heaviside.

This is not to say that Heaviside deserves credit for the conception of electromagnetic theory. Using the standard mathematics available to him, Maxwell expressed his theory in Cartesian coordinates, separately denoting the x, y, and z components of the electric and magnetic fields and writing down complicated differential equations, occupying many pages, to capture the variation of these three components with respect to each of the three coordinates. Heaviside, in the 1880s, was a pioneer of what is now called vector calculus. A vector is a quantity with magnitude and direction, such as a velocity. A scalar is simply a magnitude. In electromagnetism, electric charge is a scalar, and the electric field is a vector—because it has orientation. Moreover, it is a vector *field*, in other words a vector quantity that varies from place to place. So too the magnetic field. Maxwell's theory connects the amplitude and geometrical pattern of electric and magnetic fields to the spatial distribution of electric charge[2] and also to the time variation of the fields.

The three basic operations of vector calculus are grad (for gradient), div (for divergence), and curl (for twist or rotation), which roughly indicate the geometrical property of a vector or scalar field that the operations elucidate. Pages of repetitious equations turn into single condensed statements.

Heaviside did not invent this kind of mathematics, but he made innovative use of it in electromagnetism. Not only do the equations become simpler, but their meaning becomes more transparent. Using this compact and elegant notation, Heaviside was able to provide a more rigorous statement of the mathematical properties of the electromagnetic field than Maxwell had been able to do, and this in turn led to a more

---

[2]It is a curiosity of nature that electric charges exist but that their magnetic counterparts do not. Magnets always come in conjoined north and south poles; individual "monopoles" never occur. Why this should be remains mysterious. Some theories of elementary particles predict that there should be monopoles. At any rate, Maxwell's equations have a certain asymmetry between the electric and magnetic parts for this reason.

precise statement of the physical significance of certain aspects of the theory. Notably, Heaviside (and independently J. H. Poynting) proved that the electromagnetic field carried energy. It had been generally assumed that when electricity moved about, all the energy was carried by electric currents. This certainly was Thomson's view. But with vector notation instead of the old mess of Cartesian components, it became possible to find a mathematical definition of energy that could be followed with relative ease through complex algebraic manipulations. It became apparent that electromagnetic energy, like the fields, pervaded space and was not concentrated only in charges and currents.

These and other insights were the work, in the 1880s and 1890s, of a young group of mathematical physicists who became known as the Maxwellians. Heaviside and Lodge, along with the Irishman George Francis FitzGerald, were the leaders of this informal movement. In essence, it was the Maxwellians who not only produced Maxwell's equations as they are taught (and printed on T-shirts) today, but also illuminated and enlarged Maxwell's theory by using the new methods to apply it in much more general ways and to trace in detail the physics of electromagnetism when regarded as a form of energy.

To all this Thomson remained cool. It is not altogether easy to see why. The Maxwellians put great emphasis on the primacy of energy, a philosophy Thomson had long endorsed. Their use of a compact mathematical notation ought to have pleased him, as he could have taken it as part of his lifelong battle against "aphasia," the unaccountable inability of otherwise intelligent people to understand mathematical arguments. Yet he did not like grad, div, and curl, and preferred to stick with the cumbersome Cartesian notation of old. Perhaps he, if no one else, could discern physical meaning in arrays of equations laboriously written out component by component. Heaviside wrote to him once that the new style "save[s] letters, and *eases the memory*, an important matter when there are a great many vectors." But the point, he said, "is not just to save space, it is to simplify ideas and language, and harmonise our symbolization with Faraday's way of viewing things; components never come in them, in general investigations, and I am sure Faraday never thought of components."

To no avail. In Baltimore Thomson had declared that Faraday "did the most" to cure the "mathematical disease of aphasia from which we suffered so long. . . . The old mathematicians used neither diagrams to help people understand their work, nor words to express their ideas. It was formulas, and formulas alone. Faraday was a great reformer in that respect with his language of 'lines of force.'" Maxwell had carried through Faraday's project to completion, and Heaviside transformed it into a more accessible language. Along the way Thomson fell behind. He favored mathematical theories based on physical pictures, but the physical pictures had to be of a certain kind. Wheels and springs and pulleys he could countenance, but not an intangible vector field stretching and flexing unseen through space. Above all, the notion that these abstract entities purported to carry energy distressed him.

Perhaps too Thomson was influenced in part by his long-running battle against the quaternionic notation his friend Tait so heartily espoused. Tait himself objected with typical vehemence to the new vector notation because he regarded it as a watered-down version of his cherished quaternions. A quaternion was a particular combination of a vector and scalar, constructed so that quaternion operations always produced other quaternions. This appealed to Tait's sense of mathematical tidiness, but it went against nature. The vector electric field and the scalar electric charge have distinct and separate physical identities. The corresponding quaternion, a combination of the two, does not. Heaviside remarked in one paper that "if we put aside practical application to Physics, and look upon Quaternions entirely from the quaternionic point of view, then Prof. Tait is right, thoroughly right, and Quaternions furnishes a uniquely simple and natural way of treating *quaternions*."

At one point the dispute boiled over into the pages of *Nature*, with Tait attacking the growing number of adherents to the vectorial doctrine. But Heaviside was no Tyndall, responding with the measured distaste and veiled disdain of a Victorian gentleman. He went in for outright mockery, delivered with transparent glee: "Passing to Prof. Tait's letter, it seems to be very significant. The quaternionic calm and peace have been disturbed. There is confusion in the quaternionic citadel; alarms and excursions, and hurling of stones and pouring of boiling water upon the invading host. . . . It would appear that Prof. Tait, being unable to bring his massive intellect to understand my vectors . . . has delegated to Prof.

Knott the task of examining them, apparently just upon the remote chance that there might possibly be something in them that was not utterly despicable."

Thomson disliked both quaternions and vectors, mainly for the same reason: To him they obscured rather than illuminated physics. He referred to "Heaviside's nihilism," and this opinion extended to the philosophy of the Maxwellians in general. He thought they embraced a kind of mathematical formalism that distanced itself further from true physics the more formal it became. He hankered still after mechanical models of the ether, as he had in Baltimore. Heaviside, many years later, commented: "Lord Kelvin used to call me a nihilist. That was a great mistake, (though I did throw a bomb occasionally, to stimulate an official humbug to say something about electricity and how to apply it). He was most intensely mechanical, and could not accept any ether unless he could make a model of it. Without the model he did not consider electromagnetics to be dynamical. But I regard electrodynamics as being fully dynamical."[3]

This was the essence of Kelvin's difficulty over Maxwell. When he said he wanted a mechanical model of the ether, he meant something he could construct out of wheels and pulleys and springs and gyrostats, all embedded in some suitable jelly or wax. These were the mechanical ingredients he permitted in his theorizing. Heaviside and the other Maxwellians believed equally strongly in the existence of an ether—that is, a medium in which electromagnetic waves traveled. They believed, however, that the electric and magnetic fields that their new treatment of Maxwell revealed so clearly were, in themselves, dynamical entities with genuine physical significance. But they were *sui generis*, not reducible to jellies and pulleys.

FitzGerald had criticized Thomson's ideas as early as 1884. The Baltimore lectures had been reported in summary fashion in *Nature*, with reference to shoemaker's wax and various kinds of pitch as analogs to the

---

[3]Heaviside also remarked, on reading S. P. Thompson's *Life of Kelvin*: "Found out why he did not like 'curl.' He broke his leg when curling! Who can wonder?" (Gossick, 1976).

sort of ether Thomson imagined. FitzGerald found this highly unsatisfactory. He objected strongly to "Sir Wm. Thomson's speaking of the
ether as *like* a jelly. It is in some respects *analogous* to one, but we certainly know a great deal too little about it to say that it is *like* one. May be
Maxwell's conceptions as to its structure are not very definite, but neither
are any body's as to the actual structure of a jelly. . . . It seems very
unlikely that any jelly is at all *like* the ether that Maxwell supposes.
. . . I also think that Sir Wm. Thomson, notwithstanding his guarded
statements on the subject, is lending his overwhelming authority to a
view of the ether which is not justified by our present knowledge and
which may lead to the same unfortunate results in delaying the progress
of science as arose from Sir Isaac Newton's equally guarded advocacy of
the corpuscular theory of optics." The last phrase refers to Newton's insistence that light consisted of particles, not waves, an opinion that had its
merits at the time but retarded the later acceptance of wave theory in
England.

Coming from a man just 33 years old, this was sharp criticism of his
renowned elder. But Thomson never took personal offense in scientific
debate; indeed he embraced a blunt exchange of views. According to
Rayleigh, in fact, Kelvin admitted late in his life that "a certain amount of
opposition was good for him." He and FitzGerald embarked on a substantial though frankly useless correspondence. Thomson could never
accept certain aspects of Maxwell's theory, simply because he could find
no familiar physical analog to them. FitzGerald tried to persuade him
that these parts of the theory corresponded to real physical phenomena,
but no reconciliation came.

Rather remarkably, Thomson's views on Maxwell merited occasional
mention in the newspapers, in the way that momentous meaning was
teased out of official pronouncements from the Kremlin in the last days
of the Soviet Union. At the 1888 British Association meeting in Bath,
the correspondent from the *Times* reported with extreme circumspection
that "Sir William Thomson in one paper cautiously made what must be
regarded as a somewhat noteworthy admission with reference to Clerk-
Maxwell's fundamental theory. . . . He considered Maxwell's fundamental assumption 'not wholly tenable.' In all his previous utterances on the
subject, Sir William has described Maxwell's views on this point as com-

pletely untenable, so that the change in his position is of great importance to all interested in electro-magnetic theory." Thomson wrote to the paper to explain that he had slightly softened his wording after talking to FitzGerald, among others. But this was the full extent of FitzGerald's influence. In 1896 he was still making the same point he had tried to make after Baltimore. Responding to a letter from Kelvin, FitzGerald wrote: "You say . . . 'The luminiferous ether we must imagine to be a substance which so far as luminiferous vibrations are concerned moves as if it were an elastic solid.' Now this 'we must' is entirely unjustifiable. We *need* do nothing of the kind. . . . I cannot see how you are justified in concluding that 'we *must*' deal with the ether as if it were an elastic jelly. The electromagnetic properties of the ether are a much better key to its properties than light waves, and I cannot see, nor apparently can you, how it can be both electric and magnetic and at the same time an elastic solid."

The ether had by this time come under experimental as well as theoretical attack. In 1887, at the Case Research Institute (now Case Western Reserve University) in Cleveland, Ohio, Albert Michelson and Edward Morley performed the celebrated experiment in which they tried, and failed, to find a difference between the speed of two light beams running at right angles. This addressed an old and unresolved issue. If light propagated through an ether that filled space, and if the earth is also moving through that ether, should not light beams have slightly different velocities depending on their direction relative to the motion of the earth? From the theoretical standpoint, other issues presented themselves. How could the earth pass through a solid ether? Or, in fact, would the ether (because of friction) move with the earth in the vicinity of the planet but revert to a cosmically stationary state at great distances? Whatever the answer, there would be consequences for the way light traveled near the earth's surface.

Michelson and Morley showed, to a high degree of precision, that light moved at precisely the same speed near the earth, regardless of its direction. Kelvin took this to mean that the earth dragged the ether along with it (his old friend Stokes had made a similar proposal decades ago), and that in turn meant further complication for his models of the ether. FitzGerald, along with the Dutch physicist Hendrik Lorentz, made a more radical suggestion: Perhaps the physical dimensions of moving objects shrank slightly when they moved relative to the ether. In that case, light would move a little more slowly when it had to go upwind, so to speak, but any measuring stick would shrink by the same amount, so the apparent velocity of light would remain unchanged.

Still, there was the assumption in either case that the ether existed and that some complex interaction between ether, light, and matter would explain the result of the Michelson-Morley experiment. FitzGerald was closer to the truth than Kelvin, but he died in 1901, at the same age as Maxwell had died and apparently of a similar cause. Not until 1905 did Albert Einstein propose his special theory of relativity, which said that light always moved at the same speed and that moving objects apparently got shorter. This was not, in Einstein's proposal, an absolute effect—it depended on who was doing the measuring and was a consequence of the "relativity" of measurement for observers moving at different velocities. There was no genuine FitzGerald-Lorentz contraction of moving objects. Einstein made no mention of an ether. In his theory the ether simply vanished, and so came to an end half a century of strenuous and increasingly baroque efforts to construct mechanical models of the ether, none of which ever proved satisfactory.

***

At the British Association meeting of 1892 in Edinburgh, the presidential address was delivered by Archibald Geikie, a Scottish geologist and friend of Kelvin. He began by reminiscing about the early influence of arguments from physics restricting the lifetime of the earth, coming as they did at a time when geologists had given the matter no thought at all: "It is not a pleasant experience to discover that a fortune which one has unconcernedly believed to be ample has somehow taken to itself wings and disappeared. When the geologist was suddenly awakened by the energetic warning of the physicist, who assured him that he had enormously overdrawn his account with past time, it was but natural under the circumstances that he should think the accountant to be mistaken, who thus returned to him dishonoured the large drafts he had made on eternity."

The geologists, grumbling and uncomfortable, had nevertheless accepted the limitations imposed by physics, and with a salutary effect on their reasoning. But the physicists, Geikie complained, had still not been satisfied. "The geologist found himself in the plight of Lear, when his bodyguard of one hundred knights was cut down. 'What need you five-and-twenty, ten, or five?' demands the inexorable physicist, as he remorselessly strikes slice after slice from his allowance of geological time. Lord Kelvin is willing, I believe, to grant us some twenty millions of years, but Professor Tait would have us content with less than ten millions."

Geologists were becoming more confident of their science, however, particularly in their ability to reason quantitatively about the formation and erosion of terrestrial rocks. They now had their own calculations about age, which they were willing to put up against the numbers coming from the physicists' camp. Geikie went so far as to suggest that the physicists might not know as much as they thought they knew. "Some assumption, it seems to me, has been made, or some consideration has been left out of sight, which will eventually be seen to vitiate the conclusions," he told his audience. "After careful reflection on the subject, I affirm that the geological record furnishes a mass of evidence which no arguments drawn from other departments of Nature can explain away, and which, it seems to me, cannot be satisfactorily interpreted save with an allowance of time much beyond the narrow limits which recent physical speculation would concede."

A wary impasse reigned. Kelvin, in truth, was more inclined to allow 100 million years as a reasonable maximum, while Tait's assertion that the age could hardly exceed 10 million years was strident but lonely. If physicists' numbers rested on a handful of assumptions, however, geological arguments seemed full of guesses and speculations about weathering and erosion and sedimentation and deposition, none of which seemed to have the fundamental certitude that physical law offered.

But that certitude began to show cracks. One of the weaker arguments limiting the earth's age came from consideration of the effect of tides in slowing the planet's rotation, coupled with measurements of the departure of the planet's shape from a perfect sphere, which indicated its rotation at the time it solidified. This line of analysis had always been rife with physical uncertainties and mathematical difficulties.

By odd coincidence, the man who refined these arguments enough to extract reliable results was George Howard Darwin, the fifth child of Charles and Emma Darwin and the third to grow to adulthood. Often in poor health, he struggled to get into Trinity College, Cambridge, but then surprised himself and delighted his father by becoming second wrangler in 1868. As a fellow at Trinity he subsequently dabbled for a while in various mathematical ventures, including a sophisticated attempt to statistically analyze ill health among the offspring of marriages between first cousins. In 1877, having seen some earlier work by Thomson, he wrote a paper, "On the Influence of Geological Changes in the Earth's Axis of Rotation," in which he addressed the "wandering" of the poles due to slow viscous stirring of the earth's interior. The paper was sent to Thomson for review. Thomson, as was his habit when he saw something that struck him as possessing insight and originality, made contact with the author directly to discuss not only the work at hand but possible ramifications of it. Thus George Darwin encountered the difficult problem of analyzing tidal effects on the earth's rotation, to which he added the further complication of regarding the planet's interior as a stiff semi-liquid rather than an absolutely rigid solid.

Darwin was not an original mathematician or an especially imaginative physicist, but he was a prodigious calculator. The problems he tackled would today be programmed into a computer, but solving them "by hand" had some advantages. The success of some approximate methods and the failure of others often indicates which physical effects are important and which negligible. Darwin, said a colleague, "never hesitated to embark on the most complicated computations if he saw a chance of attaining his end," and he disparaged displays of elegant mathematics "which are in fact mere conjuring tricks with symbols." In short, he was a man after William Thomson's heart, and he had moreover the patience or perhaps monomania to mount an almost lifelong investigation of a problem Thomson could never quite find the time to properly address.

Charles Darwin was overjoyed that his son made such an impression on a matter that had so agitated him over the years. "My dear old George," he wrote, "All of us are delighted, for considering what a man Sir William Thomson is, it is most grand that you should have staggered him so quickly, and that he should speak of your 'discovery, etc.' . . . Hurrah for the bowels of the earth and their viscosity and for the moon and for the Heavenly bodies and for my son George."

Through lengthy and laborious calculation, and using the best empirical knowledge he could find of the properties of rocks making up the earth's mantle, Darwin proved (what seems not at all surprising today) that the body of the planet is amenable to small, slow changes in shape, as external forces and internal conditions vary. His work laid the foundations for understanding many aspects of tidal friction (including both ocean tides and the much smaller but still significant tidally induced flexure of the crust and mantle). He showed how variations in the rotation and figure of the earth could diagnose inaccessible physical parameters such as the viscosity of the planet's interior. Darwin also analyzed the orbit of the moon, which recedes as the earth spins slower, in order to maintain overall conservation of angular momentum. This led to models in which the moon broke off originally as a fragment of the spinning earth, and so led to another estimate of the age of the system from the time the moon would take to achieve its present orbit. The coupling of the earth-moon system with the sun's gravity induces changes in the tilt of the earth's axis. . . .

And so on. As far as Thomson's particular interest went, the main conclusion of Darwin's lifelong work was a negative one. Because the earth is not perfectly rigid, it can adapt its shape slowly as its rotation slows. There is no credible way to determine the planet's age from its present rotation period and current measures of tidal friction. Darwin could at best only establish limits. In papers published in 1879 he estimated that tidal friction operated on a timescale of perhaps 700 million years. This alone meant that tidal arguments had no ability to limit the earth's lifetime to the kind of number that Thomson had long talked about. In any case, Darwin freely admitted that there were too many uncertainties in the properties of the earth's interior to be sure even of the estimates he gave. "Under these circumstances, I cannot think that any estimate having any pretension to accuracy can be made as to the present rate of tidal friction," he concluded.

Of course, this simply meant that the tidal argument was of no use in Thomson's battle with the geologists and biologists. Heat loss from the earth continued to provide a stricter limit and a smaller allowable age. But the fact that one of the restrictions Thomson had long insisted on had now been lifted from their shoulders, and by Charles Darwin's son, gave geologists, biologists, and their sympathizers reason to think the other restrictions might turn out to have concealed flaws. In 1895 John Perry, a physicist and former student of Kelvin, went back to the original calculation of heat loss and claimed to have found just such a loophole.

Thomson, in order to obtain an answer, had assumed the earth to be uniform throughout in its thermal properties—the same conductivity and heat capacity everywhere. Perry offered a simple modification. He imagined the earth as a crust surrounding an interior and allowed the two components to have different properties. The thermal attributes of the crust, he observed, were known from direct measurement, but as to the interior there was only guesswork or assumption. By solving this more complex mathematical problem, in which he had assistance from Oliver Heaviside, Perry showed that with a not excessively outrageous choice of thermal properties for the interior, he could obtain strikingly different conclusions about the age of the earth. The starting point (a uniformly molten sphere) and endpoint (a solid earth with measured surface temperature gradient of 1°F per 50 feet of depth) were the same, but because the interior distribution and flow of heat were significantly different in the two-component model, the time from start to finish could be much longer. Perry claimed that an age as much as a few hundred times the original estimate of 100 million years was possible. Ten billion years was surely enough for any geologist or biologist.

Perry sent a draft of his paper to Kelvin but got no immediate response and was reluctant to pursue his criticism. As he explained, with excessive melodrama, "I was Lord Kelvin's pupil, and am still his affectionate pupil. . . . He has been uniformly kind to me, and there have been times when he must have found this difficult. One thing has not yet happened; I have not yet received the thirty pieces of silver."

Unwisely, Perry then approached Tait, who responded with barely coherent scorn. He wrote dismissively of his "entire failure to catch the *object* of your paper. For I seem to gather that you don't object to Lord Kelvin's mathematics. Why then drag in mathematics at all. . . ?" Tait told Perry that it was "absolutely obvious" that changing the thermal properties of the interior would alter the result: "I don't suppose Lord Kelvin would care to be troubled with a demonstration of *that*." As to what the interior properties of the earth might be, Tait simply declared, "I don't suppose anyone will ever be in a position to judge," as if that settled the matter.

Perry reiterated that his point was not to establish definitively a greater age for the earth, only to show that a greater age was distinctly possible and that Tait and Kelvin, if they disagreed, ought to supply some counterargument. "What troubles me," he told Tait, "is that I cannot see one bit that you have reason on your side, and yet I have been so accustomed to look up to you and Lord Kelvin, that I think I must be more or less of an idiot to doubt when you and he were so 'cocksure.'"[4] Tait responded by asking Perry again why he thought the interior was different from the crust and, with startling irrelevance, added, "do you fancy that any of the *advanced* geologists would thank you for 10 billion years instead of 100 million? Their least demand is for one trillion."

Kelvin, when he finally weighed in, was by contrast eminently reasonable. Perry's argument was "clearly right" but neither new nor surprising to him. His original analysis of 1862 referred explicitly to the possibility of a difference in conductivity between crust and interior, and this was one of the reasons he had allowed a range of 20 million to 400 million years for the earth's age. He observed, slyly or more likely obliviously, that "100 millions . . . is all Geikie wants," this being the figure the geologists had reluctantly accepted because it was as much as Kelvin would give them.

More pertinently, Kelvin questioned whether the difference between crust and interior could be as great as Perry suggested. He adduced some relevant bits of laboratory data to argue otherwise. He admitted "it is quite possible I should have put the superior limit a good deal higher, perhaps 4,000 instead of 400."

No solid conclusion emerged. Perry had pointed out a difficulty, but neither he nor Kelvin (and certainly not Tait) was able to come up with any sound estimate of what the heat loss argument now said about the age of the earth. It could be bigger than 100 million years, but not nearly as big as Perry imagined, unless the interior of the earth was wildly strange

---

[4]This is part of the exchange published in *Nature*, which still today remains somewhat racy by the standard of scientific publishing, but I doubt that letters like this would be printed nowadays.

and different from its crust. Still, geologists and biologists took heart from the confusion. Another of the supposedly restrictive calculations that Kelvin had so long promulgated had turned out be much shakier than he had let on.

The shackles were loosening. At the 1896 British Association meeting, E. B. Poulton of Oxford explained to his fellow biologists the uncertainties that had recently been demonstrated in the physicists' calculations. He told them Tait's oft-repeated views were "entirely indefensible. . . . The obligation is all on the other side, and rests with those who have pressed their conclusions hard and carried them far," and concluded roundly that "Natural Selection will never be stifled in the Procrustean bed of insufficient geological time."

At the same meeting George Darwin began his lecture to the physicists by saying "amongst the many transcendent services rendered to science by Sir William Thomson, it is not the least that he has turned the searching light of the theory of energy on to the science of geology," but then he went on to enumerate the mounting difficulties. The tidal argument had proved empty. "Professor Tait cuts the limit down to 10,000,000 years; he may be right, but the uncertainties of the case are far too great to justify us in accepting such a narrowing of the conclusion."

All in all, Darwin concluded, there were so many uncertainties "that we should do wrong to summarily reject any theories which appear to demand longer periods of time than those which now appear allowable. . . . It should be borne in mind that many views have been utterly condemned when later knowledge has only shown us that we were in them only seeing the truth from another side." (Privately, Darwin had said the same thing to Thomson 10 years earlier: "I do not wish to combat the fundamental proposition at all, & only wish to speak against such dogmatism as I find in Tait's writings & not in yours. It appears to me that we know far too little as yet to be sure that we may not have overlooked some important point.")

Wise words, but Kelvin did not like to heed them. The following year, in his last formal pronouncement on the subject, he fulfilled a longstanding promise to Stokes by delivering a lecture, "The Age of the Earth as an Abode Fitted for Life," to the Victoria Institute in London. He repeated his by now tired old disparagement of the geological uniformitarians, who really no longer existed. He admitted the tidal argument was probably not helpful but restated his figure of 100 million years for the age of both the sun and the earth. No real progress had been made in understanding the heat of the sun, but as the century drew to an end, radioactivity had entered the world's laboratories as a mysterious physical phenomenon awaiting experimental scrutiny and theoretical explanation. What it was no one then knew. But clearly there were things in the world of physics that went beyond the limits of established knowledge. Kelvin mentioned none of this and repeated his offer of 100 million years, no more.

He sent a copy of his lecture to Archibald Geikie, who sent thanks for this "latest blast of the anti-geological trumpet." But he was no longer willing simply to cave in to Kelvin's strictures. "The geological & biological arguments for a longer period than you would allow seem to me so strong that I do not see how they are to be reconciled with the physical demands." Geikie did not know how to get around the physics, but he was sure there must be a way.

***

Meanwhile the early promise of the vortex atom began to fade in the light of both mathematical and physical problems. Maxwell had written to Tait in 1867 endorsing what he called "worbles" (a play on the German *Wirbel*, presumably, not some obscure Scotticism), and he wrote approvingly of them in 1872, in a review for *Nature* of Thomson's collected papers on electromagnetism. He particularly noted that the billiard-ball atoms of simple kinetic theory could not explain spectroscopy: "It would puzzle one of the old-fashioned little round hard molecules to execute vibrations at all. There was no music in those spheres." He praised Thomson as well as Helmholtz for developing the theory of vortex atoms. Even so, he had some concern it was a difficult road that might lead nowhere. "But why does no one else work in the same field? Has the multiplication of symbols put a stop to the development of ideas?"

Equally, however, Maxwell saw the virtues of the coming methods for dealing with electromagnetism. At the 1870 British Association meeting he spoke favorably of helpful mathematical innovations that "can often transform a perplexing expression into another which explains its meaning in more intelligible language," and he cited vectors as a specific example. When he died in 1879 his guiding intelligence was lost. Had he lived, though, it is clear Maxwell would have been a Maxwellian, along-side FitzGerald, Heaviside, and the rest.

In 1882 the subject of the Adams Prize at Cambridge University (named for the mathematician John Couch Adams, who in 1843 had predicted the existence of Neptune from its perturbing influence on other planets) was the interaction of two vortex rings. J. J. Thomson won the prize. "Like most problems in vortex motion," he recalled in his dry but oddly humorous way, it "involved long and complicated mathematical analysis, and took a long time." Few besides William Thomson saw in the increasing complexity and difficulty of vortex analysis the prospect of a universal theory. Most, like the Maxwellians, sought to pare and simplify.

Thomson, on the other hand, continued to develop vortex models despite evident shortcomings. He had realized that the original assumption of vortex ring stability was not quite watertight: "After many years of failure to prove that the motion in the ordinary Helmholtz circular ring is stable, I came to the conclusion that it is essentially unstable, and that its fate must be to become dissipated." The total amount of rotation, as Helmholtz had proved, remains constant, but the rotating fraction of the fluid spins out into ever finer and more filigreed threads, so that the rotating and nonrotating parts of a fluid become ever more minutely intermixed.

That was the end of the vortex atom, but Thomson turned this dis-appointment into a new model of the ether, which he called the vortex sponge. It consisted of a fine-grained admixture of rotating and nonrotating elements. Regarded as a fluid, it could support waves with a form analogous to electromagnetic waves in Maxwell's theory—or approximately so. The vortex sponge was so difficult to analyze that even Thomson could only come up with inexact solutions that he hoped captured the essential physics. Undeterred, he pushed ahead, trying to pin down the exact nature of the little rotating elements in his sponge ether. He took off into realms of fluid behavior that were permissible, under Newtonian mechanics, but so far removed from the world of tangible phenomena that to most of his colleagues it seemed he had lost sight of his goal of constructing an ether model that was comprehensible because it was "mechanical." Rayleigh wrote to a physicist friend: "Sir W. is full of a *froth* theory of the ether! This will lend itself to sarcasm even better than the jelly theory."

In 1889, when Thomson spoke to the Institution of Electrical Engineers in praise of Heaviside's improved theory of telegraphy and practical electromagnetism, he spoke plainly of his unhappiness with the state of affairs as it then stood. "I may add that I have been considering the subject for forty-two years—night and day for forty-two years. I do not mean all of every day and all of every night; I do not mean some of each day and some of each night; but the subject has been on my mind all these years. I have been trying, many days and many nights, to find an explanation, but have not found it." Seven years later, writing again to FitzGerald, nothing much had changed. He could not find a satisfactory ether model, but neither would he accept the bare mathematical formalism of the Maxwellians. "It is mere nihilism, having no part or lot in Natural Philosophy, to be contented with two formulas for energy, electromagnetic and electrostatic, and to be happy with a vector and delighted with a page of symmetrical formulas. . . . I have not had a moment's peace or happiness in respect to electromagnetic theory since Nov. 28 1846 [his early works on analogies between electric fields and elasticity]. All this time I have been liable to fits of ether dipsomania, kept away at intervals only by rigorous abstention of thought on the subject."

Five years later Lord Kelvin took to a desperate assertion: "It has occurred to me that, without contravening anything we know from observation of nature, we may simply deny the scholastic axiom that two portions of matter cannot jointly occupy the same space, and may assert, as an admissible hypothesis, that ether does occupy the same space as ponderable matter." It had always been the goal to find an ether model that explained electromagnetic phenomena in their own right and also explained their interaction with matter. Now Kelvin was saying ether and matter could occupy the same portion of space and know nothing of each other. This was not a suggestion embraced by other physicists.

***

As the end of the century approached, Kelvin's growing isolation was not only intellectual. His brother James, who had been beside him at Glasgow as professor of engineering since the death of Rankine in 1872, died in 1892 at the age of 70. His sister Elizabeth died in 1896, having reached 77 years of age. Of his siblings only his brother Robert in Australia still lived, but they had not seen each other since Robert left Scotland in 1850 and would never do so again. Robert died in 1905. Kelvin had a collection of nieces and nephews. James Thomson Bottomley, son of his long-dead sister Anna, was his assistant in Glasgow, generally lived with him there or at Netherhall, performed experiments for him, and lectured frequently when Kelvin was away on business. Elizabeth had three surviving children, her daughters Elizabeth and Agnes, to whom Kelvin remained close, and a son George who figures little in family tales. (Another son, David Thomson King, had gone into the cabling business and died in a shipwreck in 1875.) Elizabeth edited her mother's memoir of Kelvin and added her own recollections. Agnes wrote her own personal reminiscence.

Friends and colleagues were beginning to disappear too. In 1885 Fleeming Jenkin died at the age of 52 after what should have been minor surgery. Their collaboration on telegraph matters had subsided over the years, but having been professor of engineering at Edinburgh since 1868 Jenkin saw Thomson from time to time and made the acquaintance of the Blackburns at their lonely, lovely house on the Moidart Peninsula, where he took up highland dancing with enthusiasm. More affecting for Kelvin was the death in 1894 of Hermann von Helmholtz, whom he had known and admired since their first meeting in the Rhine valley almost 40 years earlier. Helmholtz, at the age of 72, had decided to visit the 1893 World's Fair in Chicago. His wife, already concerned about his health, went with him, and on their tour they got as far west as Denver, which they found dull and unsophisticated. They preferred the East Coast, Boston in particular, which was "quite English. . . . Some intellectual interest attaches to this city, unlike that awful Chicago," Anna von Helmholtz reported home.

On the steamer returning to Europe, Helmholtz fell badly down a narrow stairway, lost consciousness, and never fully recovered his physical or mental strength. He suffered a stroke the following summer and died a few months later. Scientifically, Helmholtz and Kelvin had much in common. Whereas Kelvin had started as a mathematician and moved toward physics and engineering, Helmholtz had trained as a physician, moved to physics, and taught himself mathematics. Both were versatile and ingenious, makers of instruments and solvers of problems more than philosophers. Both became public figures in their own countries, though Helmholtz was an able administrator while Kelvin remained a free spirit in matters of bureaucracy and organizations. Helmholtz had served as first director of the Physikalische-Technische Reichsanstalt in Berlin—the world's first government-funded laboratory for applied science and technology, a project that Werner Siemens had both pushed for politically and underwritten financially. Kelvin saw the need for such an institute in his own country but, perhaps wearied by his old battles with the Admiralty, had exerted no great effort to bring it about. Not until 1900, due largely to the efforts of Lord Rayleigh, did the British government inaugurate the National Physical Laboratory in the London suburbs.

In Germany Helmholtz was regarded by many younger researchers as a fearsome man of stiff formality. In fact he was rather shy. Kelvin was oblivious to matters of etiquette. (Fanny once had to shush him at dinner with Queen Victoria, when he was about to correct in front of distinguished guests Her Majesty's misstatement on some nautical question.) In Kelvin's unconstrained company it was impossible to indulge in formalities, and Helmholtz gladly let go the attempt. At the Edinburgh British Association meeting in 1892, "Helmholtz and Uncle William were inseparable, and both spoke a good deal in the sections," a niece reported.[5] Helmholtz, who was not as powerful a mathematician as Kelvin, had a stronger and simpler sense of physics. He made great contributions to the physical understanding of hearing and sound, for example, and he never once tried to make an ether model. His death took away one of the few people whose opinions Kelvin could occasionally bring himself to attend to. More characteristic is a tale recounted by Lord Rayleigh's son. Kelvin

---

[5]Lord Rayleigh, a reticent man himself, reported when Helmholtz stayed with him in Cambridge for a couple of days that "there is not very much to be got out of him in conversation" (Strutt, 1968, p. 130).

was visiting the Terling estate "full of indignation" at some new electrolytic theory he had recently heard about. He pounced on a textbook from Rayleigh's library to learn more about this theory but came across some smallish error after a couple of pages and immediately put the book aside: "It is Mayer's old mistake of 1842, and here it is again in 1895!" Then he was persuaded to keep reading anyway and started to see there was something to the idea after all. "He will think before long that he discovered it himself," observed Rayleigh to his son, after Kelvin had left. Rayleigh also remarked on the difficulty of getting Kelvin to concentrate on some argument, even to the extent of reading a page of a paper: "The first line would send him off on some train of thought of his own, and his eye would wander from the printed page."

Throughout his long travails over ether theories Kelvin had corresponded with Stokes on the finer points of fluid mechanics. He remarked in passing during the Baltimore lectures how "I always consulted my great authority, Stokes, whenever I got a chance." Even so, it often took repetitious explanations from Stokes, a patient and long-suffering man, to bring a point finally to Kelvin's full attention. When they talked in person at Cambridge, J. J. Thomson recalled, "Stokes would remain silent until Kelvin seemed at any rate to pause. On the other hand, when Stokes was speaking, Kelvin would butt in after almost every sentence with some idea which had just occurred to him, and which he could not suppress." On just one occasion, when Kelvin was talking wildly about atoms, Stokes got his dander up enough to resist: "He was so much in earnest that Kelvin for once could not get a word in edgeways: as soon as he started to speak, Stokes raised his hand in a solemn way and, as it were, pushed Kelvin back into his seat."

***

If physics, or rather natural philosophy, represented by this time a source of frustration and even failure to Kelvin, he enjoyed compensations in the form of his reputation and public demand for his pronouncements. In 1897 the British Association met again in Canada, this time in Toronto. The visit of the noted savants, Kelvin prominent among them, was greeted by banner headlines and extravagant prose on the front page of the local newspaper. "The Men of Science Arriving," the Toronto *Globe* informed its readers on Monday, August 16, assuring them on Wednesday that the city was "Ready for the Men of Science." When, the next day, the Men of Science arrived, they saw news of their meeting blazoned across the entire front page of the *Globe*, with woodcuts depicting the university buildings and the emblems of the British Association, along with reports of cordial welcoming speeches from the mayor and others, which drew hearty thanks from the distinguished visitors.

Kelvin featured strongly in reports from the Friday sessions, for which the *Globe*'s valiant but struggling headline writer came up with the banner "A Day of Good Things: Extremely Interesting Proceedings at the Meeting of the British Association." The question of large-scale power production and consumption had only lately impinged on public consciousness, as the coal-based economy went from strength to strength and, crucially, electricity generation and distribution began to blossom, bringing what had been industrial matters into the domestic realm. Kelvin spoke about the world's supply of coal and, more provocatively, about the supply of oxygen in the atmosphere, which the burning of coal used up. He offered an ingenious calculation: If all atmospheric oxygen came up originally from the respiration of plants, and if all ancient plants decayed and turned into fuel of one sort or another, then he could show that the ultimate limit on terrestrial power would come not from running out of coal but from running out of oxygen to combust it with. This is not at all true, in fact, since only a tiny fraction of vegetation turns into coal, but his fearlessness in tackling enormous questions by means of a few simple scientific assumptions showed he had lost none of his bravado.

Kelvin had come to Toronto after visiting Niagara, where he had official business as a consultant on continuing efforts to generate electricity from the power of the falls. In 1890 he had been invited by an American consortium to serve as chairman of an international commission to study the feasibility of electricity generation from Niagara Falls, an idea first seriously proposed by William Siemens in 1887. As early as 1879, however, Kelvin had spoken to a British parliamentary committee on the advantages of electricity over gas for lighting and industrial purposes and said then that he "believed the Falls of Niagara would in the future be used for the production of light and mechanical power over a large area of North America." By 1893 the international commission had chosen a design offered by Westinghouse, and two years later the first power plant came into operation. When Kelvin visited in 1897, two generators were running, the nearby town of Niagara Falls was lit by electricity, and a number of industries, notably Union Carbide and American Cyanamide, had set up plants to take advantage of the newly abundant electric power.

Impressed by the scale of activity, Kelvin looked grandiosely ahead. In interviews with local reporters he talked of a time "when the whole water from Lake Erie will find its way to the lower level of Lake Ontario through machinery, doing more good for the world than even that great scene which we now possess in contemplation of the splendid scene which we have before us in the waterfall of Niagara. . . . I do not hope that our children's children will ever see the Niagara cataract." He repeated his prediction to a Toronto reporter: "As the demand goes on increasing, so the amount of horse-power developed will increase, until the whole water power of Niagara will be used for doing mechanical work."

Kelvin's attitude toward nature was a little inconsistent. He may have been happy to see Niagara Falls vanish, but he loathed motor cars and voted in the House of Lords for a bill restricting the use of cars because he didn't want to see the pristine Scottish landscape ruined. In London on one occasion his niece Agnes took him to an art exhibition and showed him a romantic painting of Glen Sannox, with mist adorning the mountains. Kelvin, she reported, was annoyed that the artist had not waited until the mist cleared because it obscured a notable geological feature.

Kelvin envisaged a time when electricity from Niagara would travel farther afield, but the transmission of electric power was a controversial scientific question. Any wire, even the best copper, had some resistance to an electric current, which generated heat and wasted power. As Joule had first shown more than half a century ago, that power loss went in proportion to the resistance times the square of the current. Higher voltages and lower currents meant more efficient transmission of power, but high voltages presented dangers and practical difficulties. Speaking to the British Parliament two decades earlier, Kelvin had talked of thick copper conductors in the form of tubes, with cooling water running down them, able to transmit power for hundreds of miles. Since then, however, he had thought more closely about the economics of power loss and formulated what has sometimes been called Kelvin's law of power transmission:

The "most economical size of the copper conductor for the electric transmission of energy . . . would be found by comparing the annual interest on the money value of the copper with the money value of the energy lost in it annually in the heat generated in it by the electric current." Guided by this principle he now estimated that electricity could travel up to 300 miles with acceptable power loss, if 20,000 volts were used, but he recoiled from such large potentials. "I would not advise manufacturers to settle farther than ten miles from Niagara Falls," he told a reporter for the *Buffalo Express*. Speaking just the next day to a Toronto reporter, he was a little more generous: He thought power might usefully travel 20 or even 30 miles from the falls.

Kelvin gave out his opinions freely during his visit and spoke easily to journalists. "A gentleman of exceedingly pleasant manners" with an "amiability of disposition," wrote one. "He is a remarkable example of a great man whose native character has remained unchanged despite . . . the elevation to a lofty social position." The Buffalo reporter reported with dry humor the contrast between this member of the British nobility and "a plain American citizen," the chief Niagara engineer Coleman Sellers. "The man with the title looks and acts like a plain citizen. The plain citizen looks and acts as if he were the autocrat of all the Russias. Lord Kelvin is approachable and affable." During the interview Sellers kept interrupting to say that Lord Kelvin was hungry and wished to go to dinner. Kelvin at first smiled indulgently and carried on talking. On the second occasion he "looked annoyed. He looked at Coleman Sellers for a moment, then turned to the reporter again." Finally Sellers dragged him away by the arm.

"Lord Kelvin is short and thin and gray and plain," the Buffalo reporter told his readers. "He is very lame, but there is something in his appearance that does not belie his youthful record as an athlete, when he won the Silver Sculls at Cambridge. . . . His head is large and his gray beard is thin and straggling. His baldness runs to the crown and his immense forehead is smooth and polished as a roc's egg. . . . His small blue eyes are kindly and genial in their expression. His clothes fit badly, after the English fashion."

Kelvin had some years earlier allied himself with the losing side of the peculiarly fierce controversy that raged, around 1890, over the relative merits of direct current and alternating current systems for power transmission. The debate took off when the eccentric Serbian engineer Nikola Tesla, brought over from the Edison laboratories in Paris to work with Thomas Edison himself at Menlo Park, had given a presentation in 1888 describing the virtues and promise of his "polyphase generator" for producing alternating currents at high voltage. George Westinghouse, a railway entrepreneur, heard the talk and promptly hired Tesla away from Edison to design power transmission systems using the new technology. Edison, a devout direct current man, started a campaign assailing the dangers of alternating current, the height of which involved proposing that a man sent to the electric chair should be said to have been "westinghoused" or "consigned to the westinghouse."

The suggestion that alternating current is fearsomely dangerous while direct current is pleasantly safe seems absurd, when a few thousand volts of either will satisfactorily kill anyone. There is a smidgen of reason at the bottom of the argument. A given alternating voltage will transmit less power down a wire than the same direct voltage, because the alternating current oscillates from some peak value, through zero, to the same peak value in the opposite direction, and so on. The numerical factor of importance here is the square root of 2, approximately 1.414. A somewhat higher voltage is needed for an alternating current system to achieve the same efficiency in transmission as a direct current system. And higher voltage means greater danger.

Although the Niagara commission that Kelvin chaired endorsed Westinghouse's alternating current system for the falls, Kelvin himself never abandoned his preference for direct current. It fit in with his idealized law about the economics of power transmission. Having calculated costs and efficiencies, he insisted it made sense to choose the better solution. This argument, however, neglected utterly the relative ease (and therefore lower cost) of producing large alternating voltages and reducing them to safe values for domestic use. The transformer—two coils of wire wrapped around a common iron core, the same device Faraday had used in 1831 to demonstrate simple electromagnetic induction—could step alternating voltages up and down with no fuss. A changing electromagnetic field induces a current in a wire; a static one does not, and this is why direct current transformers do not exist. For efficient power transmission, high voltages are essential, and direct current systems cannot do the job.

Kelvin understood all these matters perfectly well, yet insisted that the quantitative efficiency argument ought to trump all other concerns. It is an extreme example of the kind of tunnel vision he increasingly showed on technological as well as scientific matters. Having once grasped a certain point of view and justified it with an appropriately quantitative analysis, he seemed impervious to all other considerations. It was the nearest thing he possessed to a philosophy of science. In one of his most famous and memorable remarks about the power of the scientific method, he declared: "I often say that when you can measure what you are speaking about, and express it in numbers, you know something about it; but when you cannot measure, when you cannot express it in numbers, your knowledge is of a meagre and unsatisfactory kind: it may be the beginning of knowledge, but you have scarcely, in your thoughts, advanced to the stage of *science*, whatever the matter may be."

This was the wellspring of his attitude in all departments of science—indeed in all useful kinds of rational thought. It was why he favored the precise arguments of physics limiting the lifetimes of the earth and the sun over the woolly speculations of geologists and biologists. It was why, in a broader sense, he insisted on literally mechanical models of ether and atoms and would not succumb to the nihilism of the Maxwellians, who were willing to accept a mathematical structure based on no tangible model. Where in that were physical notions he could touch, assess, and quantify?

Kelvin valued mathematics not for any formal elegance or alleged esthetic qualities but because it enabled him to think and reason with confidence. Mathematics "is merely the etherealisation of common sense," he told the citizens of Birmingham in an address at the town hall there in 1883. In a similar vein he praised empirical science, bolstered by mathematical argument, over abstract theorizing. Opening new laboratories at University College in Bangor, Wales, in 1885, he said forcefully: "There is one thing I feel strongly in respect to investigation in physical or chemical laboratories—it leaves no room for shady, doubtful distinctions between truth, half-truth, whole falsehood. In the laboratory everything is found either true or not true. Every result is *true*. Nothing not proved true is a *result*;—there is no such thing as doubtfulness." And further, if merely measuring things seemed like dull work, he declared that when investigation is done with a purpose, "measurement itself becomes an object to inspire the worker with interested ardour. Dulness [sic] does not exist in science."

Reason flowing from quantitative knowledge suffused Kelvin's attitude toward life in general. When he visited Stokes in Cambridge, the two elderly men used any opportunity to engage in playfully intense scientific analysis. "For instance, the eggs were always boiled in an egg-boiler on the table, and Lord Kelvin would wish to boil them by mathematical rule and economy of fuel, with preliminary measurements by the millimetre scale, and so on," the physicist Joseph Larmor recalled. In more serious matters, Kelvin's insistence on simplicity of thought came across as naivete. At an 1887 dinner commemorating the jubilee of Britain's first telegraph line (from Euston to Camden, in London), he confidently informed the assembled guests that instant telegraphic communication between London and Dublin demonstrated "the utter scientific absurdity of any sentimental need for a separate parliament in Ireland. [This] seems to me a great contribution of science to the political welfare of the world." Oceanic telegraphy, despite the expectations of editorial writers for the *Times* of London and New York, had not banished international tension and war; it was hardly likely to quell the ancient disagreements between Ireland and England. But these disputes were, to Kelvin, irrational to begin with; they ought not, therefore, to exist in the first place, and so he imagined it would be an easy matter to dispose of them, if only people would talk plainly and stick to the facts.

***

Plain talk had not dissipated Kelvin's scientific disagreements. Some of the novel discoveries of the 1890s were more revolutionary than others. When Wilhelm Röntgen announced in 1895 the discovery of a penetrating kind of radiation that could pass through flesh and create an image of bones, Kelvin hoped at first it would bear out an aspect of his mechanical conception of the ether. In Maxwell's theory, electromagnetic radiation consisted strictly of transverse waves, or side-to-side oscillations, as of a violin string. But if the ether bearing electromagnetic radiation were some kind of solid-liquid-jelly-sponge affair, then in general it would be expected to support longitudinal waves, or alternations of compression and rarefaction along the direction of radiation, rather than perpendicular to it. Perhaps the Röntgen waves were longitudinal ether oscillations. Kelvin had held out ever since the Baltimore lectures that such waves, absent from Maxwell's theory, could not be ruled out and ought to be sought experimentally.

Quickly, though, it became apparent that X rays, as the new radiation became known, were ordinary electromagnetic waves beyond the ultraviolet, with higher energy and shorter wavelength than had been encountered before. Kelvin (whose confidence in his numerous ether models was wavering anyway) accepted this conclusion with little hesitation.

Electrons, J. J. Thomson's discovery of 1897, also fitted in at first with Kelvin's view of the physical world. That there might exist tiny objects carrying electric charge was no great surprise. What exactly they were remained mysterious. Inevitably, Kelvin's thoughts took him in his own direction. In a 1902 essay eccentrically entitled "Aepinus Atomized," he tried to revive, in modern clothing, a mid-18th-century theory of electricity due to Franz Ulrich Theodor Aepinus of Rostock, in what is now northern Germany. Aepinus proposed that there was a single kind of electric substance, of which an excess represented a positive charge and a deficit a negative charge. Kelvin proposed that "Aepinus' fluid consists of exceedingly minute equal and similar atoms, which I call electrions, much smaller than the atoms of ponderable matter." These electrions were not simply free particles reacting to electric forces alone. They interacted with atoms of ordinary matter through a complicated force law. A neutral atom contained a certain number of electrions; with too few or too many, an atom acquired an overall electric charge. The forces controlling the loss and reacquisition of electrions by atoms were meant, Kelvin explained, to account for the variety of chemical properties displayed by the atoms of the (newly formulated) periodic table. And he hoped too he could explain the geometry of crystal lattices and the regular arrangement of atoms in solids through this single force law.

This proposal, complex yet sketchy, was entirely characteristic of Kelvin's thinking. Beginning with a handful of simple assumptions (electrions, atoms, a force law between them), it quickly piled up mathematical complications, without yielding specific predictions. It was, to use another historical term, a Boscovichian theory, harking back to the ideas of Roger Boscovich, an 18th-century Serbo-Croat priest who had proposed a model in which both attractive and repulsive forces operated between atoms, depending on the distance between them. By adjusting the force appropriately, Boscovich hoped to explain chemical reactions, absorption, adhesion, and so on. Not too many years earlier Kelvin had hoped his vortex atoms, in which the only ingredients were a suitable ether and Newtonian mechanics, would provide a universal explanation of matter. But that project had failed, and Kelvin reached back to the 18th century for ideas that had been proposed as philosophical speculations but never developed into mathematical theory. The weakness of Boscovichian atomic theory, from the modern perspective, is that the force law between atoms is assumed into existence and made as complicated as it needed to be in order to explain whatever the theory was meant to explain. But this was Kelvin's way of thinking, taken to a final extreme. Forces and particles he was happy with, though the particles were of a mysterious nature and the forces complex and unproven. On the other hand, he still would not accept the disembodied field theory of the Maxwellians.

***

Kelvin celebrated his 75th birthday in 1899 and at last retired from his position as professor of natural philosophy[6] at Glasgow University. Except for five years at Cambridge and a few months in Paris, he had belonged to the university, student and teacher, for 67 years, going back to the days when, as an eight-year-old boy, he sat in on his father's math-

---

[6]In the matter of terminology Kelvin had an ally in Heaviside: "For my part I always admired the old-fashioned term 'natural philosopher.' It was so dignified, and raised up visions of the portraits of Count Rumford, Young, Herschel, Sir H. Davy, &c., usually highly respectable-looking elderly gentlemen, with very large bald heads, and much wrapped up about the throats, sitting in their studies pondering calmly over the secrets of nature revealed to them by their experiments. There are no natural philosophers now-a-days" (Heaviside, 1951, vol. 1, p. 5).

ematics lectures. He insisted on maintaining a connection and on his retirement was duly enrolled as a "research student" of the university. But he lectured no more and was free to spend his summer months in London, where he engaged in scientific or political or commercial business, or in the south of France for his own and his wife's health, returning in the winter to Netherhall. In this peripatetic life he was never without a green notebook, in which he jotted ideas and worked out bits of theory and noted experimental suggestions and corresponded with his fellow natural philosophers as furiously as ever.

Health problems occasionally slowed him down. Late in 1895 his bad leg troubled him, becoming swollen and keeping him in the house well into the following year. Forced into physical inactivity, he redoubled his demands of everyone else. Propped up in bed, wearing a bright red jacket draped over with a blue and white quilt, and with papers and notebooks scattered all about him, he issued notes and instructions and urgent demands. His recovery was interrupted by a bout of pleurisy, but he was well again by the time of his jubilee that summer. Toward the end of that year he suffered for the first time an acute attack of facial pain, diagnosed as inflammation of the fifth nerve. Such attacks, lasting a few days and disappearing as abruptly as they came, troubled him for the rest of his life. "A horrid demon of the No. 5 nerve," he sometimes called it, and it caused him to cancel lectures and miss dinner engagements. But he rebelled at being cooped up, and when the pain departed he resumed his robust and enthusiastic habits and traveled and lectured and socialized as much as ever.

In 1902 he and Lady Kelvin embarked on another trip to the United States, his fourth and last. It was a triumphal tour. Arriving in New York on April 19, he immediately went to Columbia University to attend the installation of a new president. He mingled with a distinguished crowd, including President Roosevelt and Andrew Carnegie. Kelvin, "looking very venerable, limping, and wearing a large monocle," came in on the arm of the governor of New York. He spoke to a reporter for the *New York Times*, full of enthusiasm for Marconi's wireless telegraph but dismissive of heavier-than-air flight by dirigible. Two evenings later the American Institute of Electrical Engineers gave a reception for him, where he met Edison, Tesla, Westinghouse, and many other luminaries of the burgeoning U.S. electrical industry. Traveling down to Washington, D.C., he stayed with Mr. and Mrs. George Westinghouse, who gave the noble couple a grand dinner, with numerous distinguished guests, both political and academic. On April 24 Kelvin testified before a congressional committee in favor of a metrication bill. He wished his own country would take the initiative, he said, but added that he would be glad to see the United States in the vanguard, so long as the goal was accomplished.

Then it was on to Niagara, to observe the great progress there not only in the generation of electricity but in the accompanying rise of power-hungry industries. He saw a plant that made nitric acid from atmospheric nitrogen and inspected an electric furnace. "There is practically no limit to the temperature the electric furnace can get," he mused out loud. "It ought to be easy to manufacture the diamond" from ordinary carbon. This was reported in both the *Times* of April 28 and the Rochester *Democrat and Chronicle* of April 29, but when a Rochester reporter asked him about it the following day he responded with a smile, "Oh, there is nothing practical in that."

In Rochester his host was George Eastman, founder of Kodak, of which Kelvin had been named a director (he was also vice-chairman of the Kodak Company of London). He talked of the scientific basis of photography and also of its scientific potential, particularly in astronomy. The purpose of his visit to Rochester, indeed of this whole trip to the United States, was to inspect the new camera works and offer his technical advice on the many new processes and technologies operating there. But his opinion on science past, present, and future was always in demand. The Rochester reporter found him altogether down to earth. "Before Lord Kelvin has been conversing five minutes, the visitor is beguiled into thinking that he has known Lord Kelvin as long as he has known anybody. His courteous, unaffected manner puts one at ease at once."

Kelvin held forth on electric power transmission ("a success . . . no longer an experiment") and on wireless telegraphy ("I think Marconi has got it"), but soon the conversation turned to the possibility of power generation from Rochester's river, the Genesee. Now Kelvin began to pepper his interviewer with questions about the size and course of the river, the falls running to the lake, and the numerous tributary streams. A detailed map and an engineer's survey were fetched. Before long Kelvin had picked out a plausible location for a hydroelectric plant and assured the reporter that power could satisfactorily be carried from there to the city itself, but he cautioned that these were preliminary ideas. His combination of unfettered thinking with insistence on precision impressed the reporter. "He is too exact in his methods to announce a conclusion prematurely, although he is wonderfully quick to grasp details and is as keen for information on the subject like the water power of the Genesee as a newspaper reporter."

Though the Rochester journalist could not know it, this was his attitude to ether models and atomic theories too. Once the concept was set down, it was a matter of working out all the details, and unless all the details could be got right at once, no solution was yet acceptable.

There followed Kelvin's raucous reception at the university, where an attack of the number five demon kept his remarks briefer than he had intended, except that once he warmed up to his task he carried on anyway. He described to the students his own fortunate life, as the child and product of universities, who had spent his whole existence in or around colleges. "Both as a student and as a professor, I love the college atmosphere." He urged his listeners to "acquire knowledge, and make use of every hour. As we grow to advanced age, we can look back upon the pictures formed in the college days. Fill your minds with these pictures. They are pleasant to bring to your recollection as you grow to age."

Mellow thoughts such as these were probably lost on his youthful audience, but Kelvin's nostalgia was, typically, of a pleasurable kind. Fond recollection, not regret—but then he was a man who had never wasted a moment, so that even what he came to regard as failure might seem time well spent, in a worthy endeavor.

On a tour of the university Kelvin eagerly looked over the physics and chemistry laboratories but insisted also on seeing the boiler rooms and the heating and ventilation systems, which he "commended . . . in high terms." A day or two later he left Rochester for a reception at Cornell, then continued on to Yale to take an honorary degree.

Lord and Lady Kelvin traveled by private railroad carriage, furnished by George Eastman and filled by his wife with orchids and violets from their conservatory. The eager reporter from the Rochester *Democrat and*

Chronicle joined the train for the first part of the journey, anxious to hear more about Kelvin's ideas for electricity generation from the Genesee River. They were grand plans indeed, if the newspaper account is to be believed. This "greatest of living scientists, whose simple dictum is law in matters electrical, whose achievements on physical lines have deservedly carried him to the highest rank of England's nobility . . . presented Rochester with the formula by which its greatness is to be achieved and its dream of practically unlimited power for industrial purposes realized."

The reporter noted that Kelvin went against standard practice by proposing direct rather than alternating currents for transmission. He explained that with the multiphase system pioneered by Tesla and Westinghouse, voltages on neighboring high-tension cables could differ enormously if their currents were out of phase. "A cat may make a connection across the wires, and in its death disable the system," he said. Kelvin proposed instead a system of 20 direct current generators at 2,000 volts each, connected in series to produce 40,000 volts, which would satisfactorily transmit power up to 100 miles with acceptable losses. As Kelvin elaborated on his plan, the reporter "covertly pinched himself, to make sure that he was not dreaming."

After an hour of these thrilling revelations, Kelvin talked briefly of his fascination with the camera works he had seen in Rochester. As the reporter made to leave, Kelvin immediately took up page proofs that needed his urgent attention. They represented a version, to be published finally in book form, of the Baltimore lectures he had given almost 20 years earlier. "I work on these proofs whenever I get even fifteen minutes' time," Kelvin told the starry-eyed reporter.

Two days later the Rochester newspaper ran another story, with careful comments from "one gentleman, who is in a position to speak with authority," scotching rumors that a company had already been formed to put Kelvin's plan into effect. Discussions would take place, it was said, but a good deal of preliminary investigation would be needed. It hardly needs saying that Kelvin's extravagant plan for a system of direct current generators and transmission lines spreading across the Genesee valley to Rochester never came to pass.

***

In March 1903 Pierre Curie and Albert Laborde announced that the radioactive decay of a radium salt released heat. This was seven years after Becquerel's discovery of radioactivity, and the transmutation of an atom of one element into an atom of another was an established fact. The nature of the transmutation was baffling, however, not least because the nature of atoms remained mysterious.

Curie and Laborde's finding caused a stir in the physics world. At the British Association meeting in September of that year there was a demonstration of a little piece of radium making the mercury rise in an ordinary thermometer. One scientist commented that this phenomenon "can barely be distinguished from the discovery of perpetual motion, which it is an axiom of science to call impossible, [and] has left every chemist and physicist in a state of bewilderment." It seemed like energy from nowhere, appearing out of an otherwise inert mineral.

Whatever the explanation, this new source of energy had implications for the age of the earth and the sun. In a very short note published in *Nature* on July 9, 1903, an amateur astronomer by the name of William Wilson calculated that the entire output of energy by the sun could be accounted for if the sun contained 3.6 grams of radium in every cubic meter. This simplistic but telling bit of arithmetic attracted little notice. But at the end of September, shortly after the BA meeting, George Darwin weighed in. He cited recent measurements by the young New Zealand physicist Ernest Rutherford, then at McGill University in Montreal, of the heat released by radium, and reckoned that if the sun as a whole were made of some such material, its age could be many times greater than the age calculated long ago by Kelvin. This was all speculative, he admitted, but "knowing, as we now do, that an atom of matter is capable of containing an enormous store of energy in itself, I think we have no right to assume that the sun is incapable of liberating atomic energy, to a degree at least comparable to that which it would do if made of radium." Unlike the commentator at the BA meeting, Darwin did not hesitate to conclude that any energy released by radioactive atoms must have been stored up somehow beforehand. Whatever an atom might be, it could not create energy from nothing.

A week later J. Joly of Dublin applied the same thinking to the age of the earth. Kelvin had always assumed that the planet was a passively cooling body, slowly losing whatever original heat it had possessed. But if radioactive minerals constantly generated interior heat, the old argument fell apart. Joly concluded confusedly that "the hundred million years which the doctrine of uniformity requires may, in fact, yet be gladly accepted by the physicist." The uniformitarians, of course, had originally assumed unlimited time; it was the physicists who had beaten the geologists down to 100 million years.

Kelvin responded at the 1903 BA meeting with off-the-cuff remarks, later written up in more elaborate form, to the effect that he thought radioactive heat must come not from within the decaying atoms but ultimately from the surrounding ether. Atoms took in energy, stored it up temporarily, and released it when they decayed. The heat, he said, comes "from without the atoms, where it exists in a form we have not yet found the means of detecting." In papers published in the next couple of years, and in presentations at succeeding BA meetings, he reached back again to the old ideas of Aepinus and Boscovich to devise mechanical models of the atoms, with components held in place by forces whose form he concocted for the purpose. These atoms could be, so to speak, spring loaded by absorbing energy from the ether. Then they would burst apart, shooting "electrions" and other fragments into space, possibly at speeds exceeding the speed of light. These models were inventive and ingenious, as ever, but increasingly detached from the developing ideas about atoms. Rutherford, along with other young men such as Frederick Soddy, were measuring precisely the heat released in radioactive transformations and determining the identity of the particles released. Lord Rayleigh's son, Robert Strutt, who became the fourth Lord Rayleigh in 1919 when his father died, was among those who established the possibility of dating rocks by assaying the radioactive decay products they contained. Soon, ages of hundreds of millions of years were spoken of for perfectly ordinary minerals in the earth's crusts.

One old combatant who did not speak in this final round of the debate over the age of the earth was P. G. Tait, who had died in 1901. His last year was miserable. His third son, Freddie, had taken up golf as soon as he could walk and became one of the great players of his day. He won the British amateur championship twice, in 1896 and 1898. Tait took fierce pride in his son's achievements. One of Tait's lasting contributions to physics was his 1891 proof that backspin on a golf ball, via the agency of a fluid mechanical phenomenon called the Magnus effect, imparts lift and thus allows the ball to fly much farther than if it were not spinning. Tait's ferocity in debates over physics was matched by an equally intense patriotism. When the Boer War erupted in 1899 between British forces and rebellious Afrikaaners, Tait rejoiced that his son signed up and went to South Africa to fight. Young Tait was brave in a way that tends to excite mockery today—a good-hearted, good-looking, cheery sporting fellow, by no means an intellectual, sailing out eagerly and unquestioningly to the fringe of the empire to defend British pride and power. His father doted on him. Freddie Tait, a lieutenant in the Black Watch, shipped out on October 24, 1899. On December 11 he was wounded by a bullet in his leg at the Battle of Magersfontein. By the end of the month he had recovered. Early in February he was sent to Koodoosberg and on the seventh he was hit by a bullet to the chest and died where he fell.

Tait, who had lost never an atom of self-confidence in all his long-running scientific controversies, was vanquished utterly by the death of his beloved son. He continued to teach, but listlessly, with none of his former vigor. He went to St. Andrews as usual in the summer of 1900, but the golf course now held only painful memories. He retired from teaching that winter, simply leaving one day and never coming back to the lecture room. John Low, a golfing friend who published a life of Freddie Tait in 1900, recorded that "the Professor seemed very depressed as though afraid to enter into any conversation which might become reminiscent of the golf which had Freddie for its central figure. . . . I do not think that he ever got back into his true gait after Freddie's death; the light seemed to have left the eyes which in repose often wore an expression of weariness."

The following summer he had the use of a friend's house and sat in the garden clutching his copy of Low's book, reading over and over again the accounts of Freddie's tournaments and victories. He died on July 4 at the age of 70. In his obituary Kelvin recalled Tait's "rough gaiety . . . cheerful humour . . . this was a large factor in the success of our alliance for heavy work, in which we persevered for eighteen years. . . . We had keen differences (much more frequent agreements) on every conceivable subject,—quaternions, energy, the daily news, politics, *quicquid agunt homines*, etc., etc. We never agreed to differ, always fought it out. But it was almost as great a pleasure to fight with Tait as to agree with him. His death is a loss to me which cannot, as long as I live, be replaced."

Barely 18 months later, on February 1, 1903, Kelvin had to contend with the death of his oldest scientific friend and confidante, George Gabriel Stokes, his lifelong adviser and consultant in mathematical physics. Stokes was by then 83 years old and died quietly after a peaceful retirement. He had lived in Cambridge with his daughter's family since the death of his wife in 1899. Stokes was a firm, taciturn man. He had served for some years in Parliament, under the old system by which Cambridge University nominated a member. Isaac Newton had been M. P. likewise in earlier times, and it is said he spoke only once, to ask for a window to be shut. Stokes surpassed Newton, dutifully attending every session and saying never a single word.

J. J. Thomson recalled that Stokes had great powers of analysis but was "exceedingly cautious about coming to a conclusion." The voluminous correspondence between Kelvin and Stokes—some 650 letters spanning 55 years—is almost wholly technical. Occasionally there are personal remarks in the gruff style of Victorian men, such as when each announces to the other that he is about to be married. "My principal intelligence must belong to the non-scientific head which is that I am engaged to be married to Miss Robinson daughter of Dr Robinson," Stokes informed William Thomson in 1856, in an unpunctuated rush.

In 1879 Stokes assembled a selection of his papers for publication in book form and consulted Thomson on many points he wished to revise or refine. In one letter, after describing some changes he proposed in order to make the allotment of credit for original ideas scrupulously fair and unambiguous, he signed off by remarking that "it is curious how these things bring back our work together in 1847." In their entire correspondence, this small plangent sentence is the closest either Stokes or Thomson came to a confession of intimacy. Kelvin's obituary notice of Stokes was conventional enough, describing his numerous original researches and the help he freely offered over the years to other researchers, including himself. But Arthur Schuster, a young physicist, saw a deeper response at Stokes's funeral: "I shall always remember Lord Kelvin, as he stood at the open grave, almost overcome by his emotion, saying in a low voice: 'Stokes is gone, and I shall never return to Cambridge again.'"

***

Exactly 20 years after he had delivered his celebrated series of lectures in Baltimore, Kelvin published in 1904 the version he had been tinkering with off and on ever since. In 1900 Max Planck's first intimation of quantum theory had appeared; in 1905 Einstein's four remarkable papers on quantum theory and relativity ushered in a new era of physics. The appearance, between those milestones, of Kelvin's fantastically elaborated mechanical models of the ether, supplemented with new materials such as his curious essay "Aepinus Atomized," was a bizarre anachronism. Details had changed to accommodate new discoveries, but the intellectual foundation of this project remained what it had been in Baltimore—a mechanical universe, in the old-fashioned Newtonian clockwork sense.

The lectures were politely received. Kelvin was the most publicly known scientist of the day, and everyone liked him, though they might mock his thinking. Commenting in *Nature*, the young physicist Joseph Larmor couched his reservations in carefully respectful terms. Kelvin's "expression of distrust of 'the so-called electro-magnetic theory of light' stands as in the original. . . . In this chain of simple, yet brilliant and attractive ideas, Lord Kelvin has gradually forged a reconciliation between fact and theory that would probably have been received with universal acclaim thirty years ago. Nowadays, as regards most people, the need has ceased to be so strongly felt. . . . [Kelvin] would perhaps say that [Maxwell's electromagnetism] is a successful description rather than an explanation, and he would probably desire to modify the terms of the description in order to bring it closer to the train of dynamical ideas in which he would search for the explanation. And here we are at the parting of the ways."

Confusion over the nature of radioactivity rumbled on. In August 1906 Kelvin wrote to the London *Times* arguing against the idea, by then widely accepted, that radioactive decay involved the transmutation of one element into another. His line of thinking was as much semantic as physical. He proposed that heavier elements were compounds, in a molecular sense, of lighter ones, and split into their various components when they disintegrated. Radium, in other words, was a compound of helium and other lighter elements, not a true element in its own right. Nowadays, knowing that atomic nuclei are built from protons and neutrons, we say that the different elements are all combinations of the same ingredients. One might almost suggest that Kelvin was reaching in this direction, but since no one at that time had any clear idea of what atoms were made of, the debate had no real substance. Other physicists wrote in to disagree. The *Times* itself weighed in with an editorial asking for Kelvin's views to be taken seriously, on account of his great reputation and experience.

But young scientists have little respect for seniority. Frederick Soddy wrote to the paper on August 31, putting the case against Kelvin, and concluding that "it would be a pity if the public were misled into supposing that those who have not worked with radio-active bodies are as entitled to as weighty an opinion as those who have. . . . Atomic disintegration is based on experimental evidence, which even its most hostile opponents are unable to shake or explain in any other way."

Summarizing these inconclusive exchanges a few weeks later in *Nature*, Soddy's veiled tribute to Kelvin came close to condescension. "Whatever opinion may be formed of the merits of the controversy, all must unite on admiration for the boldness with which Lord Kelvin initiated his campaign, and the intellectual keenness with which he conducted, almost single-handed, what appeared to many from the first almost a forlorn hope against the transmutational and evolutionary doctrines framed to account for the properties of radium. The weight of years and the almost unanimous opinion of his younger colleagues against him have not deterred him from leading a lost cause, if not to a victorious termination, at least to one from which no one will grudge him the honours of war."

Even Ernest Rutherford, the great pioneer of radioactivity and atomic theory, who had written to his mother years ago of his admiration for Kelvin, could not help but think of the aging natural philosopher as a child. They had met at a scientific party at Terling, Lord Rayleigh's estate. Rutherford described the proceedings to his wife: "Lord Kelvin has talked radium most of the day, and I admire his confidence in talking about a subject of which he has taken the trouble to learn so little. I showed him and the ladies some experiments this evening, and he was tremendously delighted and has gone to bed happy with a few small phosphorescent things I gave him."

Long afterward, Rutherford recalled Kelvin with fond indulgence. In 1904 he had given a public lecture at the Royal Institution on radioactivity, in which he intended to touch on the question of the earth's age. Kelvin was in the audience. In his much-retold account, Rutherford recalled that "to my relief, Kelvin fell fast asleep, but as I came to the important point, I saw the old bird sit up, open an eye, and cock a baleful glance at me! Then a sudden inspiration came, and I said Lord Kelvin had limited the age of the earth, *provided no new source was discovered*. That prophetic utterance refers to what we are now considering tonight, radium! Behold! the old boy beamed upon me."

EPILOGUE

Lord Kelvin died on December 17, 1907, at Netherhall, in the house he had built some 30 years earlier. That year he had often appeared frailer and lamer to those who knew him well, and he had lost the sight in one eye, probably because of a detached retina. But at other times he was as lively as ever. At the British Association meeting in July at Leicester he had participated in a memorable discussion with Rutherford, Soddy, and others on radioactivity. His views were sharp but hard to fathom. At the 1904 BA meeting he had reportedly acknowledged that radioactive decay released energy locked up within the atom. But in March 1907, in one of his last published papers, he argued again for a sort of rechargeable atom, which took up energy from its surroundings then released it in a sudden burst of disintegration. "Lord Kelvin preferred to regard the atom as a big gun loaded with an explosive shell," reported *Nature* from the BA. "The impossibility of the transmutation of one element into any other he declared to be almost absolutely certain."

Even so, he expressed his opinions with vigor and with his old delight in argument. He had lost none of his simple joy in the enlightenment that only science could provide. After a lecture on recent progress in astronomy, he had spoken up to praise the speaker and his message.

"In proposing a vote of thanks," one observer recalled, "Lord Kelvin burst into a sort of rhapsody, in which, with unaffected enthusiasm, he declared that we had been taken on a journey far more wonderful than that of Aladdin on the enchanted carpet; we had been carried to the remotest stars, and well-nigh round the Universe, and brought back safely to Leicester on the wings of science, and the most marvellous thing about it all was that it was true!"

From Leicester, Lord and Lady Kelvin went to the south of France for their habitual month of rest, but after the long journey home Lady Kelvin collapsed at Netherhall on September 15, apparently from a stroke. Kelvin stayed with her, as she began to improve. He himself stayed mostly well, but toward the end of November he caught a chill, took to his bed, seemed to be on the mend, but by the middle of December he relapsed and died quietly a few days later. He was 83.

Following memorial services at St. Columba's Episcopal Church in Largs and at Glasgow University's Bute Hall, Kelvin's coffin went by train to London, arriving at Westminster Abbey on the morning of December 22. Lady Kelvin, still unwell, remained in Scotland. Kelvin's funeral at Westminster on the 23rd, a dark, cold, midwinter day, manifested that combination of pomp and humility characteristic of high church Anglicanism. Distinguished scientists joined with representatives of the Crown and the government, foreign ambassadors, and other dignitaries. The Duchess of Argyll, a personal friend, was there; so was the First Lord of the Admiralty, and Kelvin's old nautical ally Admiral Sir John Fisher. Scientific societies from around the world sent representatives, as did countless universities and scientific and technological associations from Great Britain and elsewhere. The pall bearers included Lord Rayleigh, the geologist Archibald Geikie, and George Darwin (Sir George by now), who had been Kelvin's closest scientific confidante of the rising generation.

Into this solemn and magnificent assembly in Westminster Abbey Kelvin's coffin was borne to the accompaniment of Purcell's Music for Queen Mary and Chopin's Funeral March. After the hymn "Brief Life Is Here Our Portion" came readings from Psalm 90 ("The days of our years are threescore years and ten; and if by reason of strength they be fourscore years, yet is their strength labour and sorrow; for it is soon cut off, and we fly away") and 1 Cor. xv ("O death, where is thy sting? O grave, where is thy victory?").

Kelvin was interred beside Isaac Newton, beneath a plain slab inscribed "William Thomson, Lord Kelvin, 1824-1907." As he was laid to rest the dean of the Abbey intoned the old words of the burial service, singularly inappropriate in this case: "Man that is born of woman hath but a short time to live, and is full of misery."

***

In writing about scientists of previous centuries and the mysteries they spent their lives trying to resolve, I have often wanted to speak back into the past, or bring these scientists forward to the present day, to show them how it all worked out. Of course, science solves one set of problems only to present new ones for the next generation, but I can't help thinking that pioneers such as Faraday and Maxwell, Joule and Clausius, and especially such unrecognized explorers as Carnot would have been gratified to see how their difficult endeavors and half-perceived ideas laid the foundation for what we now understand.

In Kelvin's case I am not so sure. His early thinking on thermodynamics and electromagnetism was profound and influential, yet as he grew old he grew unhappy with the developments that others, working from his clues, had pursued. J. J. Thomson recalled a conversation in which Kelvin said he thought the most valuable work of his career had been the long struggle to limit the ages of the earth and the sun. Few modern scientists would agree. The Baltimore lectures, representing the summation of Kelvin's mechanical picture of the universe, now seem impossibly antique.

Obituary notices of Kelvin reflected this delicate judgment. In newspapers around the world he was "one of the greatest scientists and ablest men of the age," "the foremost scientist of Great Britain," and "the most distinguished British man of science." His scientific prowess was noted, but the popular press mostly reminded readers of his contributions to electrical technology, especially in the Atlantic cable and its instrumentation. Such achievements characterized the astonishing transformation of the world by science and technology in the course of the 19th century, a transformation of which Kelvin was a prominent symbol. It was not for science alone that Kelvin became famous but because of the way he brought science into ordinary life.

Scientific assessments of his life, on the other hand, emphasized his contributions to the understanding of energy and electromagnetism as the core of his legacy, along with his essential work in establishing scientific units for the new sciences, and a great deal of valuable investigation of fluid mechanics, elasticity of matter, and other more mundane but important works. Of his later scientific activity, S. P. Thompson observed in *Nature*, "it is less easy to speak." The Baltimore lectures "remain a witness to his extraordinary fertility of intellectual resource," while of his final ideas on radioactivity and "electrions," Thompson carefully concluded, "it would be entirely premature to evaluate [their] ultimate importance."

Also writing in *Nature*, Joseph Larmor described Kelvin as "a main pioneer and creator in the all-embracing science of energy, the greatest physical generalisation of the last century," but conceded that even there "his fragmentary and often hurried writings on this subject" left to others the task of rationalizing the new ideas into rigorous and whole science. In a longer notice written for the Royal Society, Larmor's reservations stood out more clearly still. Kelvin's brilliance was not in doubt. "What a happy strenuous career his must have been. . . . New discoveries and new aspects of knowledge crowding in upon him faster than he could express them to the world. . . . In the first half of his life, fundamental results arrived in such volume as to leave behind all chance of effective development." Kelvin had inspired Maxwell to start work on electromagnetism, Larmor noted, but Maxwell's "genius was as systematic as Thomson's was desultory." There was the problem: Kelvin thought fast, so fast, in fact, that he never stopped to think.

And there was another misfortune: "One is at times almost tempted to wish that the electric cabling of the Atlantic . . . had never been undertaken by him." Because after that his powers of original invention in science waned, and he channeled his admittedly enormous energy into too many distracting enterprises. A slightly silly story emphasizes the point. At Netherhall, Kelvin had been irked by a dripping tap. Naturally, he set out to design a better one, and his niece Agnes King told of happy hours running up and down stairs to turn the water on and off while her uncle designed and experimented. He devised a valve and took out a patent. In 1894 Kelvin visited his niece at rented accommodations in London and, finding no one home, left his card. The landlord, seeing this, remarked, "Lord Kelvin is rather a clever man. He invented a water-tap." Delighted, his niece told the story later at a dinner party for Kelvin, and "everyone laughed except himself; he could not see the joke! . . . He was quite satisfied to be thought 'rather clever,' and wondered at our amusement."

If family members were tickled at the thought of the great philosopher of nature being known for a piece of plumbing, scientists such as Larmor tut-tutted sadly. Kelvin was less of a physicist than he could have been, the thinking goes, because he dabbled in domestic engineering and other inconsequential matters. But laying aside the old snobbery that an intellectual cannot have practical talents, where is the truth in this? Kelvin was busy his entire life. Helmholtz, writing to his wife after a visit, at first thought that "Sir Wm might do better than apply his eminent sagacity to industrial undertaking" but quickly changed his mind: "I did Thomson an injustice in supposing him to be wholly immersed in technical work; he was full of speculations as to the original properties of bodies, some of which were very difficult to follow; and as you know he will not stop for meals or any other consideration."

When he was inventing his tap and refining his compass, in other words, Kelvin was also working up his Baltimore lectures, elaborating his vortex models of atom and ether, corresponding with FitzGerald on electromagnetism, and so on. If he had not invented the tap, would he somehow have been led to change his mind about Maxwell's theory? Of course not. The fact that he pursued an intellectual course largely alone, long after everyone else had abandoned it, and which turned out to be a historical dead end, has nothing to do with his skills of practical invention.

Kelvin clung to his outdated view of a strictly mechanical universe because he sincerely believed it was the right line to take. In his last years he must have been dismayed that he would not live to see how all the confusing questions thrown up at the beginning of the new century would turn out. Would he, brought back a century later, react to the modern physicists' view of the world with amazement or horror? It is easy to imagine that quantum mechanics would have befuddled and further dismayed him, since it took away forever the elementary picture of absolute cause and effect, of microphysics that could be interpreted as mechanical systems writ small.

But then I wonder. As much as he stuck to his established conceptions, Kelvin was capable of abandoning ideas with little regret when it was clear they would not do. By the end of his life he had discarded his old faith in vortex atoms and acknowledged that not one of his ether models was adequate. He did not know what would come in their place, but as he and others said, he needed a model in order to understand a theory. Quantum mechanics, in its pure form, might well have seemed elusive to him, but surely he would have loved particle physics. The giant machines, a mile or more around, accelerating tiny fragments of matter and smashing them into each other to see what smaller fragments would fly out—this is a science Kelvin could embrace. The strange subatomic forces that keep quarks bundled up inside atomic nuclei: This is Kelvin's physics! It's Boscovich! It's Aepinus Atomized! Elementary particles as constructions of tiny ingredients held together by peculiar but necessary forces—Kelvin would have convinced himself, I am sure, that this was exactly the kind of thing he had in mind when he was preaching against the nihilism of the Maxwellians and their mathematically austere electric and magnetic fields.

Even the vacuum, according to modern physics, is not a void but a seething medium of virtual particles coming and going. Electromagnetic fields, likewise, are portrayed now not as continuous entities, elastic but abstract, pervading space, but as the manifestation of forces transmitted by quantum particles, specifically photons. Many physicists today who probe the universe at its most fundamental level of construction speak of superstrings and their offspring, lines and loops wiggling around in multidimensional spaces and creating for us, in our limited three-dimensional world, the appearance of point particles and continuous forces. It is (as S. P. Thompson said of Kelvin's late thinking) entirely premature to judge whether superstrings and their ramifications will serve as the foundation of a final theory of physics, or lead the way to some other as yet unconceived theory, or, like the vortex atom and the sponge ether, go the way of the dinosaurs. But I think Kelvin, with his ability to dream up

endlessly ingenious pictures based on elementary principles and his fond-
ness for adding mathematical complication to explain hard empirical phenomena, would after being taken aback by the dizzying scope of modern
theoretical physics decide that, after all, it was exactly what he had been
trying to say.

BIBLIOGRAPHY

The basic biographical source for Kelvin is the two-volume work by Sylvanus P. Thompson, published in 1910. Concerning this, R. J. Strutt (p. 422, note to p. 253) recounts an amusing tale told to him by George Forbes, who among other things reported the Baltimore lectures for *Nature*: "At the British Association at York [in 1881], Thompson covered the blackboard with a mathematical calculation which did not commend itself to Lord Kelvin. The latter, as a protest, as soon as it was finished, silently took a duster and wiped it all out! Forbes many years afterwards reminded Thompson of the incident and said he thought it very generous of him to have written so laudatory a biography of Lord Kelvin. Thompson indicated that he attached no importance to it, that he had never given it another thought."

Thompson's life of Kelvin is, at any rate, not at all critical but is filled with detail, most of which stands up when compared to information from other sources. Thompson emphasizes Kelvin's science rather than his technological work, and writing as a physicist at the beginning of the 20th century, he naturally cannot judge a number of scientific issues that were unresolved at the time.

The memoirs by Kelvin's nieces Elizabeth Thomson King and Agnes

317

Gardner King overlap a good deal with each other and with Thompson but add interesting perspectives on Kelvin's youth and daily life. (Thompson evidently obtained most of his account of Kelvin's childhood and family background from the nieces.) Elizabeth's book is mainly her edited account of the recollections of her mother Elizabeth, Kelvin's sister, along with her own thoughts and memories. By remarkable coincidence, my copy of her book is dedicated by George King, presumably the brother of Elizabeth T. and Agnes G., to Captain John Gibb "in recollection of a delightful voyage in R.M.S. *Makura* [I think], Sydney to Vancouver, 28 Augt to 19 Sept 1911." Captain Gibb, sad to report, never read it. I bought it (on the Internet) from a used-book store in Sydney, Australia, and when it arrived I found its pages entirely uncut.

A more recent scholarly account of Kelvin is the book by Crosbie Smith and M. Norton Wise. This is a thorough but, to me anyway, excessively ideological work. Smith and Wise deconstruct Kelvin's science in a painstaking and mostly convincing way, but everything they find they fit into a sociopolitical straitjacket. They are enamored of a model of science couched in the terminology of economics. A discovery or idea has value in the scientific community according to the extent to which other scientists buy it. By generating consistently saleable products, a scientist can boost his intellectual credit rating. A winning theory outsells competing ideas and corners its market sector. Or something of the sort. Of course, Kelvin is then an accomplished scientific capitalist as well as an old-fashioned money-making one. How clever!

This sociological approach tends to strip individual scientists of originality or idiosyncrasy or psychology, turning them into anonymous actors responding helplessly to social forces. No doubt scientists are creatures of their times, but then so are historians of science, unless they claim a special exemption. Regardless of all that, I have tried to tell Kelvin's story as the saga of a man equipped with a particular set of talents and a particular cast of mind. Where these things come from I don't pretend to know.

The hardest part of researching this book was untangling the origins of thermodynamics. Scientific progress in any area tends to be a steady refinement of qualitative ideas into analytical laws, but in thermodynamics this journey was unusually tortuous, with the result that the academic literature today contains many discrepant opinions on what the several pioneers of the subject said, what they thought they were saying, and how important their contributions were. I found no single account that provided a thorough and measured estimation of the whole subject. Of particular note is Clifford Truesdell's exceedingly strange book, *The Tragicomical History of Thermodynamics*. Truesdell is bombastic, judgmental, and sesquipedalian, and a terrible scold to boot. He writes in a constant state of exasperated wonderment that men such as Carnot, Joule, Kelvin, Clausius, and Rankine were unable to see clearly all the things that are so apparent to him. He takes the line that no law is a good law until it has been given rigorous mathematical expression and for this reason tends to underestimate the value and difficulty of coming up with new physical concepts. His judgments as a result are idiosyncratic and questionable, but his detailed dissection of the many papers contributing to the foundation of thermodynamics is nonetheless illuminating.

In addition to the works cited here, I gleaned numerous odds and ends of information, too many and too minor to merit individual reference, from scanning the *Reports* of the British Association, the *Philosophical Magazine*, and *Nature*.

Celebration of Jubilee of Lord Kelvin, 15th, 16th and 17th June, 1896. 1896. Glasgow: MacLehose.

Kelvin Centenary: Oration and Addresses Commemorative. 1924. London: P. L. Humphries.

The New Amphion: Being the Book of the Edinburgh University Union Fancy Fair. 1886. Edinburgh: T. & A. Constable.

The North Atlantic Telegraph via the Faröe Isles, Iceland, and Greenland: Miscellaneous Reports, Speeches, and Papers on the Practicability of the Proposed North Atlantic Telegraph: The Results of the Surveying Expedition of 1859. 1861. London: Edw. Stanford.

The Practical Applications of Electricity: A Series of Lectures Delivered at the Institution of Civil Engineers. 1884. London: Institution of Civil Engineers.

Verses by M. T. 1874. Glasgow: Maclehose and Macdougall. (These are Margaret Thomson's poems, published posthumously and without explicit identification of the author.)

Babbage, C. 1971. *Reflections on the Decline of Science in England and on Some of Its Causes.* Shannon, Ireland: Irish University Press. Reprint of the original 1830 edition.

Bacon, R. H. 1929. *The Life of Lord Fisher of Kilverstone.* 2 vols. London: Hodder and Stoughton.

Baldwin, N. 2001. *Edison: Inventing the Century.* Chicago: University of Chicago Press.

Barrie, J. M. 1889. *An Edinburgh Eleven: Pencil Portraits from College Life.* London: British Weekly.

Berton, P. 1998. *Niagara: A History of the Falls.* New York: Penguin.

Bottomley, J. T. 1872. Physical Science in the University of Glasgow. *Nature* 6:29-32.

————. 1882. Scientific Worthies XX: James Joule. *Nature* 26:617-619.

Brett, J. W. 1858. *Brett's Submarine Telegraph.* London: Author.

Bright, C. 1889. *Submarine Telegraphs: Their History, Construction, and Working.* London: Crosby Lockwood & Son.

Buchwald, J. Z. 1977. William Thomson and the Mathematization of Faraday's Electrostatics. *Historical Studies in the Physical Sciences* 8:101-136.

Burchfield, J. D. 1990. *Lord Kelvin and the Age of the Earth.* Chicago: University of Chicago Press.

Campbell, L., and W. Garnett. 1884. *Life of James Clerk Maxwell.* 2nd ed., abridged. London: Macmillan.

Cardoso Dias, P. M. 1996. William Thomson and the Heritage of Caloric. *Annals of Science* 53:511-520.

Cardwell, D. S. L. 1972. *The Organisation of Science in England.* London: Heinemann.

————. 1989. *James Joule.* Manchester: Manchester University Press.

Carter, S. 1968. *Cyrus Field: Man of Two Worlds.* New York: G. P. Putnam's Sons.

Clark, J. L. 1868. *An Elementary Treatise on Electrical Measurement for the Use of Telegraph Inspectors and Operators.* London: E. & F. N. Spon.

Coates, V. T., and B. Finn. 1979. *A Retrospective Technology Assessment: Submarine Telegraphy: The Transatlantic Cable of 1866.* San Francisco: San Francisco Press.

Colvin, S., and J. A. Ewing, eds. 1887. *Fleeming Jenkin: Papers, Literary, Scientific &c., with a Memoir by Robert Louis Stevenson.* 2 vols. London: Longmans, Green, and Co. (The memoir can also be found in many editions of Stevenson's works.)

Cotter, C. H. 1976. George Biddell Airy and his Mechanical Correction of the Magnetic Compass. *Annals of Science* 33:263-274.

————. 1977. The Early History of Ship Magnetism: The Airy-Scoresby Controversy. *Annals of Science* 34:589-599.

Crosland, M., and C. Smith. 1978. The Transmission of Physics from France to Britain: 1800-1840. *Historical Studies of the Physical Sciences* 9:1-61.

Crowther, J. G. 1935. *British Scientists of the 19th Century.* London: Routledge & Kegan Paul.

Daiches, D. 1977. *Glasgow.* London: Andre Deutsch.

Darwin, G. H. 1916. *Scientific Papers,* vol. 5. Cambridge: Cambridge University Press.

Daub, E. E. 1967. Atomism and Thermodynamics. *Isis* 58:293-303.

Davie, George. 1964. *The Democratic Intellect: Scotland and Her Universities in the 19th Century,* 2nd ed. Edinburgh: University of Edinburgh Press.

de Cogan, D. 1985. Dr. E.O.W. Whitehouse and the 1858 trans-Atlantic cable. *History of Technology* 10:1-17.

Dibner, B. 1959. *The Atlantic Cable.* Norwalk, Conn.: Burndy Library.

Eve, A. S. 1939. *Rutherford.* Cambridge: Cambridge University Press.

Eve, A. S., and C. H. Creasey. 1945. *Life and Work of John Tyndall.* London: Macmillan.

Ewing, J. A. 1933. *An Engineer's Outlook.* London: Methuen.

Fairley, R. 1988. *Jemima: The Paintings and Memoirs of a Victorian Lady*. Edinburgh: Canongate.

Fanning, A. E. 1986. *Steady as She Goes: A History of the Compass Department of the Admiralty*. London: Her Majesty's Stationery Office.

Feldenkirchen, W. 1994. *Werner von Siemens: Inventor and International Entrepreneur*. Columbus: Ohio State University Press.

Fergusson, A. 1887. *Chronicles of the Cumming Club and Memories of Old Academy Days 1841-1846*. Edinburgh: T. & A. Constable.

Fisher, J. A. (Admiral of the Fleet Lord Fisher). 1919. *Memories*. London: Hodder and Stoughton.

————. 1919. *Records*. London: Hodder and Stoughton.

Fleming, J. A. 1934. *Memories of a Scientific Life*. London and Edinburgh: Marshall, Morgan and Scott.

Fourier, J. (A. Freeman, trans.). 1955. *The Analytical Theory of Heat*. New York: Dover.

Gooch, D. (R. B. Wilson, ed.). 1972. *Memoirs and Diary*. Newton Abbot, U.K.: David and Charles. Revised and expanded version of the 1892 edition.

Gooding, D. 1980. Faraday, Thomson, and the concept of the magnetic field. *British Journal of the History of Science* 13:91-120.

————. 1982. A Convergence of Opinion on the Divergence of Lines: Faraday and Thomson's Discussion of Diamagnetism. *Notes and Records of the Royal Society* 36:243-259.

Gordon, J. S. 2002. *A Thread Across the Ocean: The Heroic Story of the Transatlantic Cable*. New York: Walker.

Gossick, B. R. 1976. Heaviside and Kelvin: A Study in Contrasts. *Annals of Science* 33:275-287.

Gray, A. 1908. *Lord Kelvin: An Account of His Scientific Work*. London: J. M. Dent.

————. 1897. Lord Kelvin's laboratory in the University of Glasgow. *Nature* 55:487-492.

Harman, P. M. 1982. *Energy, Force, Matter*. Cambridge: Cambridge University Press.

————, ed. 1985. *Wranglers and Physicists*. Manchester: Manchester University Press.

Heaviside, O. 1951. *Electromagnetic Theory*. London: E. & F. N. Spon. (Single-volume reprint, with an introduction by E. Weber, of three volumes published earlier.)

Hiebert, E. N. 1962. *Historical Roots of the Principle of Conservation of Mechanical Energy*. Madison: State Historical Society of Wisconsin.

Hitchins, H. L., and W. E. May. 1952. *From Lodestone to Gyro-Compass*. London: Hutchinson's Scientific and Technical Publications.

Hunt, B. J. 1994. The Ohm Is Where the Art Is: British Telegraph Engineers and the Development of Electrical Standards. *Osiris* 9:48-63.

James, F. A. J. L., ed. 1996. *The Correspondence of Michael Faraday*, vol. 3. London: Institution of Electrical Engineers.

Kahl, R., ed. 1971. *Selected Writings of Hermann von Helmholtz*. Middletown, Conn.: Wesleyan Press.

Kargon, R., and P. Achinstein, eds. 1987. *Kelvin's Baltimore Lectures and Modern Theoretical Physics: Historical and Philosophical Perspectives*. Cambridge, Mass.: MIT Press.

King, A. G. 1925. *Kelvin the Man*. London: Hodder and Stoughton.

King, E. T. 1910. *Lord Kelvin's Early Home*. London: Macmillan.

Knight, W. 1896. *Memoir of John Nichol*. Glasgow: Maclehose & Sons.

Knott, C. G. 1911. *Life and Scientific Work of Peter Guthrie Tait.* Cambridge: Cambridge University Press.

Knudsen, O. 1972. From Lord Kelvin's Notebook: Ether Speculations. *Centaurus* 16:41-53.

———. 1976. The Faraday Effect and Physical Theory, 1845-73. *Archive for the History of the Exact Sciences* 15:235-281.

Königsberger, L. (F. A. Welby, trans.) 1906. *Hermann von Helmholtz.* Oxford: Clarendon.

Kremer, R. L., ed. 1990. *Letters of Hermann v. Helmholtz to His Wife, 1847-1859.* Stuttgart: Franz Steiner Verlag.

Kuhn, T. S. 1977. Energy Conservation as an Example of Simultaneous Discovery. Pp. 321-356 in *The Essential Tension: Selected Studies in Scientific Tradition and Change.* Chicago: University of Chicago Press.

Larmor, J., ed. 1902. *The Scientific Writings of the Late George Francis FitzGerald.* Dublin: Dublin University Press.

———, ed. 1907. *Sir George Gabriel Stokes, Memoir and Scientific Correspondence.* 2 vols. Cambridge: Cambridge University Press.

———. 1936. The Origins of Clerk Maxwell's Electric Ideas, as described in Familiar Letters to W. Thomson. *Proceedings of the Cambridge Philosophical Society* 32:695-750.

Lindley, D. 2001. *Boltzmann's Atom.* New York: Free Press.

Lloyd, J. T. 1970. Background to the Joule-Mayer Controversy. *Notes and Records of the Royal Society* 25:211-225.

Low, J. L. 1900. *F. G. Tait: A Record.* London: Nisbet & Co.

Lynch, A. C. 1985. History of the Electrical Units and Early Standards. *Proceedings of the Institute of Electrical and Electronic Engineers A* 132:564-573.

Mackie, J. D. 1954. *The University of Glasgow 1451-1951: A Short History.* Glasgow: Jackson, Son & Co.

Maxwell, J. C., and F. Jenkin. 1863. On the Elementary Relations Between Electrical Measurements. BA *Reports* 1863; also *Philosophical Magazine* 29(1865):436, 507.

May, W. E. 1973. *A History of Marine Navigation.* Henley-on-Thames, U.K.: G. T. Foulis & Co.

———. 1979. Lord Kelvin and His Compass. *Journal of Navigation* 32:122-134.

MacDonald, D. K. C. 1964. *Faraday, Maxwell, and Kelvin.* London: Heinemann.

Mendoza, E., ed. 1960. *Reflections on the Motive Power of Fire and Other Papers on the Second Law of Thermodynamics.* New York: Dover. (A translation of Carnot's *Réflexions,* along with translations of the 1834 paper by Clapeyron and the 1850 paper by Clausius, as well as a brief biographical sketch of Carnot.)

Millar, W. J., ed. 1881. *William John Macquorn Rankine: Miscellaneous Scientific Papers. With a Memoir of the Author by P. G. Tait.* London: Charles Griffin.

Murray, D. 1924. *Lord Kelvin as Professor in the Old College of Glasgow.* Glasgow: Maclehose & Jackson.

Olson, R. 1975. *Scottish Philosophy and British Physics, 1750-1880.* Princeton, N.J.: Princeton University Press.

Rolt, L. T. C. 1989. *Isambard Kingdom Brunel.* London: Penguin.

Rothblatt, S. 1981. *The Revolution of the Dons: Cambridge and Society in Victorian England.* Cambridge: Cambridge University Press.

Rowlinson, J. S. 1971. The Theory of Glaciers. *Notes and Records of the Royal Society* 26:189-204.

Russell, A. 1908. *Lord Kelvin.* London: Blackie.

Russell, W. H. 1866. *The Atlantic Telegraph.* London: Day & Son.

Schuster, A. 1932. *Biographical Fragments.* London: Macmillan.

Searle, G. F. C. (I. Catt, ed.). 1987. *Oliver Heaviside, the Man.* St. Albans, U.K.: C.A.M. Publishing.

Sharlin, H. I. 1979. *Lord Kelvin: The Dynamic Victorian.* State College: Pennsylvania State University Press.

Siemens, W. von. 1966. *Inventor and Entrepreneur: Recollections of Werner von Siemens.* London: Lund Humphries.

Silliman, R. 1963. William Thomson: Smoke Rings and 19th Century Atomism. *Isis* 54:461-474.

Smith, C. 1998. *The Science of Energy.* Chicago: University of Chicago Press.

————, and M. Norton Wise. 1989. *Energy and Empire.* Cambridge: Cambridge University Press.

Smith, W. 1891. *The Rise and Extension of Submarine Telegraphy.* London: J. S. Virtue & Co.

Strutt, R. J. (J. N. Howard, ed.). 1968. *Life of John William Strutt, Third Baron Rayleigh.* Madison: University of Wisconsin Press. (Reprint with additional material of the 1924 edition.)

Sviedrys, R. 1976. The Rise of Physics Laboratories in Britain. *Historical Studies of the Physical Sciences* 7:405-436.

Tait, P. G. 1866. *On the Value of an Edinburgh Degree.* Edinburgh: Maclachlan and Stewart.

————. 1885. *Lectures on Some Recent Advances in Physical Science (with a special lecture on force),* 3rd ed. London: Macmillan.

Thompson, S. P. 1910. *Life of Lord Kelvin.* 2 vols. London: Macmillan.

Thomson, J. J. 1936. *Recollections and Reflections.* London: Macmillan.

Thomson, W. (Lord Kelvin). 1872. *Reprint of Papers on Electrostatics and Magnetism.* London: Macmillan.

————. 1882-1911. *Mathematical and Physical Papers.* 6 vols. Cambridge University Press. (First three volumes were edited by Thomson, last three by J. Larmor.)

————. 1889-1894. *Popular Lectures and Addresses.* 3 vols. London: Macmillan.

Truesdell, C. A. 1980. *The Tragicomical History of Thermodynamics, 1822-1854.* New York/Heidelberg/Berlin: Springer.

Tunbridge, P. 1992. *Lord Kelvin: His Influence on Electrical Measurements.* London: Peter Peregrinus for the Institution of Electrical Engineers.

Watson, E. C. 1939. College Life at Cambridge in the Days of Stokes, Cayley, Adams, and Kelvin. *Scripta Mathematica,* 6:101-106.

Whittaker, E. T. 1989. *A History of the Theories of Aether and Electricity.* New York: Dover. (Single-volume reprint of two volumes originally published 1951 and 1953.)

Williams, L. P. 1961. The Royal Society and the Founding of the BAAS. *Notes and Records of the Royal Society* 16:221-233.

————. 1965. *Michael Faraday.* London: Chapman and Hall.

Wilson, D. 1910. *William Thomson, Lord Kelvin: His Way of Teaching.* Glasgow: John Smith and Son.

Wilson, D. B. 1987. *Kelvin and Stokes: A Comparative Study in Victorian Physics*. Bristol, U.K.: Adam Hilger.

————. 1990. *The Correspondence Between Sir George Gabriel Stokes and Sir William Thomson, Baron Kelvin of Largs*. 2 vols. Cambridge: Cambridge University Press.

Yavetz, I. 1995. *From Obscurity to Enigma: The Work of Oliver Heaviside, 1872-1889*. Basel/Boston/Berlin: Birkhäuser.

Notes

Quotations are not referenced here if their origin is clear from the text. In particular, extracts from Kelvin's scientific papers can be found by consulting his collected works, and remarks by scientists at British Association meetings are taken from the single-volume (but eccentrically paginated) BA *Reports* for that year. Likewise, extracts from newspapers are not cited here if the source and date are identified in the text.

Correspondence indicated as B128, etc., comes from the Kelvin collection at Cambridge University Library (CUL), with the exception of the correspondence between Kelvin and Stokes, where reference is to Wilson (1990). In these references WT is William Thomson, and JT is his father, James Thomson. Siblings are referred to by their first names only, for brevity.

Abbreviations
SPT I and II: *Life of Kelvin*, Thompson (1910)
MPP I-VI: *Mathematical and Physical Papers*, Thomson (1882-1911)
PLA I-III: *Popular Lectures and Addresses*, Thomson (1889-1894)
E&M: *Papers on Electrostatics and Magnetism*, Thomson (1872)
AGK: *Kelvin the Man*, King (1925)
ETK: *Lord Kelvin's Early Home*, King (1910)

INTRODUCTION

p. 1, **"not . . . in robust health":** This, the reception at Columbia University, and later interview extracts are from the *New York Times*, April 20, 1902.

p. 2, **"broke forth such a cheer . . .":** This and subsequent remarks by Rhees are from Rochester (N.Y.) *Democrat and Chronicle*, May 2, 1902.

p. 2, **In Washington D.C.:** *Washington Post*, April 23, and *Washington Star*, April 24, 1902.

p. 3, **" . . . eminent electrician":** *Toronto Globe*, August 19, 1897.

p. 9, **"I wish I could have made it more clear . . .":** from the Q&A following Kelvin's lecture "Electrical Units of Measurement," in *Practical Applications of Electricity* (1884). PLA I includes the text of the lecture but not the remarks.

1
CAMBRIDGE

p. 12, **"Since he has left . . .":** This and following remarks to Anna are from B128, October 23, 1841.

p. 12, **"I had no idea . . .":** WT to Elizabeth, n.d.; ETK 196.

p. 12, **"I have got no time to be dull . . .":** WT to Elizabeth, n.d.; ETK 201.

p. 13, **"asked your age . . .":** T180, JT to WT, October 28, 1841.

p. 14, **"It was certainly a great honour . . .":** T185, WT to JT, November 21, 1841.

p. 14, **"delightful young man":** ETK 162.

p. 15, **" . . . we accordingly determined to wait . . .":** from Notebook 7 at CUL.

p. 16, **"The mathematics is very difficult . . ."** and **"in the first half of the month of May . . .":** Lord Kelvin and his first teacher in natural philosophy, *Nature,* 68(1903):623-624.

p. 17, **"On the 1st of May . . .":** SPT I 14.

p. 17, **"Primary causes are unknown to us . . .":** Preliminary Discourse, Fourier (1955).

p. 18, **"shocked to be told . . ."**: SPT I 17.

p. 18, **"Papa! Fourier is right, and Kelland is wrong!"**: ETK 183.

p. 19, **"As to the insertion of the paper . . ."**: K4, Kelland to JT, March 4, 1841.

p. 20, **"sole object is to establish what is true . . ."**: K5, JT to Kelland, March 6, 1841.

p. 20, **"the flippant manner . . ."**: G182, Gregory to JT, March 6, 1841.

p. 20, **"I am very much pleased . . ."**: K6, Kelland to JT, March 8, 1841.

p. 20, **The paper appeared . . . May 1841:** On Fourier's Expansions of Functions in Trigonometric Series, MPP I.

p. 22, **"You know my views . . ."**: T180, JT to WT, October 28, 1841, SPT I 29.

p. 22, **"Recollect my maxim . . ."**: T186, JT to WT, December 6, 1841, SPT I 32.

p. 22, **"Never forget to take every care . . ."**: T234, JT to WT, April 9, 1843, SPT I 53.

p. 22, **"Healthful and innocent exercise . . ."**: T197, JT to WT, February 21, 1842.

p. 22, **"farther, that it will not lead . . ."**: T186, JT to WT, December 6, 1841.

p. 23, **"With regard to wine parties . . ."** and **"I always row by myself . . ."**: T187, WT to JT, December 12, 1841.

p. 23, **"an idle and extravagant set"**: T181, WT to JT, October 30, 1841.

p. 23, **"built of oak, and as good as new"**: T196, WT to JT, February 19, 1842.

p. 23, **"You are quite right in anticipating . . ."**: T197, JT to WT, February 21, 1842.

p. 23, **"I hope [the boat] is to your liking . . ."**: T496, John to WT, March 1, 1842.

p. 24, **"so favourably and so kindly"**: T203, JT to WT, March 27, 1842.

p. 24, **"Good-hearted . . ."**: Knight (1896) 20.

p. 24, **"the fine old stock of Scottish Covenanters"**: AGK 3.

p. 26, **"frightened to see my beautiful father . . ."** and following extract: ETK 71.

p. 27, **"both father and mother to us . . .":** ETK 87.

p. 27, **"There was something in the very disamenities . . .":** G. G. Ramsay, quoted by Gray (1908) 11.

p. 28, **"dingy old place":** Knight (1896) 18.

p. 28, **"both clever, good talkers and sketchers":** Knight (1896) 19.

p. 28, **"Do, papa, let me answer!":** ETK 101.

p. 29, **"P'itty b'ue eyes . . .":** ETK 27.

p. 29, **"William was a great pet with him . . .":** ETK 87.

p. 29, **"A most engaging boy . . .":** recollection by Canon Grenside, reported by AGK 11.

p. 29, **"I have been reading moderately . . .":** T206, WT to JT, April 1842.

p. 30, **" . . . I can read with much greater vigour . . .":** T209, WT to JT, April 14, 1842.

p. 30, **recorded his weight:** CUL notebook NB29, entry for February 13, 1843.

p. 30, **"I am sure you will perfectly approve . . .":** T213, WT to JT, May 6, 1842.

p. 30, **"to relieve my head from the seediness . . .":** NB29, March 31, 1843.

p. 32, **Hopkins charged £72 per student:** Rothblatt (1981) 68n, 200.

p. 32, **"College Tutors and Lecturers take but small part . . .":** Tait (1866).

p. 33, **"at an appointed time . . .":** Fleming (1934) 57.

p. 34, **Hopkins . . . coached 17 Senior Wranglers:** Rothblatt (1981) 202.

p. 36, **"Fourier made Thomson":** Knott (1911) 191.

p. 36, **One of William's fellow undergraduates:** Canon Grenside, reported by SPT I 25.

p. 37, **William noted on February 15:** These and subsequent extracts are from NB29.

p. 37, **"Fischer does not get on quite so well . . .":** T230, WT to JT, March 12, 1843.

p. 40, **"What mortal in the world . . .":** From the 1824 translation of *Wilhelm Meister* by Thomas Carlyle.

p. 40, **"I hope most intensely . . .":** T250, WT to JT, May 5, 1843.

p. 40, **" . . . I have been much gratified . . .":** T243, JT to WT, May 23, 1843.

p. 41, **"I am practising now everyday . . .":** B204, WT to Anna, November 30, 1843.

p. 41, **"containing all your reasons . . .":** Anna to WT, October 1843, SPT I 63.

p. 41, **"better than winning in an examination.":** AGK 9.

p. 43, **"This is a very pleasant place . . .":** T265, WT to JT, June 13, 1844.

p. 43, **"Your lodgings are surely unnecessarily fine . . .":** T266, JT to WT, June 21, 1844.

p. 43, **"I think you might write a *little* oftener . . .":** T553, Robert to WT, June 3, 1844.

p. 43, **" . . . he has given me entire satisfaction . . .":** H122, Hopkins to JT, August 7, 1844.

p. 44, **"Your son is going on extremely well . . .":** C135, Cookson to JT, December 19, 1844.

p. 44, **"I do not feel at all confident . . .":** G11, WT to Aunt Gall, December 11, 1844.

p. 44, **"The prospect is of course rather terrible . . .":** T278, WT to JT, December 29, 1844.

p. 46, **" . . . Dr Meikleham has a second attack . . .":** T228, JT to WT, December 7, 1843.

p. 46, **"I felt . . . I ought to mention . . .":** T236, JT to WT, April 1843.

p. 47, **"What you have to do, therefore . . .":** T239, JT to WT, May 4, 1843.

p. 47, **"Dr. Meikleham has had another attack . . .":** T246, JT to WT, August 1843.

p. 47, **"seems to be getting on very well . . .":** T255, WT to JT, April 4, 1844.

p. 48, **"three years of Cambridge drilling . . .":** T253, WT to JT, March 6, 1844.

p. 48, **"For the project we have . . .":** T257, WT to JT, April 22, 1844.

p. 48, **"greatly changed . . .":** T262, JT to WT, May 13, 1844.

p. 48, **"A Cambridge education did not *always* give . . .":** T271, JT to WT, September 22, 1844.

p. 48, **William dashed off a brief note:** T279, WT to JT, January 1, 1845.

p. 49, **" . . . with vigour and cheerfulness . . .":** H123, Hopkins to JT, January 5, 1845.

p. 49, **"I think that he cannot fail . . .":** C138, Cookson to JT, January 6, 1845.

p. 49, **"I have been getting on very well . . .":** T281, WT to JT, January 5, 1845.

p. 49, **"the Johnians are talking confidently . . .":** T286, WT to JT, January 14, 1845.

p. 49, **"This present year, however . . .":** extracts from Charles A. Bristed (1852), *Five Years in an English University* (New York: Putnam), from Watson (1939).

p. 50, **"You see I was right . . .":** T287, WT to JT, January 17, 1845.

p. 50, **"I hope by this time . . .":** T288, WT to JT, January 18, 1845.

p. 50, **"The place you have got . . .":** T289, JT to WT, January 19, 1845.

p. 51, **"I must confess . . .":** K74, Elizabeth to WT, January 22, 1845.

p. 51, **" . . . your son's not being senior wrangler . . .":** H124, Hopkins to JT, January 18, 1845.

p. 51, **"Hopkins's letter has done you *great* good . . .":** T290, JT to WT, January 22, 1845.

p. 52, **The exam took place at Stokes's house:** Larmor (1907) 11.

p. 52, **"I have seen your son . . .":** C140, Cookson to JT, January 24, 1845.

p. 54, **"Since the days of Newton, however . . .":** *Belfast Magazine and Literary Journal,* 1(1825):270-271.

p. 55, **"despotic Whewell":** T319, WT to JT, October 10, 1845.

p. 56, **"he asked me to write a short paper . . .":** T303, WT to JT, March 30, 1845.

p. 57, **" . . . the Alma Mater of my scientific youth . . .":** *Comptes Rendus de l'Académie des Sciences,* 121(1895):582 (my translation).

p. 58, **"Dr W Thomson [the medical man] and Dr Nichol . . .":** T296, JT to WT, February 12, 1845.

p. 58, **"Dr W.T. is much pleased . . .":** T305, JT to WT, April 8, 1845.

p. 58, **"very much contrary to my expectations":** T314, WT to JT, June 28, 1845.

p. 58, **" . . . about fit to mend his pens":** SPT I 98; said to be the words of R. L. Ellis.

p. 58, **"forward so far at so early an age!":** T315, JT to WT, July 1, 1845.

p. 59, **"important matters in consideration at present":** T317, WT to JT, August 17, 1845.

p. 59, **"as many pupils as I would wish":** T326, WT to JT, November 1, 1845.

p. 59, **"I am afraid I should have to give up . . .":** T329, WT to JT, February 11, 1846.

p. 59, **"said not to bring down your instructions . . .":** T333, JT to WT, May 2, 1846.

p. 60, **"The idea, if there is such an idea . . .":** E62A, Ellis to JT, May 11, 1846.

p. 60, **"timidity and want of effective locution":** T341, JT to WT, May 16, 1846.

p. 60, **"took me quite by surprise . . .":** T336, WT to JT, May 10, 1846.

p. 61, **"Could you 'get at them'? . . .":** T334, JT to WT, May 7, 1846.

p. 61, **"double your efforts to procure testimonials . . .":** T342, JT to WT, May 17, 1846.

p. 61, **"I am afraid you are resting too quietly . . .":** T349, JT to WT, June 13, 1846.

p. 61, **"get a beard fast . . .":** T558, Robert (relaying Anna's words) to WT, July 8, 1845.

p. 61, **"Cookson &c are right in their views . . .":** T355, JT to WT, June 21, 1846.

p. 62, **"hoped I do not intend . . .":** T357, WT to JT, June 29, 1846.

p. 62, **"He is already blessed . . .":** Cookson's testimonial in CUL item PA34.

p. 63, **"I believe M. William Thomson . . .":** Liouville, PA35 (my translation).

p. 63, **"a countenance more expressive of delight . . .":** David King to Elizabeth, September 12, 1846; ETK 232.

p. 63, **"every now and then . . .":** David King to Elizabeth, September 14, 1846; ETK 232.

p. 63, **"William does not look in the slightest degree elated.":** Elizabeth to Agnes Gall, September 1846; ETK 233.

2
CONUNDRUMS

p. 66, **"*Caino?*":** On the dissipation of energy, PLA II; also SPT I 133.

p. 67, **"the production of motive power . . ."** and other subsequent remarks from Carnot's *Réflexions*: Mendoza (1960).

p. 70, **"The preliminary part . . .":** T415, James to WT, February 22, 1846.

p. 70, **"Of the sons I liked James . . .":** Knight (1896) 20.

p. 70, **"a level-headed fellow . . .":** Königsberger (1906) 222.

p. 70, **"It was also, sometimes, difficult . . .":** Ewing (1933) 171.

p. 71, **"I have a good many warnings . . .":** T382, James to WT, 1842.

p. 71, **"I wish my apprenticeship was as nearly done . . .":** T404, James to WT, December 23, 1844.

p. 72, **"blister over my heart . . ."**: T297, James to WT, February 15, 1845.

p. 72, **"really a most painful and distressing thing"**: T425, James to WT, May 30, 1847.

p. 73, **"had the courage to say . . ."**: Cardwell (1989) 6.

p. 74, **"I could imagine . . ."**: Schuster (1932) 201.

p. 75, **"felt strongly impelled at first . . ."**: WT quoted in Bottomley (1882).

p. 75, **"I gained ideas . . ."**: Address delivered on the unveiling of Joule's statue in Manchester, December 7, 1893, PLA II.

p. 75, **"Joule is I am sure wrong . . ."**: T367, WT to JT, July 1, 1847.

p. 75, **"I enclose Joule's papers . . ."**: T429, WT to James, July 12, 1847.

p. 75, **"I certainly think [Joule] has fallen into blunders . . ."**: T433, James to WT, July 24, 1847.

p. 76, **"Whom should I meet walking up but Joule . . ."**: Bottomley (1882).

p. 76, **"Before leaving the St Martin road . . ."**: T373, WT to JT, September 5, 1847.

p. 77, **"I must say I am not at all satisfied . . ."**: F95, Fischer to WT, October 26, 1847.

p. 77, **"Your most grave & sober counsel . . ."**: D124, Dykes to WT, June 1847.

p. 77, **"Mind you don't get married . . ."**: F295, Frederick Fuller to WT, September 14, 1846.

p. 77, **"There is a tremendous report . . ."**: F297, Fuller to WT, November 14, 1846.

p. 77, **"I was asked to go to balls . . ."**: Jemima Blackburn's recollection in Fairley (1988) 33.

p. 77, **"professed utter scorn"**: ETK 141.

p. 78, **"regular drudgery"**: T505, John to WT, May 10, 1844.

p. 78, **"He burst out rather faintly . . ."**: WT to David King, January 12, 1849; ETK 241.

p. 78, **"Elizabeth! Elizabeth Thomson! . . ."**: Agnes Gall to Elizabeth, January 30, 1849; ETK 243.

p. 79, **starting on £20 a year:** T560, Robert to WT, May 9, 1846.

p. 79, **William bought stock:** T366, Robert to WT, June 20, 1847; T563, WT to JT, July 6, 1847.

p. 80, **Going to Hopkins' rooms:** SPT I 113.

p. 80, *"Ah! Voilà mon affaire!"*: T298, WT to JT, February 23, 1845; also SPT I 119.

p. 81, **"My education was of the most ordinary description . . .":** Williams (1965) 7, quoting directly from the 1870 biography of Faraday by Bence-Jones.

p. 83, **"I do not think I could work in company . . .":** Williams (1965) 99.

p. 84, **"In all kinds of knowledge . . .":** Williams (1965) 105.

p. 86, **"inoculated with Faraday fire":** SPT I 19.

p. 86, **"in w^h Faraday and Daniell . . . got (abused)²":** CUL notebook NB 29, entry for March 16, 1843.

p. 87, **"merely as actual truths . . .":** On the Mathematical Theory of Electricity in Equilibrium, E&M.

p. 87, **"What I have written is merely a sketch . . .":** WT to Faraday, June 11, 1847, James (1996); also quoted in SPT I 203.

p. 89, **"straightforward course is, to decline . . .":** S35, Stokes to WT, February 12, 1845.

p. 89, **"once or twice or three times . . .":** S36, WT to Stokes, February 24, 1845.

p. 90, **"Ye'll no lach when ye're in hell!":** ETK 120.

p. 91, **"no case can prove the noxiousness . . .":** S39, WT to Stokes, February 20, 1845.

p. 91, **"When I consider thy heavens . . .":** from Psalm 20; SPT I 251.

p. 91, **"According to his own account . . .":** Elizabeth to David King, November 3, 1846; ETK 233.

p. 91, **"the lecturer was greatly downhearted at its conclusion":** SPT I 191.

p. 92, **"an enthusiastic and inspiring teacher . . .":** Murray (1924).

p. 92, **"Explanation . . . was never his forte":** Wilson (1910).

p. 92, **" . . . Thomson soared to heights . . .":** Russell (1938) 35.

p. 92, **"Now, Mr. Macintosh . . .":** Murray (1924).

p. 92, **"will not answer when questioned . . ."** and further remarks about his lab at Glasgow: From the Bangor Laboratories, PLA II.

p. 94, **"had none of the air or manner of a superior":** Murray (1924).

p. 94, **"What I liked best . . .":** Wilson (1910).

p. 94, **"I suppose it is out of the question . . .":** F99, Fischer to WT, February 18, 1850.

p. 95, **"I think will please you . . .":** WT to Elizabeth, July 13, 1852, SPT I 232.

p. 95, **"sometime, probably early in September . . .":** S96, WT to Stokes, July 31, 1852.

p. 95, **"We have one interest in common . . .":** Margaret Crum to Elizabeth, n.d., SPT I 233.

p. 96, **"The day is somewhat dark and cold . . .":** WT to Elizabeth, September 19, 1852; SPT I 234.

p. 96, **"a rather pretty woman . . .":** letter 39, August 7, 1855, Kremer (1990) (my translation).

p. 96, **"give my best regards to . . . Thomson . . .":** Fairley (1988) 43.

p. 96, **"*They have sung to thee, O grave!*":** all poetical selections are from *Verses by MT* (1874).

p. 97, **"surgical nursing":** SPT I 238.

p. 97, **"in a wretched state . . .":** letter 39, Kremer (1990).

p. 98, **"she looks much better . . .":** WT to Elizabeth, August 1854; SPT I 305.

p. 99, **the Frenchman Guillaume Amontons:** For the early history of these temperature-scale calibrations, see Truesdell (1980), ch. 1.

p. 101, **"As you have taken so much trouble . . .":** F194, Forbes to WT, April 20, 1848.

p. 101, **"I write to remind you . . .":** F198, Forbes to WT, November 27, 1848.

p. 102, **" . . . without the expenditure of mechanical effect":** F199, WT to Forbes, December 7, 1848.

p. 104**: "I believe we should not be daunted . . .":** Clausius (1850), in Mendoza (1960).

p. 105, **"wherever motive power is destroyed . . ."**: Carnot's later notes, in Mendoza (1960).

p. 112, **"she suffers much . . ."**: T445, WT to James, March 17, 1855.

p. 112, **"From Edinburgh I traveled for a couple of hours . . ."**: letter 34, September 22, 1853, Kremer (1990) (my translation).

p. 113, **" . . . one of the leading mathematical physicists in Europe . . ."**: letter 39, Kremer (1990).

3
CABLE

p. 114, **"expressed their great astonishment . . ."**: Brett (1858) x.

p. 114, **"What a mad scheme!"**: W. Smith (1891) 5.

p. 115, **"from one continent to another"**: Brett (1858) x.

p. 115, **Faraday . . . sent a short note**: *Philosophical Magazine*, 32(1848):165-167.

p. 115, **"some few, more or less incoherent, letters . . ."**: Bright (1898), ch. 1.

p. 115, **"the jest or scheme of yesterday . . ."**: London *Times*, August 24, 1850.

p. 115, **"a man in London might sign a bill . . ."**: *Spectator*, August 21, 1850, quoted by Brett (1858) 34.

p. 116, **"oceanic and subterranean inland electric telegraphs"**: Brett (1858) includes a facsimile of the original telegraph.

p. 119, **"the simplicity of Morse's apparatus . . ."**: Siemens (1966) 82.

p. 120, **"not ten minutes after . . ."**: Brett (1858) xiii.

p. 122, **Faraday published his analysis**: *Philosophical Magazine*, 7(1854):297-208.

p. 122, **The problem finally came to Thomson's attention:** WT recounts the story in Ether, Electricity, and Ponderable Matter, MPP III.

p. 122, **"devoting myself as much as possible . . ."**: S115, WT to Stokes, October 28, 1854.

p. 123, **"the remedy for the anticipated difficulty . . ."**: S119, WT to Stokes, December 1, 1854.

p. 123, **"suggested the plan . . .":** T442, WT to James, January 13, 1855 (incorrectly dated 1854).

p. 124, **"distinctness of the utterance":** S116, WT to Stokes, October 30, 1854.

p. 125, **Edward Orange Wildman Whitehouse:** A few details about him are in a short obituary notice, *The Electrician*, 24(1890):319.

p. 125, **"has been able to show most convincingly . . .":** *Athenaeum*, August 30, 1856.

p. 126, **"coiled in a large tank . . .":** Whitehouse's contribution in BA *Reports*, 1856.

p. 126, **"depends on the nature of the electric operation . . .":** *Athenaeum*, November 8, 1856.

p. 127, **"like every *theory* . . .":** *Athenaeum*, October 4, 1856; also MPP III.

p. 131, **"was surprised to find differences . . .":** On Electric Conductivity of Commercial Copper of Various Kinds, MPP III.

p. 132, **"It was not until practical testing . . .":** footnote added June 27, 1883, to Analytical and Synthetical Attempts to Ascertain the Cause of the Differences of Electrical Conductivity Discovered in Wires of Nearly Pure Copper, MPP II. (Despite the title WT never did satisfactorily ascertain the cause.)

p. 133, **a hair plucked from his dog:** AGK 55.

p. 133, **"the frantic fooleries of the Americans . . .":** Carter (1968) 147.

p. 134, **"experienced mariners gazed in apprehension . . .":** Russell (1866) 22.

p. 136, **"in a fearful state of excitement . . .":** From an eyewitness account in the Sydney *Morning Herald*, quoted in SPT I 360-364.

p. 136, **"The electrical signals sent and received . . .":** *New York Post*, August 5, 1858.

p. 138, **"True, the Queen's message . . .":** *New York Post*, August 17, 1858.

p. 138, **"Glorious Recognition . . .":** *New York Herald*, September 2, 1858.

p. 138, **"It is the harbinger of an age . . .":** *New York Post*, September 2, 1858.

p. 139, **"near the shore, and remediable"**: *New York Post*, September 21, 1858.

p. 139, **at least in some accounts**: SPT I 369, but his account is only partly corroborated by others.

p. 140, **"Instead of telegraphic work . . ."**: WT to Joule, September 25, 1858, SPT I 378.

p. 140, **"I should like much to know . . ."**: A111, Field to WT, June 29, 1859.

p. 141, **"the interior of the jar lit up . . ."**: Bright (1898) 52.

p. 141, **"I must not hide from you . . ."**: L9, C. M. Lampson to WT, October 22, 1858.

p. 141, **thought about proposing Whitehouse**: see S163, WT to Stokes, November 7, 1857.

p. 142, **"The foundation of a real and lasting success . . ."**: SPT I 390.

p. 145, **"the gentlemen who constitute the Committee . . ."**: quoted in Lynch (1985).

p. 147, **"At the least sign of unrest . . ."** and following extracts: *Memoir of Fleeming Jenkin*, Colvin and Ewing (1887).

p. 150, **"I have had the opportunity . . ."**: Feldenkirchen (1994) 35.

p. 150, **Debate over the resistance standard came to a stalemate**: Account is from Tunbridge (1992), ch. 4, who takes it from *La Vie et les oeuvres de E. Mascart* (Paris: P. Janet, 1910):32-39.

p. 151, **Names for the units**: Tunbridge (1992), ch. 5; also SPT I 418.

p. 151, **"I object to Galvad . . ."**: Tunbridge (1992), ch. 5; the letter is at Glasgow University.

p. 153, **"I was not mathematical enough . . ."**: C91, Clark to WT, May 3, 1883.

p. 153, **" . . . an aged and severe philosopher . . ."** and following extract: Ewing (1933) 172; letter to him from Annie Jenkin.

p. 156, **"Cyrus Field, from the other side of the Atlantic . . ."**: On the Early History of Submarine Cable Enterprise, MPP V.

p. 156, **"With the perseverance characteristic of the English . . ."**: Siemens (1966) 119.

p. 157, **"A line of two thousand miles . . .":** *The North Atlantic Telegraph* (1861).

p. 158, **"a submarine telegraph cable would be designed . . .":** WT's remark is from a discussion following presentations by H. C. Forde and C. W. Siemens on the Malta and Alexandria Telegraph Submarine Telegraph Cable, *Proceedings of the Institute of Civil Engineers*, 1861-1862 session (London: Wm Clowes and Sons, 1863).

p. 158, **"There is your ship":** Rolt (1989) 396n.

p. 159, **" . . . the appearance of a dead forest . . .":** From a magazine article *A summer trip across the Atlantic: A reminiscence* 'by one who helped to lay the cable' above the nom de plume (well, I presume!) Henry Plantagenet Dynamometer, from a collection of pamphlets bound together at the Bodleian Library, Oxford, BOD 247931 e.22. I have been unable to find where this was originally published.

p. 159, **"was ordered by his board . . .":** *Cornhill Magazine*, September 1865.

p. 160, **"What we had taken for assassination . . .":** Russell (1866) 75.

p. 160, **"steady as a Thames steamer":** Russell (1866) 61.

p. 161, **"I will never forget this hour . . .":** Gooch (1972) 100.

p. 161, **"sad and dreadful discouragement . . .":** On the Early History of Submarine Cable Enterprise, MPP V.

p. 161, **"I remember well a night . . .":** WT interviewed in Rochester (N.Y.) *Democrat and Chronicle,* April 28, 1902.

p. 161, **"I am very pleased to learn . . .":** WT to Field, January 1, 1866, in Cyrus Field collection at the Library of Congress.

p. 163, **"in no other branch of engineering . . .":** Presidential Address to the Society of Telegraph Engineers, January 14, 1874, PLA II.

4
CONTROVERSIES

p. 165, **"undoubtedly meteoric":** On the Mechanical Energies of the Solar System, MPP II.

p. 168, **"For eighteen years it has pressed on my mind . . .":** On the Secular Cooling of the Earth, MPP III.

p. 170, **"guess at the half . . ."**: F. Jenkin (although anonymously published) in *North British Review*, June 1867, 277-318.

p. 171, **Thomson accosted the geologist Andrew Ramsay:** anecdote and remarks from Age of Earth as an Abode Fitted for Life, MPP V.

p. 172, ***Beautiful Round! Superbly played:*** Knott (1911) 55.

p. 172, **"The small twinkling eyes . . ."**: Barrie (1889).

p. 173, **"A great reform in geological speculation . . ."**: On Geological Time, PLA II.

p. 174, **"Thomson's views . . ."** and **"then comes Sir W. Thomson . . ."**: Darwin to A. R. Wallace, April 14, 1869; Darwin to St. G. Mivart, July 12, 1871; both quoted by E. B. Poulton, BA *Reports* (1896).

p. 174, **"The rotation of the earth *may* be diminishing . . ."**: Huxley, *Quarterly Journal of the Geological Society of London*, 25 (1869):xxxviii-liii.

p. 175, **"so many geologists are contented . . ."**: Of Geological Dynamics, PLA II.

p. 175, **a lengthy review:** Tait, *North British Review*, July 1869, 406-439.

p. 177, **"We cannot give more scope . . ."**: Tait (1885).

p. 177, **Darwin thought his views "monstrous":** Burchfield (1990) 110.

p. 178, **"a German of the name of Mayer . . ."**: J64, Joule to WT, December 9, 1848.

p. 178, **"I have not the slightest wish . . ."**: J66, Joule to WT, March 8, 1849.

p. 178, **"I have not pursued the controversy further . . ."**: J77, Joule to WT, March 17, 1851.

p. 179, **"I suffer in a righteous cause . . ."**: Eve and Creasey (1945) 10.

p. 179, **"I am not stubborn . . ."**: Eve and Creasey (1945) 11.

p. 180, **"Thomson completely backed out . . ."**: Eve and Creasey (1945) 55.

p. 181, **"*most absolutely honest* man . . ."**: Sharlin (1979), ch. 11. The remarks are from Tait's notebook, shown to Sharlin by Tait's granddaughter.

p. 181, **"peremptory, abrupt and dogmatic . . ."**: Eve and Creasey (1945) 71.

p. 182, **"astonished at the multitude of . . ."**: Eve and Creasey (1945) 94.

p. 182, **"To whom, then, are we indebted . . ."**: Tyndall, *Philosophical Magazine*, 24(1862):57.

p. 183, **"We were certainly amazed . . ."** and other extracts: Energy, *Good Words*, September 1862; not included in the collected papers.

p. 186, **"to give to Mayer . . ."**: Joule, *Philosophical Magazine*, 24(1862):121.

p. 186, **"Water Babies, Sunken Rocks, and Women of Italy"**: Tait, *Philosophical Magazine*, 25(1862):263.

p. 186, **"were so discordant . . ."** and **"What you have the hardihood to affirm . . ."**: Tyndall, *Philosophical Magazine*, 25(1862):368-378.

p. 187, **"Allow me to say . . ."**: Thomson, *Philosophical Magazine*, 25(1862):429.

p. 188, **the first incontrovertible . . . statement of a true conservation law:** see Truesdell (1980), ch. 9.

p. 189, **"I have commenced trying . . ."**: WT to Elizabeth, July 3, 1868, SPT I 26.

p. 190, **"my wife has been feeling much better . . ."**: WT to Helmholtz, July 24, 1868, SPT I 527.

p. 191, **"the days of signalling by the 'spot of light' . . ."**: WT to Jessie Crum, July 21, 1870, SPT I 577.

p. 192, **"The signals are, comparitively . . ."** and **"I am sorry to say however . . ."**: Items E2A at CUL: from George Stacey at Aden, September 28, 1872; from C. H. Reynell in Bombay, September 20, 1872.

p. 192, **"At Malta, the Mirror is a thing of the past . . ."**: E10, from R. Portelli in Malta, November 21, 1873.

p. 193, **"You quite take away my breath . . ."**: S201, Smith to WT, September 12, 1870.

p. 193, **"It is the *Lalla Rookh* . . ."**: T480, WT to James, September 21, 1870.

p. 194, **"Full of impatience and excitement . . ."**: Knott (1911) 32.

p. 195, **As long ago as 1845:** T324, JT to WT, October 21, 1845.

p. 196, **"the great advantages I have here . . .":** WT to Cookson, December 1, 1870, SPT I 563.

p. 197, **"I am very glad Maxwell is standing . . .":** S273, WT to Stokes, March 3, 1871.

p. 198, **"The lady was neither pretty, nor healthy, nor agreeable . . .":** Fairley (1988) 107.

p. 198, **" . . . terrible wife":** Whittaker (1989) vol. 1, 246n.

p. 198, **"James, it's time you went home . . ."** and **"Mrs. Maxwell, although . . .":** McDonald (1964) 20, 21. Second recollection is by G. P. Thomson, son of J. J. Thomson.

p. 198, **should not call on him at home:** Strutt (1968) 407 (footnote to p. 44).

p. 199, **"Suppose a man . . .":** Maxwell to WT, February 20, 1854; Larmor (1936).

p. 199, **"been rewarded of late . . .":** Maxwell to WT, November 30, 1854, Larmor (1936).

p. 199, **"I would be much assisted . . .":** Maxwell to WT, September 13, 1855; Larmor (1936).

p. 201, **"How few understand the physical lines of force!":** from diary of Faraday's niece, November 7, 1855; Williams (1965) 507.

p. 201, **"one of the most valuable . . .":** from Maxwell's review of E&M, *Nature*, 7(1873):218-221.

p. 202, **"as far as I know you are the first person . . .":** Maxwell to Faraday, November 9, 1857, Williams (1965) 511.

p. 202, **"I have been most happy in your kindness . . .":** Williams (1965) 494.

p. 202, **"I am, I hope, very thankful . . .":** Faraday to Auguste de la Rive, 1861, Williams (1965) 500.

p. 203, **"Do you know of any elementary work . . .":** F101, Fischer to WT, October 20, 1855.

p. 203, **"I fancy that we might easily give . . .":** T6B, Tait to WT, December 12, 1861.

p. 204, **"Let us apportion our work . . .":** T6C, Tait to WT, December 25, 1861.

p. 204, **the *expense to the students* . . .**": T6D, Tait to WT December 28, 1861.

p. 204, **"I will shortly send you . . .**": T6G, Tait to WT, January 15, 1862.

p. 205, **"I wish you would send back . . .**": T6L, Tait to WT, January 30, 1862.

p. 205, **"*at all events act speedily*"**: T6M, Tait to WT, January 31, 1862.

p. 205, **"Dʳ T, Do look alive . . .**": T6W, Tait to WT, May 5, 1864.

p. 205, **"I wish you would go ahead . . .**": T6X, Tait to WT, June 20, 1864.

p. 205, **"You are a terrible fellow . . .**": S159, Stokes to WT, January 20, 1857.

p. 206, **"the making of the first part . . .**": Obituary of Tait, MPP VI.

p. 206, **"better known in my year . . .**": Barrie (1889).

p. 206, **"The credit of breaking up the monopoly . . .**": Maxwell's review of Thomson and Tait, *Nature*, 10(1879).

p. 207, **"Three *pages* of formulae can easily . . .**": Tait to Cayley, October 22, 1888, Knott (1911) 159.

p. 207, **"remarkable condensation not to say coagulation . . .**": Knott (1911) 153.

p. 207, **"We have had a thirty-eight year . . .**": C87, Kelvin to G. Chrystal, July 13, 1901; also in Knott (1911) 185.

p. 207, **"Oh! That the Cayleys . . .**": WT to Helmholtz, July 31, 1864, SPT I 432.

p. 208, **"the art of reading mathematical books . . .**": Gray (1908) 294.

p. 209, **" . . . no proof at all . . .**": Tait to WT, January 18, 1868, Knott (1911) 220.

p. 209, **"Is it fair to ask you . . .**": Tait to Helmholtz, March 27, 1867; Knott (1911) 217.

p. 209, **"I enclose a letter just rec'd . . .**": Quoted by J. T. Lloyd (1970), who does not identify the author of the letter.

p. 209, **"For my part I must say . . .**": Helmholtz to Tait, April 30, 1867, Knott (1911) 217.

p. 209, **"in all its individuality"**: Knott (1911) 217.

p. 210, **a supposedly personal letter:** *Philosophical Magazine,* 7(1879):344-346; also Art. 85 in MPP V.

5
COMPASS

p. 215, **"has, after anxious consideration . . ."**: WT to Mrs. Tait, March 29, 1871, SPT II 586.

p. 216, **"married one of the sisters . . ."**: Knott (1911) 14.

p. 216, **"Thomson met me in the Kinnaird Hall . . ."**: Eve and Creasey (1945) 124.

p. 216, **"There will be a splendid row . . ."**: Tait to Tyndall, March 18, 1872, and Tyndall's reply; Eve and Creasey (1945) 162.

p. 217, **"the flow of word-painting and righteous indignation . . ."**: This and the following remarks are from *Nature,* 8(1873):381, 399, 431.

p. 217, **"especially inappropriate"**: Undated remark to David King, SPT II 649.

p. 218, ***In the very beginnings of science . . .***: Campbell and Garnett (1884) 415.

p. 218, **"for surely [they] did not hold council . . ."**: Book 1 of *De Rerum Natura,* in the translation by A. M. Esolen (Baltimore: Johns Hopkins University Press, 1995).

p. 219, **assured Mrs. Tait:** WT to Mrs. Tait, May 8, 1871, SPT II 592.

p. 219, **"You will not rest . . ."**: SPT II 595.

p. 219, **he amusingly recounted to a friend:** WT to G. Darwin, July 25, 1882, SPT II 784.

p. 220, **"It was a strange reunion . . ."**: WT to Jessie Crum, undated, SPT II 597.

p. 220, **"learn (at its headquarters) . . ."**: Tait to Helmholtz, May 20, 1871, Knott (1911) 196.

p. 220, **"Mr. Tait knows nothing here besides golf."**: Helmholtz to his wife, August 29, 1871; German original in Knott (1911) 197n (my translation). SPT II 612-624 has a longer English version.

p. 220, **"an indescribably sad impression . . ."** and subsequent extracts: Helmholtz to his wife, August/September 1871; given in English by SPT II 613-614. Königsberger (1906), ch. 10, includes many of the same letters, in slightly different translations.

p. 221, **"a husband who is no longer in his first youth . . .":** Königsberger (1906), ch. 10; not included by SPT.

p. 222, **"Now, mind, Helmholtz . . .":** SPT II 614.

p. 222, **". . . immense intellectual strength . . .":** Crowther (1935) 201.

p. 223, **"much ashamed":** On Deep-Sea Sounding by Pianoforte Wire, PLA III.

p. 225, **"No harm was done":** *Memoir of Fleeming Jenkin*, ch. 3, Colvin and Ewing (1887).

p. 225, **"Goodbye, goodbye, Sir William Thomson":** SPT II 639; also AGK 33.

p. 226, **His new wife was in her mid-30s:** Smith and Wise (1989), ch. 21, give Fanny's year of birth as ca. 1838; provenance unstated.

p. 226, **"When I came to Madeira . . .":** WT to Elizabeth, May 12, 1874, SPT II 645.

p. 227, **"as I have so many engagements . . .":** WT to Charles A. Smith, April 28, 1874, SPT II 643.

p. 228, **"Oh, I'll tell you what you should do.":** Knott (1911) 31.

p. 229, **"utterly surprised":** S356, Stokes to WT, June 5, 1879.

p. 229, **"That is the very thing for me":** S362, Stokes to WT, July 9, 1979; see also S361, an unsent draft of 362.

p. 233, **"evil so pregnant with mischief":** Fanning (1986), introduction.

p. 235, **"to see 'the Retribution' swing . . .":** S70, WT to Stokes, July 19, 1850.

p. 236, **"When I tried to write on the mariner's compass . . .":** Terrestrial Magnetism and the Mariner's Compass, PLA III.

p. 237, **"By a happy coincidence":** Obituary of Archibald Smith, MPP VI.

p. 239, **" . . . so dangerous a tool as a moveable magnet . . .":** Fanning (1986) 69.

p. 239, **"a process of 'Artificial Selection' . . ."**: Terrestrial Magnetism and the Mariner's Compass, PLA III.

p. 239, **"between 1850 and 1880 . . ."**: Hitchins and May (1952) 79.

p. 240, **"innovation is very distasteful to sailors . . ."**: Sounding by Pianoforte wire, PLA III.

p. 241, **Some naval historians:** Cotter (1976, 1977) disparages WT's work and (in my estimation) overstates Airy's contributions.

p. 241, **"enunciated no new principles . . ."**: Hitchins and May (1952) 82.

p. 241, **"It won't do"**: SPT II 710.

p. 242, **thlipsinomic, platythliptic . . .**: Kargon and Achinstein (1987) 131.

p. 243, **"marvellously distinct"**: SPT II 671, from WT's report at the exhibition.

p. 244, **"the originality, the inventiveness . . ."**: WT in BA *Reports* (1876); also *New York Times,* October 4, 1876.

p. 244, **"To see such men is a privilege . . ."**: *Montreal Gazette,* September 4, 1876.

p. 244, **" . . . England's great mathematician and electrician"**: *Philadelphia Inquirer,* September 4, 1876.

p. 244, **"the great event in the year's work . . ."**: *Baltimore Sun,* September 20, 1884.

p. 245, **"would give a strong impulse . . ."**: Gilman to WT, August 13, 1882, SPT II 811.

p. 245, **"the very best and most effective . . ."**: G75, Gilman to WT, January 8, 1884; enclosure from W. Gibbs.

p. 247, **"the lecturer is a man tall . . ."**: *Baltimore Sun,* October 2, 1881.

p. 247, **"What an extraordinary performance that was! . . ."**: Strutt (1968) 145.

p. 248, **" . . . the usual Thomsonian style . . ."**: Rayleigh to his mother, October 19, 1884, Strutt (1968) 147.

p. 248, **"he has been known to lecture for an hour . . ."**: J. J. Thomson (1936) 424.

p. 248, **the "wiggler"**: SPT II 832.

p. 251, **" . . . cinematics instead of kinematics . . ."**: Kargon and Achinstein (1987) 129, 148.

p. 251, **"I never satisfy myself until I can make a mechanical model . . ."**: WT reported in *Nature*, 31(1885):508.

p. 252, **"stripped of the scaffolding . . ."**: Whittaker (1989) vol. 1, 255.

p. 252, **"writing out the Lord's Prayer . . ."**: Bacon (1929) 6.

p. 253, **"We still have ancient Admirals . . ."**: Fisher (1919), *Memories*, 99.

p. 253, **"the most suitable number Captain Fisher could think of"**: Bacon (1929) 50.

p. 253, **"No, thank you, I am quite warm."**: Fisher (1919), *Memories*, 251.

p. 254, **" . . . the Incarnation of Revolution"**: Fisher (1919), *Records*, 20.

p. 254, **"He diagnosed the matter . . ."**: Bacon 77.

p. 254, **"Well," Fisher asked . . .**: Fisher (1919), *Memories*, 142.

p. 255, **"We fight God . . ."**: Fisher (1919), *Records*, 71.

p. 255, **"It was an immense difficulty . . ."**: Fisher (1919), *Records*, 63.

p. 255, **"pig-headed and self-opinionated . . ."**: May (1979).

p. 256, **"I can state from long experience . . ."**: F105, Fisher to WT, October 16, 1885.

p. 257, **"'My Lord, what has this to do with the case? . . ."**: J. J. Thomson (1936) 386.

p. 258, **"much mean and underhand work . . ."**: Elizabeth to her daughters, November 23, 1889, AGK 89.

p. 258, **Fanning tells a different story**: Fanning (1986), ch. 4.

p. 258, **"I may single out . . ."**: C168, Creak to WT, December 3, 1883.

p. 258, **"When the Thomson compass was first introduced . . ."**: Fanning (1986) 154.

6
KELVIN

p. 261, **Lord Cable! Lord Compass!:** AGK 106.

p. 262, **"I am afraid it cannot be . . .":** WT to G. Darwin, November 20, 1884, SPT II 840.

p. 262, **"Some day you will be proud . . .":** J. J. Thomson (1936) 10.

p. 262, **"to my great surprise . . .":** J. J. Thomson (1936) 98.

p. 262, **"pre-eminent service in promoting arts . . .":** SPT II 968-973.

p. 262, **"friends and comrades . . .":** SPT II 984.

p. 263, **"seemed to ring through the hall . . .":** SPT II 988; AGK 116.

p. 264, **"Naturalist. A person well versed . . .":** On the Rigidity of the Earth, MPP III.

p. 264, **"magnificent display of smoke-rings . . .", "the clash of atoms"** and other extracts: On Vortex Atoms, MPP IV.

p. 265, **"pungent and disagreeable":** WT to Helmholtz, January 22, 1867, SPT I 514.

p. 268, **The following story is true . . .:** Waves from moving sources, Heaviside (1951), vol. 3, 1.

p. 268, **"What would Edison say . . . ":** Gossick (1976).

p. 269, **his address as inaugural president of the IEE:** Ether, Electricity and Ponderable Matter, MPP III.

p. 270, **" . . . the cart-men shouted abuse . . .":** Searle (1987) 10.

p. 270, **"I have to give you my best thanks . . .":** H53, Heaviside to WT, February 27, 1889.

p. 270, **"You may judge of the intensity . . .":** Searle (1987) 77.

p. 272, **"save[s] letters, and *eases the memory* . . ."** H53, supra.

p. 272, **Faraday "did the most":** Kargon and Achinstein (1987) 148.

p. 273, **"if we put aside practical application to Physics . . .":** Heaviside (1951), vol. 1, 301.

p. 273, **"Passing to Prof. Tait's letter . . .":** Heaviside (1951), vol. 3, 509.

p. 274, **"Heaviside's nihilism":** Kelvin to FitzGerald, April 29, 1896, SPT II 1070n.

p. 274, **"Lord Kelvin used to call me a nihilist . . .":** Heaviside (1951), vol. 3, 479.

p. 275, **"Sir Wm. Thomson's speaking of the ether . . .":** *Nature,* 32(1885):4.

p. 275, **"a certain amount of opposition . . .":** Strutt (1968) 252.

p. 275, **"Sir William Thomson in one paper . . .":** London *Times,* September 14, 1888.

p. 276, **"You say . . . 'The luminiferous ether . . .'":** F127, FitzGerald to Kelvin, April 17, 1896.

p. 279, **"never hesitated to embark on . . .":** E. W. Brown in Darwin (1916).

p. 279, **" . . . mere conjuring tricks with symbols . . .":** Darwin's Inaugural Plumian Lecture, Darwin (1916).

p. 279, **"My dear old George . . .":** From the biographical sketch by F. Darwin, Darwin (1916).

p. 280, **"Under these circumstances . . .":** G. Darwin, *Philosophical Transactions of the Royal Society,* 170(1879):529.

p. 281, **"I was Lord Kelvin's pupil . . .":** This and subsequent extracts, including the letters to Perry from Tait and Kelvin, *Nature,* 51(1895):224-227.

p. 283, **"I do not wish to combat . . .":** D35, G. Darwin to WT, June 4, 1886.

p. 284, **"latest blast of the anti-geological trumpet . . .":** G48, Geikie to Kelvin, February 25, 1898.

p. 284, **"worbles":** Maxwell to Tait, November 13, 1867, Knott (1911) 106.

p. 284, **"It would puzzle . . ."** and **"But why does no one . . .":** Maxwell's review of E&M, *Nature,* 7(1873):218-221.

p. 285, **"Like most problems in vortex motion . . .":** J. J. Thomson (1936) 95.

p. 285, **"After many years of failure . . .":** SPT II 1047; footnote to a paper of 1904, but referring to work done sometime after 1887.

p. 286, **"Sir W. is full of a *froth* theory of the ether!":** Rayleigh to A. Schuster, October 4, 1888; Strutt (1968) 243.

p. 286, "**. . . I have been considering the subject for forty-two years . . .**": Ether, Electricity and Ponderable Matter, MPP III.

p. 286, "**It is mere nihilism . . .**": Kelvin to Fitzgerald, April 9, 1896, SPT II 1064.

p. 286, "**It has occurred to me . . .**": Nineteenth-Century Clouds over the Dynamical Theory of Heat and Light, *Philosophical Magazine,* 2(1901):1-40; not in the collected works but appears as Appendix B to the 1904 Baltimore Lectures.

p. 287, "**. . . unlike that awful Chicago**": Königsberger (1906), ch. 11.

p. 288, "**Helmholtz and Uncle William were inseparable . . .**": SPT II 926.

p. 288, **a tale recounted by Lord Rayleigh's son:** Strutt (1968) 240.

p. 289, "**The first line would send him off . . .**": Strutt (1968) 243.

p. 289, "**I always consulted my great authority . . .**": Kargon and Achinstein (1987) 129.

p. 289, "**Stokes would remain silent . . .**": J. J. Thomson (1936) 50.

p. 290, "**believed the Falls of Niagara . . .**": *Nature,* 20(1879):110.

p. 291, "**when the whole water from Lake Erie . . .**": A statement to the press reprinted by SPT II 1002.

p. 291, "**As the demand goes on increasing . . .**": *Toronto Daily Mail and Empire,* August 19, 1897.

p. 292, "**most economical size . . .**": On the Economy of Metal in Conductors of Electricity, MPP V.

p. 292, "**I would not advise manufacturers . . .**": *Buffalo Express,* August 18, 1897.

p. 292, "**A gentleman of exceedingly pleasant manners . . .**": *Toronto Daily Mail and Empire,* August 19, 1897.

p. 292, "**a plain American citizen . . .**":*Buffalo Express,* August 16, 1897.

p. 293, "**westinghoused**": Baldwin (2001) 202.

p. 294, "**I often say . . .**": Electrical Units of Measurement, PLA I.

p. 294, "**. . . the etherealisation of common sense**": The Six Gateways to Knowledge, PLA I.

p. 294, **"There is one thing I feel strongly . . ."**: The Bangor Laboratories, PLA II.

p. 295, **" . . . the eggs were always boiled . . ."**: Larmor (1907) 36.

p. 295, **"the utter scientific absurdity . . ."**: SPT II 870.

p. 296, **Aepinus Atomized:** *Philosophical Magazine*, 3(1902):257-283; not in the collected works but appears as Appendix E to the 1904 Baltimore Lectures.

p. 298, **Propped up in bed, wearing a bright red jacket:** letter from ETK, March 29, 1896, AGK 130.

p. 298, **"A horrid demon of the No. 5 nerve"**: Kelvin to SPT, October 14, 1899, SPT II 1149.

p. 298, **"looking very venerable, limping . . ."**: *New York Times*, April 20, 1902.

p. 299, **"Oh, there is nothing practical in that"** and following extracts: *Rochester Democrat and Chronicle*, April 30, 1902.

p. 300, **"Both as a student and as a professor . . ."**: *Rochester Democrat and Chronicle*, May 2, 1902.

p. 301, **"greatest of living scientists . . ."**: *Rochester Democrat and Chronicle*, May 3, 1902.

p. 301, **"one gentleman, who is . . ."**: *Rochester Democrat and Chronicle*, May 5, 1902.

p. 302, **"can barely be distinguished . . ."**: *Nature*, 68(1903):447; remark is by C. V. Boys.

p. 302, **"knowing, as we now do . . ."**: *Nature*, 68(1903):496.

p. 303, **"the hundred million years . . ."**: *Nature*, 68(1903):526.

p. 303, **"from without the atoms . . ."**: Kelvin reported in *Nature*, 68(1903):611.

p. 304, **"the Professor seemed very depressed . . ."**: Knott (1911) 63.

p. 304, **"rough gaiety . . ."**: Obituary of Tait, MPP VI.

p. 305, **"exceedingly cautious . . ."**: J. J. Thomson (1936) 50.

p. 305, **"My principal intelligence . . ."**: S153, Stokes to WT, September 24, 1856.

p. 305, **"it is curious how these things . . ."**: S371, Stokes to WT, September 23, 1879.

p. 305, **"I shall always remember . . ."**: Schuster (1932) 242.

p. 306, **"expression of distrust . . ."**: *Nature*, 70(1904) supplement, iii-v.

p. 307, **"Whatever opinion may be formed . . ."**: *Nature*, 74(1906):516-518.

p. 307, **"Lord Kelvin has talked radium . . ."**: Rutherford to his wife, May 22, 1904, Eve (1939) 108.

p. 308, **"to my relief, Kelvin fell fast asleep . . ."**: Eve (1939) 107.

EPILOGUE

p. 309, **"Lord Kelvin preferred . . ."**: *Nature,* 76(1907):457.

p. 310, **"In proposing a vote of thanks . . ."**: E. Ray Lankester, quoted by Wilson (1910).

p. 311, **"one of the greatest scientists . . ."**: *New York Times*, December 18, 1907.

p. 311, **"the foremost scientist . . ."**: *Washington Star*, December 18, 1907.

p. 311, **"the most distinguished . . ."**: London *Times*, December 18, 1907.

p. 312, **"it is less easy to speak . . ."**: *Nature*, 77(1907):175-177.

p. 312, **"a main pioneer and creator . . ."**: *Nature*, 77(1908):199-200.

p. 312, **"What a happy strenuous career . . ."**: *Proceedings of the Royal Society* A 81(1908):iii-lxxvi (appendix).

p. 313, **"Lord Kelvin is rather a clever man."**: AGK 80.

p. 313, **"Sir Wm might do better . . ."**: Helmholtz to his wife, 1884, SPT II 805.

Index

thermal properties and composition, 281, 282-283
tidal friction, 278, 279, 280, 283, 284
East India Company, 116
Eastern Cable Company, 260
Eastman, George, 2, 299, 300
Echo-sounding sonar devices, 224
Edinburgh Academy, 197, 205
Edison, Thomas, 4, 243-244, 253, 268, 293, 298
Edward VII, 255
Einstein, Albert, 82, 277, 306
Electric fields, mathematics of, 35, 86. *See also* Electromagnetic fields
Electric meters, 269
Electrical measurement, 9, 142-144, 146-153, 269
Electrician (journal), 145
Electricity, 3, 4, 8
  ac/dc controversy, 292-294, 301
  early theories, 296-297
  generation and transmission, 2, 73, 74-75, 269, 290-294, 299-300, 301
  industry and residential, 227, 260, 269, 291, 299
  on ships, 253-254
  standardization of units, 9, 142-153, 197
  and submarine telegraph, 121-125
  transformers, 293
Electricity (Murphy), 199
Electrochemistry, 84
Electromagnetic fields, 73-74, 80, 84-87, 93, 200, 202, 245-246, 251-252, 271-272, 314
Electromagnetic induction, 85
Electromagnetic spectrum, 245-247, 248-249, 267
Electromagnetic theory, 3, 5, 6, 7, 73, 79, 80, 84-85, 86-88, 91, 93, 105, 106, 112, 127, 155, 199-203, 245-246, 251-252, 262-263, 264, 267, 269-271, 284, 286, 306, 312

Electrons, 262, 263, 296
Ellis, R. L., 49, 50, 60
Ely, Bishop of, 62
Encyclopedia Britannica, 82
Energy
  internal, 188
  mathematical definition of, 272
Energy and Empire (Smith and Wise), 94 n.5
England
  industrial revolution, 64-65
  mathematics and science in, 54-55
  and submarine telegraph, 116-118
  telegraph line, 295
Entropy, 110-111, 154, 208-211, 242, 265
Epicurus, 217
Essay on the Application of Mathematical Analysis to the Theories of Electricity and Magnetism (Green), 79
Ether models, 264, 266, 267, 285-286, 289, 295-296, 300
Euclid, 53
Euler, Leonhard, 54
European and American Telegraph Company, 118
Evans, Frederick, 238, 239, 241, 242, 243, 255, 258
Evelina (Burney), 38-39, 47, 95
Evolution, theory of, 6, 8, 174, 211-214
Ewing, J. A., 70-71

F

Fahrenheit, Gabriel, 98
Fanning, A. E., 258
Faraday, Michael, 3, 35, 55, 56, 61, 73, 79, 80, 82-88, 105, 115, 121-122, 151-152, 155, 157, 178, 179, 180, 181, 199, 200, 201, 202, 223, 272-273, 293, 311
Feilitzsch (German astronomer), 180
Field, Cyrus West, 120, 121, 127-130, 131, 134, 135, 136, 138, 140, 142, 156, 157, 158, 159, 161, 162